Green Energy and Technology

More information about this series at http://www.springer.com/series/8059

M. Masud K. Khan · Ashfaque Ahmed Chowdhury
Nur M. Sayeed Hassan
Editors

Application of Thermo-fluid Processes in Energy Systems

Key Issues and Recent Developments for a Sustainable Future

Springer

Editors
M. Masud K. Khan
School of Engineering and Technology
Central Queensland University
Rockhampton, QLD
Australia

Nur M. Sayeed Hassan
School of Engineering and Technology
Central Queensland University
Cairns, QLD
Australia

Ashfaque Ahmed Chowdhury
School of Engineering and Technology
Central Queensland University
Gladstone, QLD
Australia

ISSN 1865-3529 ISSN 1865-3537 (electronic)
Green Energy and Technology
ISBN 978-981-10-0695-1 ISBN 978-981-10-0697-5 (eBook)
https://doi.org/10.1007/978-981-10-0697-5

Library of Congress Control Number: 2017947862

Printed on acid-free paper

This Springer imprint is published by Springer Nature
The registered company is Springer Nature Singapore Pte Ltd.
The registered company address is: 152 Beach Road, #21-01/04 Gateway East, Singapore 189721, Singapore

Preface

The demand for clean energy technologies is increasing because of its environmental benefits, future energy security and diversity in securing energy sources. Research on current and future application of thermo-fluid processes associated with clean energy generation, distribution and conservation is viewed fundamental for seeking solutions for clean energy and a sustainable future.

This edited book volume introduces research works and their findings on how energy efficient technologies and thermo-fluid processes are analysed and applied in engineering practices. It presents a number chapters focusing on cutting edge research work on key issues and recent developments of the thermo-fluid processes, including but not limited to, energy technologies in process industries, application of thermo-fluid processes in mining industries, application of electrostatic precipitator in thermal power plant, biofuel, energy efficiency in building systems.

These chapters will be a valuable resource to researchers, practising engineers, technologist and students and will help the reader to develop intuitive understanding of the relevant concept and solution to the global issues to achieve sustainability in medium- and large-scale industries.

The chapters of this book have been carefully selected, and they include both the relevant technical and social issues that have a significant impact on the society and the stakeholders. This edited book comprises two parts. Part I contains five chapters focusing on energy technologies in process industries, and Part II contains seven chapters emphasizing on the application of thermo-fluid processes in energy systems and the key issues and recent developments for a sustainable future. A brief summary of each part is given below.

Part I Energy Technologies

Chapter "Utilization of Nanofluid in Various Clean Energy and Energy Efficiency Applications" focuses on energy efficiency and utilization of alternative clean energy technologies by replacing fossil fuel as primary source of energy. It also reviews the uses of nano-fluid in various clean energy and energy efficiency applications such as solar energy, solar thermal collectors and solar water heaters for utilization of nano-fluid for improving solar to thermal energy conversion.

Chapter "Gaseous and Particle Emissions from a Compression Ignition Engine Fueled with Biodiesel-Diesel Blends" investigates the sustainability of rice bran biodiesel from environmental point of view. In this study, 5 and 20% biodiesel was tested in a naturally aspirated four stroke multi-cylinder diesel engine at different load and speed conditions. It was found that all biodiesel-blended fuel reduces the brake power (BP) and increases brake-specific fuel consumption (BSFC) slightly than diesel fuel. Engine emission results indicated that blended fuel reduces the average particulate matter (PM), carbon monoxide (CO) and hydrocarbons (HC) except nitric oxides (NO) emissions than diesel fuel. Finally, the chapter concludes that up to 20% rice bran biodiesel could replace diesel fuel to help in controlling the air pollution to a great extent without sacrificing engine power significantly.

Chapter "Correlation Between Physicochemical Properties and Quality of Biodiesel" introduces biodiesel feedstocks, production process, chemical compositions, standards, physicochemical properties and in-use performance to investigate the relationships between biodiesel properties and chemical composition using a principal component analysis (PCA). The PCA analysis is presented graphically and described in this chapter. Although individual biodiesel properties have a complex relationship with the parameters of chemical compositions, the PCA analysis determines the dominant relationships which were the average number of double bonds and polyunsaturated fatty fractions. This chapter will help the reader to better understanding of the physicochemical properties of biodiesel.

Chapter "A Review of Microalgal Biofuels, Challenges and Future Directions" introduces biofuels from micro-algae that have the potential to provide a sustainable and carbon-neutral energy source, complementing the shortfall of fossil fuels and enhancing the mitigation of global warming. Coupling micro-algae cultivation with wastewater and CO_2 from power plants is considered a promising route for the production of bioenergy and bio-based by-products. This chapter presents a review of current status, challenges and future of biofuel from algae as a renewable source.

Chapter "Performance Assessment of an Electrostatic Precipitator of a Coal Fired Power Plant—A Case Study for Collecting Smaller Particles" presents the collection efficiency of particles affected by different flow distribution and recommends the possible modification in physical model to increase the collection capacity of smaller particles in the existing electrostatic precipitators (ESPs) used in power plants.

Part II: Thermo-fluid Process Applications

Chapter "Experimental Investigation and Molecular-Based Modeling of Crude Oil Density at Pressures to 270 MPa and Temperatures to 524 K" investigates new experimental density data for crude oil sample obtained from the Gulf of Mexico region, and these density data were measured at pressures to 270 MPa and temperatures to 524 K. These conditions simulate those encountered from ultra-deep formations to platforms. These density data points are then used to validate both empirical-based and molecular-based equation of state models. Results show that the molecular-based perturbed chain statistical associating fluid theory (PC-SAFT) models, without the use of any fitting parameters, produced crude oil density predictions within 1% of the experimental data. These results represent an improvement over high-temperature, high-pressure density predictions from volume-translated cubic equations of state.

Chapter "Heat Transfer Enhancement in a Baffled Attic-Shaped Space" numerically investigates the natural convection heat loss in an attic-shaped enclosure introducing a single baffle under the top tip, which is a cost-effective approach. The chapter examines a wide range of governing parameters such as Rayleigh number, aspect ratio, baffle length. It is observed that the heat transfer due to natural convection in the enclosure reduces when the baffle length is increased. The chapter also discusses the effects of other parameters on heat transfer and flow field in this study.

Chapter "Enhanced Thermo-fluid Dynamic Modelling Methodologies for Convective Boiling" presents the logical development of novel thermal and dynamical enhancement approaches that overcome existing modelling limitations, and creates a precise and realistic foundation for advanced boiling design methodology with applicability over the entire boiling flow regime. The effectiveness and advantages of these model enhancements are highlighted through three special cases of boiling processes, namely convective boiling typically occurring in a straight pipe, convective boiling in a curved pipe influenced by secondary flow and pool-to-convective transitional boiling. Finally, the potential for energy saving by these techniques is identified that contributes to cleaner thermal energy generation.

Chapter "A Method of Three-Dimensional Thermo-fluid Simulation of the Receiver of a Standard Parabolic Trough Collector" develops a three-dimensional (3D) computational conjugate heat transfer (CCHT) model of a bare receiver of Luz Solar 2 (LS2) PTC. The chapter also describes the method of this CCHT modelling and its verification. The CCHT model was developed applying finite volume (FV) technique of the state-of-the-art computational fluid dynamics (CFD). The solar irradiance profile (IP) around the receiver surface of the collector was calculated using the Monte Carlo ray tracing (MCRT) technique. Moreover, the MCRT calculated-IP functions specific to the Luz Solar 2 (LS2) collector are given for facilitating further CCHT modelling of the collector system.

Chapter "Enhancement of Confined Air Jet Impingement Heat Transfer Using Perforated Pin-Fin Heat Sinks" examines the effects of fin perforations on the thermal performance of pin fin heat sinks. Results show that thermal resistance decreases and fin efficiency increases with the increase of Reynolds number due to perforation and also reduces cooling power consumption rate.

Chapter "Multiphase Flow in Porous Media: Cake Formation During Extreme Drilling Processes" emphasizes on multiphase flow in porous media to closely mimic the actual drilling fluid composed of fine particles and viscous fluid rather than focused only on single-phase flow phenomena in porous media and also simulates the fluid flow and cake formation in extreme drilling processes.

Chapter "Optimising Pyrolysis Conditions for Thermal Conversion of Beauty Leaf Tree (*Calophyllum inophyllum* L.) Press Cake" focuses on biodiesel production from beauty leaf tree (BLT) as it can thrive well in degraded soils and produces up to 3600 l of non-edible oil that can be readily converted into biodiesel. The current study tested thermal conversion of BLT press cake using a batch reactor. This study showed that up to 93% of the energy contained in the BLT press cake be recovered as biochar, bio-oil, bioliquor and syngas. The results also show that the additional products (biochar, bio-oil) from BLT press cake can make a significant contribution to the economic viability of BLT biodiesel production. It is suggested that the use of a portable and continuous feeding auger reactor could be conveniently used to convert BLT whole fruits, press cake or husks into biofuels.

We hope the selected chapters will help in enhancing your understanding and practicing of current and future application of thermo-fluid processes associated with clean energy generation, distribution and conservation and sustainability and the environment.

Rockhampton, Australia M. Masud K. Khan
Gladstone, Australia Ashfaque Ahmed Chowdhury
Cairns, Australia Nur M. Sayeed Hassan

Scope

The demand for energy technologies is increasing because of its environmental benefits, future energy security and diversity in securing energy sources. The objectives of this edited book volume are to provide the industries and the academia with an update on current and future application of thermo-fluid processes associated with clean energy generation, distribution and conservation. The book will provide resources and case studies in the field of energy technology, clean energy, energy efficiency, sustainability and the environment to academics, researchers, practising engineers, technologist and students. It will be a valuable resource to undergraduate, honours and postgraduate research students in the field of thermo-fluid engineering.

The chapters of the book will present cutting edge research work on key issues and recent developments of the thermo-fluid processes, including but not limited to, energy technologies in process industries, application of thermo-fluid processes in mining industries, application of electrostatic precipitator in thermal power plant, biofuel, energy efficiency in building systems. These chapters will help to develop and intuitive understanding of the relevant concept and solution to the global issues to achieve sustainability in medium- and large-scale industries.

Contents

Part I
Energy Technologies

Utilization of Nanofluid in Various Clean Energy and Energy Efficiency Applications

Sayedus Salehin, M. Monjurul Ehsan, Syed Rafat Faysal and A.K.M. Sadrul Islam

Abstract Policy makers around the globe, both at the national and international level, have been emphasizing on energy efficiency and utilization of alternative clean energy technologies by replacing fossil fuel as a primary source of energy. This interest was initiated when the adverse effects of fossil fuel on the environment such as greenhouse gas emission and global warming were unfolded. Multidisciplinary researches are being carried out in the research laboratories to provide effective clean energy solutions. Nanofluid, a colloidal mixture of nanoparticles in the base fluid, e.g., water, ethylene glycol, oil, offers efficiency improvements in many clean energy applications due to the improvement in its thermophysical property. In this chapter, use of nanofluid in various clean energy and energy efficiency applications is reviewed. Solar energy, the dominant clean energy source, is a potential field of application where nanofluid can be employed. Solar thermal collectors and solar water heaters are the ideal candidates for utilization of nanofluid for improving solar to thermal energy conversion. Thermal storage system employs phase change material (PCM) for storing thermal energy, and nanofluid may be added to the PCM for enhanced performance. Nanofluid can be used in carefully designed heat exchangers for extracting energy from geothermal resources. This mixture can also be used in waste heat collector to improve efficiency for its thermophysical property causing a heat transfer enhancement. Adding nanoparticles to the refrigerants improves the heat transfer characteristics of refrigerants and improves the performance of the refrigeration system from an energy efficiency perspective. However, there are challenges associated with nanofluid which are needed to overcome for optimal performance of the working fluid.

S. Salehin · M. Monjurul Ehsan · S. Rafat Faysal · A.K.M. Sadrul Islam (✉)
Department of Mechanical and Chemical Engineering,
Islamic University of Technology, Gazipur 1704, Bangladesh
e-mail: sadrul@iut-dhaka.edu; sadrul05@gmail.com

M.M.K. Khan et al. (eds.), *Application of Thermo-fluid Processes in Energy Systems*,
Green Energy and Technology, https://doi.org/10.1007/978-981-10-0697-5_1

1 Introduction

To meet the ever-increasing energy demand, petroleum-based fossil fuels have been extensively used in all sectors irrespective of geographic locations. However, this energy source is associated with adverse environmental effects such as greenhouse gas emission and global warming. In this aspect, the regulatory authorities across the globe, both at government and non-governmental level, are emphasizing on reducing the use of fossil fuel and replace it with clean and renewable energy resources and utilize energy efficiency to reduce energy consumption. The research laboratories in academia and the industry alike are working relentlessly on finding effective clean energy solutions.

Nanofluid is a comparatively new kind of synthesized fluid which is a colloidal mixture of nanoparticles, usually metals or metal oxides, in a base fluid, with enhanced thermophysical property as compared to the base fluid. This type of fluid can be used in many clean energy and energy efficiency application because of its enhanced thermal, optical, and rheological properties. In this chapter, the utilization of nanofluid in various clean energy and energy efficiency applications is reviewed. Focus has been given to the application of nanofluid in solar thermal energy, thermal energy storage, heat exchangers, refrigeration and air-conditioning, electronics cooling, and transportation.

2 Nanofluid: Definition

In last few decades, numerous researches have been carried out in a search of more suitable heat transfer fluid in thermal engineering science due to its great demand in various industrial and engineering applications such as heat exchangers, electronic devices, refrigerators, solar thermal storage, nuclear system, biomedical applications, engine cooling, and transmission oil. [85, 104]. Heat transfer properties of the working fluid can be substantially augmented by the implementation of newly emerged promising method called nanotechnology. In sustainable energy, it is one of the effective, inexpensive, and innovative ways of intensifying the heat transfer characteristics of working fluid by the stable and homogeneous dispersion of nanosized solid particles in traditional base fluid such as water, oil, ethylene glycol so called nanofluid. By the addition of nanoparticles of size below 100 nm in base fluid, the overall thermal and hydrodynamic behavior of working fluid can be significantly improved due to its improvement in thermophysical properties such as specific gravity, density, viscosity, convective heat transfer coefficient, electrical conductivity, Seebeck coefficient, and optical scattering coefficients [19, 106, 108]. The solid nanoparticles have higher thermal conductivity compared to base fluid which potentially assists in enhancing heat transfer characteristics. The fluid flow characteristics of nanofluid flowing through different channels are altered due to recirculation, vortex generation, interruption of thermal boundary layer, and flow separation. The implementation of nanofluid within a preferred volume fraction offers reduced pumping power in order to accomplish equivalent heat transfer enhancement [65, 66]. In many engineering applications, nanofluids are a novel approach for significant improvement in thermal efficiency due to its superior thermophysical properties [16, 44, 61, 80, 101, 107, 108].

Some of the examples of solid nanoparticles are copper, silver, nickel, gold, and metal oxides, and magnetic and ferromagnetic types are Al_2O_3, TiO_2, ZnO, SiO_2, Fe_2O_3, MgO, etc. Various forms of carbon are diamond, graphite, carbon nanotubes, etc. The nanoparticles could be of different size and shape and are mixed within a desired volume concentration in base fluid such as water, engine oil, ethylene glycol, ethanol, car engine coolant, terpineol, diesel oil and polydimethylsiloxane, propylene glycol, ethanol–isopropanol, or any other mixture of two homogeneous liquids.

3 Synthesis of Nanoparticles

The synthesis of nanofluid can be made by two mechanisms: one-step method and two-step method. In one-step method, the preparation and synthesis of nanoparticles to the desired size, shape, and volume fraction are performed by a combined process with minimized agglomeration [53]. In this method, nanoparticles are synthesized by physical vapor deposition (PVD) technique or a liquid chemical method and limited to small-scale production and low-pressure fluids [79, 114]. In two-step method, firstly the nanoparticles are synthesized to preferred size and shape and finally the particles are dispersed to base fluid with desired volume fraction by the addition of some additives for better stabilization [92]. Here, the nanoparticles' manufacturing and preparation is performed separately. In order to obtain better stability, minimum sedimentation, and controlling p^H to a desired level, surfactant and additives are employed during the synthesis process [74, 103]. In this process, the synthesis of nanoparticles is performed by grinding, milling, sol–gel, wet chemical methods, laser ablative technology, hydrothermal technique, gas-phase synthesis, etc., depending upon the types, size, shape of nanoparticles, and their applicability [87].

4 Thermophysical Properties of Nanofluid

4.1 Thermal Conductivity

The Maxwell model for [57] thermal conductivity for solid–liquid mixtures of relatively large particles (micro-/mini-sized) is good for low solid concentrations. The effective thermal conductivity, k_{nf}, is given by

$$k_{nf} = \frac{k_p + 2k_{bf} + 2(k_p - k_{bf})\varphi}{k_p + 2k_{bf} - (k_p - k_b)\varphi} k_{bf} \tag{1}$$

Hamilton and Crosser [29] proposed a model for liquid–solid mixtures of non-spherical particles. They introduced a shape factor, n, to account for the effect of the shape of the particles. The thermal conductivity, in which the ratio of conductivity of the solid and fluid phases is larger, than 100 ($k_p/k_{bf} > 100$), can be expressed as follows:

$$k_{nf} = \frac{k_p + (n-1)k_{bf} + (n-1)(k_p - k_{bf})\varphi}{k_p + (n-1)k_{bf} - (k_p - k_b)\varphi} k_{bf} \tag{2}$$

where "n" is the empirical shape factor given by $n = 3/\psi$, and ψ is the particle sphericity, defined as the ratio of the surface area of a sphere with volume equal to that of the particle, to the surface area of the particle. Yu and Choi [111] proposed a modified Hamilton–Crosser model to include the particle liquid interfacial layer for non-spherical particles. The effective thermal conductivity was expressed as

$$k_{nf} = (1 + \frac{n\varphi_{nf}A}{1 - \varphi_{nf}A})k_{bf} \tag{3}$$

where A is defined by

$$A = \frac{1}{3}\sum_{j=a,b,c} \frac{k_{pj} - k_{bf}}{k_{pj} + (n-1)k_{bf}})$$

And, $\varphi_{nf} = \varphi\sqrt{\frac{(a^2+t)(b^2+t)(c^2+t)}{\sqrt{abc}}}$ is the equivalent volume concentration of complex ellipsoids, which is an imaged structure of elliptical particles ($a > b > c$) with surrounding monolayers, with a general empirical shape factor n ($n = 3\psi^{-\alpha}$, here α is an empirical parameter and ψ is the particle sphericity).

Xue [109] developed a model for the effective thermal conductivity of nanofluids. His model is based on the average polarization theory and includes the effect of the interface between the solid particles and the base fluid. The derived equation for the effective thermal conductivity is

$$9\left(1 - \frac{\varphi}{\lambda}\right)\frac{k_{nf} - k_{bf}}{2k_{nf} + k_{bf}} + \frac{\varphi}{\lambda}\left[\frac{k_{nf} - k_{c,x}}{k_{nf} + B_{2,x}(k_{c,x} - k_{nf})} + 4\frac{k_{nf} - k_{c,y}}{2k_{nf} + (1 - B_{2,x})(k_{c,y} - k_{nf})}\right] = 0 \tag{4}$$

where $\lambda = \frac{abc}{(a+t)(b+t)(c+t)}$ with half radii (a, b, c) of the assumed elliptical complex nanoparticles, which consist of nanoparticles and interfacial shells between particles and the base fluids. $k_{c,j}$ is the effective dielectric constant, and $B_{2,x}$ is the depolarization factor along the x symmetrical axis, which is derived from the polarization theory.

Koo and Kleinstreuer [40] developed another model for nanofluids, which includes the effects of particle size, particle volume fraction, and temperature dependence as well as properties of the base fluid and the particle subject to Brownian motion. The resulting formula is

$$k_{nf} = \frac{k_p + 2k_{bf} + 2(k_p - k_{bf})\phi}{k_p + 2k_{bf} - (k_p - k_{bf})\phi} k_{bf} + 5 \times 10^5 \beta\phi\rho_{bf} C_{pbf} \sqrt{\frac{K_B T}{\rho_p D}} f(T,\phi) \quad (5)$$

where $f(T,\phi) = (-6.04\phi + 0.4705)T + (1722.3\phi - 134.63)$.

The first part of the equation is obtained from the Maxwell model, while the second part accounts for Brownian motion, which causes the temperature dependence of the effective thermal conductivity. β is related to the particle motion. Based on the investigation of pressure gradients, temperature profiles, and Nusselt numbers, Koo and Kleinstreuer also claimed that addition of 1–4% CuO nanoparticles and high Prandtl number base fluid such as ethylene glycol and oil could significantly increase the heat transfer performance of microheat sinks.

4.2 Viscosity

Neilsen proposed a generalized power law-based model for the relative elastic moduli of composite materials, which is also widely used for relative viscosity. The simplified form of this model is given as [64]

$$\mu_{nf} = (1 + 1.5\varphi_p)e^{\varphi_p/(1-\varphi_m)}\mu_b \quad (6)$$

where φ_m is the maximum packing fraction. For randomly dispersed spheres, the maximum close packing fraction is approximately 0.64. In order to find out the effect of Brownian motion of particles on the viscosity of a statistically homogenous suspension, Batchelor derived the following equation [5]:

$$\mu_{nf} = \left(1 + 2.5\varphi_p + 6.2\varphi_p^2\right)\mu_b \quad (7)$$

Maiga et al. studied the forced convection flow of water–Al_2O_3 and ethylene glycol–Al_2O_3 nanofluids inside a uniformly heated tube that is submitted to a constant and uniform heat flux at the wall and proposed a correlation of the effective viscosity of the nanofluid which is given as follows [55]:

$$\mu_{nf} = \left(1 + 7.3\varphi_p + 123\varphi_p^2\right)\mu_b \tag{8}$$

Koo and Kleinstreuer proposed the following correlation to calculate the effective viscosity of the nanofluid [40].

$$\mu_{Brownian} = 5 \times 10^4 \beta \rho_{bf} \varphi_p \sqrt{\frac{K_B T}{2\rho_p r_p}((-134.63 + 1722.3\varphi_p) + (0.4705 - 6.04\varphi_p)T} \tag{9}$$

where the particle motion is related to empirical parameter

$$\beta = 0.0137\left(100\varphi_p\right)^{-0.8229} \varphi_p < 0.01$$

And $\beta = 0.0011\left(100\varphi_p\right)^{-0.7272} \varphi_p > 0.01$

Kulkarni et al. proposed the following correlation to calculate the effective viscosity of nanofluid where the dependence on the temperature is shown [41].

$$\ln \mu_{nf} = -\left(2.8751 + 53.548\varphi_p - 107.12\varphi_p^2\right) + \left(1078.3 + 15857\varphi_p + 20587\varphi_p^2\right)\left(\frac{1}{T}\right) \tag{10}$$

The dynamic viscosity of nanofluid is given by the following empirical correlation derived by Corcione with 1.84% of standard deviation [14].

$$\frac{\mu_{nf}}{\mu_{bf}} = \frac{1}{1 - 34.87\left(\frac{d_p}{d_{bf}}\right)^{-0.3} \phi^{1.03}} \tag{11}$$

where dbf is the equivalent diameter of the base fluid particle and is given by

$$d_{bf} = \left[\frac{6M}{N\pi\rho_{fo}}\right]^{1/3}$$

Here, M is the molecular weight of the base fluid, N is the Avogadro number, and ρfo is the mass density of the base fluid calculated at room temperature T = 293 K.

4.3 Density and Specific Heat

The calculation of the effective density and specific heat of nanofluid are calculated in a straightforward way, using the physical principle of the mixture rule,

Density of nanofluid, $\rho_{nf} = (\frac{m}{V})_{nf}$

$$= \frac{m_{bf} + m_p}{V_{bf} + V_p}$$

$$= \frac{\rho_{bf} V_{bf} + \rho_p V_p}{V_{bf} + V_p}$$

$$\text{So,} \quad \rho_{nf} = (1 - \varphi_p)\rho_{bf} + \varphi_{bf}\rho_p \tag{12}$$

Specific heat of nanofluid is calculated as follows:

$$\begin{aligned}(\rho c_p)_{nf} &= \rho_{nf}\left(\frac{Q}{m\Delta T}\right)_{nf} \\ &= \rho_{nf}\frac{Q_{bf} + Q_p}{(m_{bf} + m_p)\Delta T} \\ &= \rho_{nf}\frac{(mC_p)_{bf}\Delta T + (mC_p)_p\Delta T}{(m_{bf} + m_p)\Delta T} \\ &= \rho_{nf}\frac{(\rho C_p)_{bf}V_{bf} + (\rho C_p)_p V_p}{\rho_{bf}V_{bf} + \rho_p V_p}\end{aligned}$$

$$\text{So,} \ (\rho c_p)_{nf} = (1 - \varphi_p)(\rho C_p)_{bf} + \varphi_p(\rho C_p)_p \tag{13}$$

5 Utilization of Nanofluid in Solar Thermal Energy Applications

Arthur et al. have reported the experimental works showing the enhancement of thermophysical and rheological properties, e.g., specific heat, thermal conductivity, and viscosity of different molten salt nanofluids that are used in solar thermal energy systems. Majority of the work on high-temperature heat transfer fluid employing nanofluid focused on two main molten salts often referred as "solar salt," lithium carbonate + potassium carbonate (Li_2CO_3–K_2CO_3) and sodium nitrate + potassium nitrate salts ($NaNO_3$–KNO_3). Other works are focused on a wide range of ionic fluids. The data on the literature suggest an increase of 10–30% of enhancement of specific heat for the molten salt nanofluids. As for the thermal conductivity, molten salts with particles having high conductivity and larger specific surface area (e.g., MWCNT, graphene, Al_2O_3 whiskers) show significant enhancement in the property. On the other hand, increasing the concentration of the nanoparticles in the base fluid enhances the thermal conductivity as well as the viscosity [4].

Ebrahimnia-Bajestan et al. have investigated water-based TiO_2 nanofluid heat transfer characteristics both numerically and experimentally for possible application in solar heat exchangers. A maximum of 21% of average heat transfer coefficient was observed for TiO_2 water nanofluids. The results indicate that with the increase of nanoparticle concentration and Reynolds number, the heat transfer coefficient increases, whereas inverse effect was observed for particle size in the nanofluid. The authors have suggested to use smaller nanoparticles with higher thermal conductivity in nanofluids to be employed in solar thermal collectors [20].

Chen et al. have conducted experimental investigation on the effect of gold nanoparticles in nanofluids for enhancing photothermal conversion in direct solar absorption solar collector. Gold nanoparticles enhance the solar light absorption due to the localized surface plasmon resonance effect as compared to the base fluids. The experiment used synthesized gold nanoparticles obtained through seed-mediated method. The use of gold nanoparticle at a low mass fraction ($\sim$0.000008% weight) in water increases the photothermal conversion efficiency in direct absorption solar collector. For a cube-shaped direct absorption solar collector, the use of gold nanoparticles augmented the efficiency by 19.9%, whereas for a flat-shaped solar collector, the value is 21.3%. The size of the gold particles has a negative effect on the photothermal conversion efficiency of flat-shaped solar collectors. However, the effect was not as significant in the case of cube-shaped collectors [9] (Fig. 1).

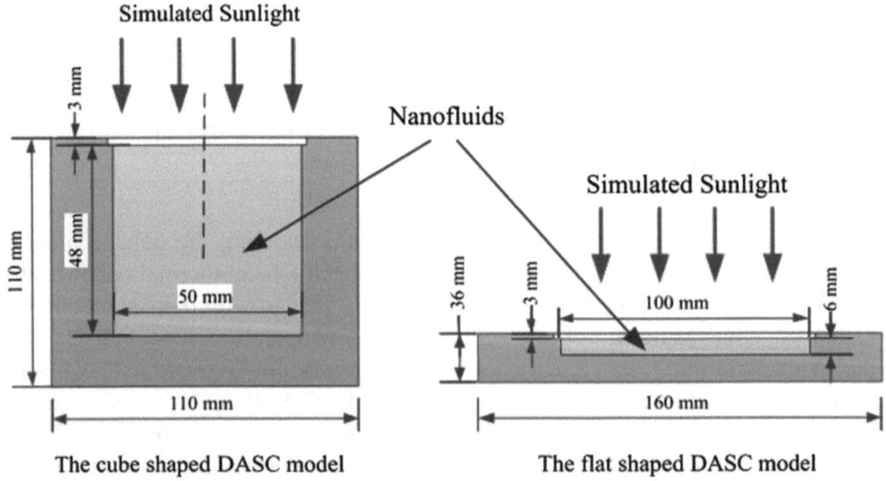

Fig. 1 Schematics of photothermal conversion test equipment (*DASC* direct absorption solar collector) [9]

Colangelo et al. have studied different properties of Al_2O_3–Therminol nanofluids for using as a heat transfer fluid in high-temperature solar energy systems. The authors have investigated the stability, viscosity, thermal conductivity, cluster size, and FT-IR spectra for the nanofluid. They have used oleic acid as a surfactant and studied their effect on the stability of nanofluids. The study revealed that nanofluids prepared without surfactants show greater sedimentation as compared to the nanofluids prepared with surfactant since the presence of surfactants creates bonds between nanoparticles and base fluid resulting in better stability. However, the addition of surfactant has little or no effect on the thermal conductivity of the nanofluid. The viscosity of the nanofluid prepared has increased with the increase of volume fraction of the nanoparticles [13].

Delfani et al. have conducted numerical and experimental investigation on nanofluid-based solar absorption solar collector used for water heating. The authors have presented thermo-optical properties of carboxyl (COOH)-functionalized multiwalled carbon nanotube (MWCNT) in water and ethylene glycol mixture (70%:30% in volume) at different volume fractions. The study reveals that the thermal conductivity of the nanofluids prepared has increased as the volume fraction and the temperature increased. The enhancement in thermal conductivity will increase the heat transfer, which can lead to increased heat transfer and hence an enhancement in collector efficiency. The authors have built prototype of MWCNT nanofluid-based direct absorption solar collector for application in domestic solar heating systems. The effect of internal emissivity, flow rate, and MWCNT nanoparticle volume fraction on the efficiency of the collector is presented in the paper. The arrangement with nanofluid has shown an increase of 29% efficiency as compared to the arrangement with base fluid only with the same flow rate [17] (Fig. 2).

Faizal et al. have reported energy and economic and environmental analysis of four different nanofluids (CuO, SiO_2, TiO_2, and Al_2O_3) for solar thermal collectors. Size reduction of the solar collector is possible due to the higher thermal conductivity. Compared to water, the area of the solar collector can be reduced up to 25.6%, 21.6%, 22.1%, and 21.5% for CuO, SiO_2, TiO_2, and Al_2O_3, respectively. Embodied energy saving is possible due to employing nanofluid in solar collector. The size reduction of the solar collector by using nanofluid leads to cost savings as well [23].

Gorji et al. have reported optical characterization of carboxyl (COOH)-functionalized carbon nanotube aqueous nanofluids for possible application as direct solar absorbers [25]. Gupta et al. have developed a low-temperature Al_2O_3–water nanofluid-based flat plate direct absorption solar collector. By employing alumina nanofluid as a direct absorbing medium, collector efficiency enhancement of 39.6% and 22.1% was observed for 0.005 vol.% and 0.001 vol.%, respectively [27].

Karami et al. have presented the performance study of a prototype of CuO nanofluid-based solar direct absorption collector for residential application of water heating. The prototype solar collector measures 60×60 cm^2 with a channel depth of 1 cm. The solar collector efficiency with CuO nanofluid has increased by 17% as compared to the base fluid at a similar flow rate [35]. In another study, Karami et al.

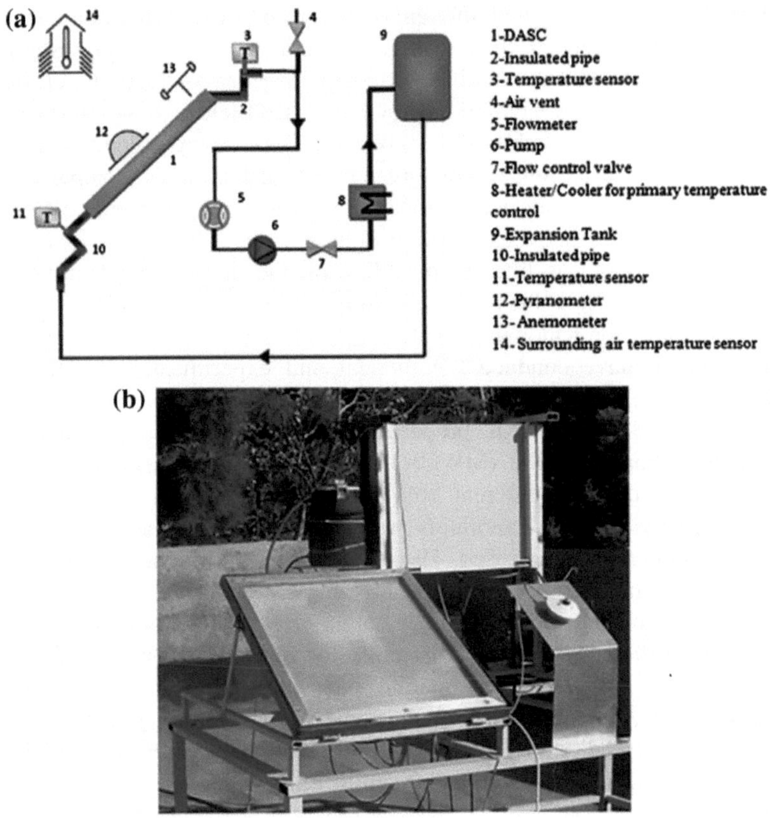

Fig. 2 DASC outdoor performance test **a** schematic of the standard test loop; **b** the photograph of the experimental setup [17]

have investigated the thermo-optical properties of CuO nanofluid (water + EG as base fluid) as the potential working fluid in low-temperature direct absorption solar collector. Even with a low nanoparticle volume concentration, the nanofluid shows greater absorption as compared to the base fluid. The experiments reveal that the dynamic viscosity of the CuO nanofluid decreases with the increase of temperature and increases with the increase in nanoparticle volume concentration. Thermal conductivity was dependent on the nanofluid temperature and volume concentration [36].

Tyagi et al. have investigated the performance of an alumina–water nanofluid-based direct absorption solar collector. In their study, the collector efficiency has increased with the increase of nanoparticle size. Under the same operating conditions, the collector efficiency was 10% higher as compared to collectors using water as the working fluid [98]. Otanicar et al. have studied the effect of different nanoparticles, e.g., carbon nanotubes, silver, graphite, in base fluid on the direct absorption solar collector performance using a microsolar thermal collector. The collector efficiency decreases with the increase of nanoparticle size [69].

Khullar et al. have quantitatively compared nanofluid-based volumetric solar collector to conventional surface-based solar collector by comparing amorphous carbon nanoparticle dispersion in ethylene glycol and multiwalled carbon nanotube (MWCNT) dispersion in distilled water with commercial material as selective solar absorber. The study suggests that the volumetric solar collector with nanoparticle dispersion shows the best performance at an optimum volume fraction [37]. Kim et al. have reported the thermal performance of U-tube solar collector by employing various nanoparticles (MWCNT, Al_2O_3, CuO, SiO_2, and TiO_2) dispersed in PG (propylene glycol)–water as base fluid. All the nanofluids used in the solar collector have increased the thermal efficiency in the following order: MWCNT, CuO, Al_2O_3, TiO_2, and SiO_2 nanofluids [39].

Meibodi et al. have experimentally investigated thermal performance of a SiO_2/EG–water nanofluid-based flat plate solar collector. Despite the low thermal conductivity of SiO_2, this nanoparticle-based nanofluid has the potential to improve the thermal performance of solar collectors. While employing the nanofluids, it is preferable to use low volume fraction since it will reduce the preparation cost and instability problem [58]. Menbari and Alemrajabi have reported the optical properties of a new class of binary nanofluid (Al_2O_3–CuO) as a potential working fluid in direct absorption solar collector. CuO and γAl2O3 are dispersed in water + ethylene glycol. The study suggests that optical coefficients are direct functions of the volume fractions [59].

Rajeb et al. have performed numerical and experimental study on the performance of a photovoltaic thermal (PV-T) nanofluid-based collector. The influence of concentration, types of different nanoparticles (Al_2O_3 and Cu), and different base fluids (water and ethylene glycol) on the electrical and thermal performance of the collector have been studied. The numerical results, validated by the experiments, show that pure water exhibits better performance than ethylene glycol as base fluid. Cu/water nanofluid shows better thermal and electrical efficiency as compared to Cu/ethylene glycol, alumina/water, and alumina/ethylene glycol. The results showed that the thermal and electrical efficiency increases with the increase of nanoparticle concentration [81] (Fig. 3).

Sabiha et al. have performed experimental study to study the thermal performance of an evacuated tube solar collector using single-walled carbon nanotube (SWCNT)-based nanofluid using water as a base fluid. The thermal efficiency of the solar collector was found to be increased with the use of nanofluid. The maximum efficiency was found to be 93.43% for 0.2 vol.% of SWCNT nanofluids at a mass flow rate of 0.025 kg/s. The thermal performance was enhanced with the increase of volume fraction of the nanoparticles and flow rate [82] (Fig. 4).

Said et al. have studied the energy and exergetic efficiency of flat plate solar collector experimentally using pH-treated alumina nanofluid. The study suggested an increase of 83.5% in energy efficiency for 0.3% volume fraction of nanoparticles and mass flow rate of 1.5 kg/min and an increase of 20.3% in exergy efficiency for 0.1% volume fraction of nanoparticles and a mass flow rate of 1 kg/min [84]. In

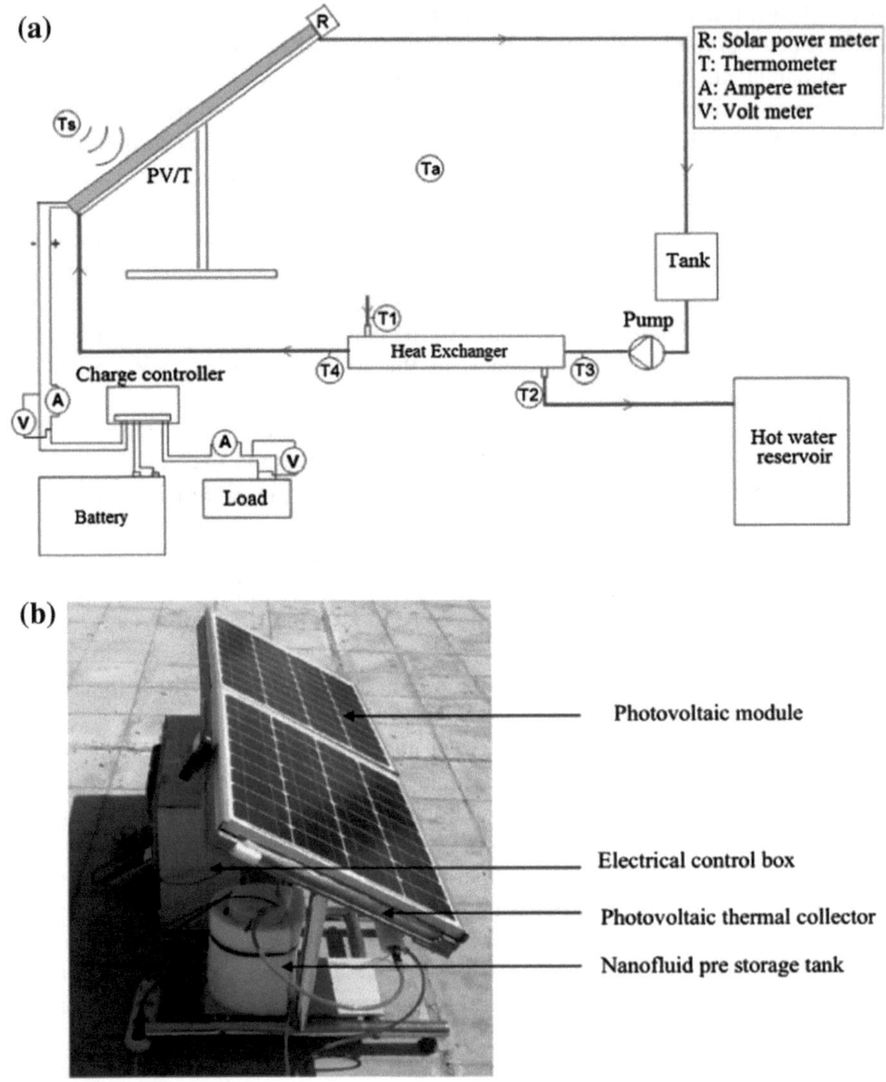

Fig. 3 **a** Schematic diagram and **b** experimental setup of PV/T system used in Rajeb et al. [81]

another study, Said et al. have reported the thermal performance of a TiO_2-based flat plate solar collector with volume fractions of 0.1 and 0.3% and mass flow rate ranging from 0.5 to 1.5 kg/min. PEG 400 dispersant was used reducing the sedimentation of nanoparticles, and its effect on the thermal performance was studied. The energy efficiency of the solar collector employing the nanofluid is increased by 76.6% for 0.1% volume fraction and mass flow rate of 0.5 kg/min, whereas the

Fig. 4 Evacuated solar collector experimental setup used in [82]

highest exergy efficiency obtained is 16.9% for 0.1% volume fraction and mass flow rate of 0.5 kg/min as compared to water as working fluid [83].

Shende and Sundara have investigated the use of N-(rGO-MWNTs), a synthesized mixture of nitrogen-doped hybrid structures of reduced grapheme oxide (rGO) and multiwalled carbon nanotubes (MWNTs) dispersed in water and ethylene glycol as base fluid in direct absorption solar collectors. The study showed that the absorption increases with the increase of particle concentration. The thermal conductivity of the nanofluid increases with the concentration and temperature [89]. Taylor et al. have studied the potential use of different nanoparticles (aluminum, copper, graphite, and silver nanoparticles) dispersed in Therminol VP-1 heat transfer fluid in power tower solar collectors. The results suggest an increase of 10% efficiency as compared to surface-based collectors for solar concentration ratio range of 100–1000 [93] (Fig. 5).

Tong et al. have reported the construction of an enclosed type evacuated U-tube solar collector employing multiwalled carbon nanotube (MWCNT) nanofluid as the working fluid. The experimental results show an increase of 4% efficiency by incorporating the nanofluid in the solar collector [95]. Michael and Iniyan studied the effect of using copper oxide/water (CuO/H_2O) nanofluid as the working fluid on the performance of a 2.08 × 1.05-m flat plate solar collector. A maximum increase of 6.3% in the efficiency was achieved using the CuO/H_2O nanofluid [60].

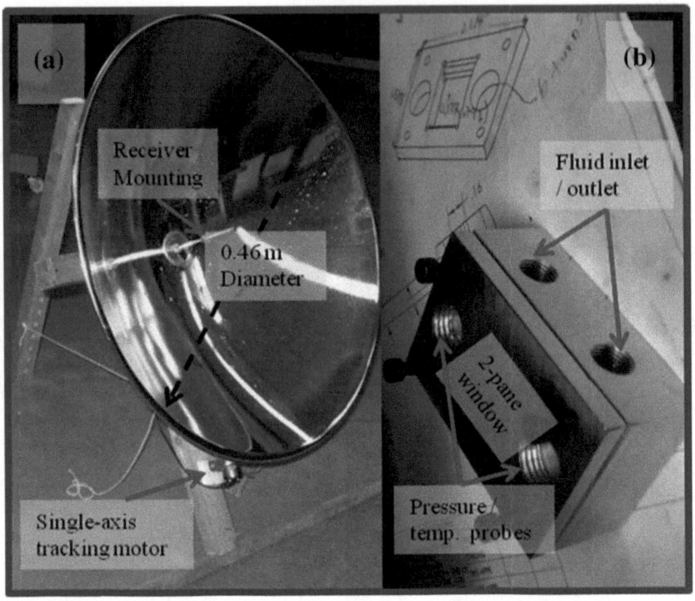

Fig. 5 Experimental scale reflective dish used in the experiments [93]

Tiwari et al. have investigated the effect of Al_2O_3 nanofluid as the working fluid on the performance of a 1 × 2-m flat plate solar collector. The results revealed that there was a maximum increase of 31.64% in the efficiency of the collector using a flow rate of 2 l/min and 1.5% volume fraction of Al_2O_3 as compared to water [94]. Lu et al. analyzed the thermal performance of an open thermosyphon using water-based CuO nanofluids for high-temperature evacuated tubular solar collectors. The study reported that optimum heat transfer augmentation was obtained with 1.2% particle concentration [50].

Liu et al. constructed an evacuated tubular solar air collector integrated with simplified compound parabolic concentrator and open thermosyphon. The authors compared the collector performance with CuO nanofluid or water as working fluid in the thermosyphon. They have remarked that solar collector integrated with open thermosyphon produces better performance [49]. Polvongsri et al. studied the thermal performance of three identical closed-loop flat plate solar collectors employing silver nanofluids as the working fluid. The experiments suggested that the convective heat transfer coefficient of silver nanofluids at 10,000 ppm was twice the value of the water. The results revealed that at high inlet temperatures, nanofluid enhances the thermal efficiency of the solar collectors [78].

Cui and Zhu presented the effect of employing MgO–water nanofluid for a PV/T system. The observations revealed that the output power from the solar cells in PV/T system was reduced with the increase of particle volume fraction and nanofluid film thickness. The results that exhibit the electrical output for a PV system is higher in comparison with PV/T systems; however, the overall efficiency of PV/T systems is more than PV systems [15].

6 Nanofluid Utilization in Thermal Energy Storage

Thermal energy storage is an imperative concept for thermal energy utilization when demand and supply period does not match. The storage allows the energy generation and consumption to be decoupled in terms of time and location. In solar thermal power plants, often the electricity generation and electricity demand are in different periods and thermal energy storage plays a vital role. Also in intermittent cooling applications, e.g., central air-conditioning in buildings, various industrial cooling processes, and electronic cooling applications, cool thermal energy storage is essential due to the intermittent nature. Phase change materials are used as thermal energy storage medium due to their high heat capacity which is stored in the form of latent heat. Nanofluids are often incorporated with these PCMs to improve their heat transfer properties and enhance their specific heat capacity.

Altohamy et al. have experimentally studied a spherical capsule with water-based nanofluid (50 nm Al_2O_3) phase change material during charging process for cool thermal energy application. The results of this study show that there is a substantial effect of alumina nanoparticle concentration on thermal properties of PCM which consecutively decreases complete charging time for all heat transfer fluid volume flow rates and also at different heat transfer fluid inlet temperatures. Also, the addition of nanofluid in the PCM increases the heat transfer coefficient rate [3] (Fig. 6).

Chandrasekaran et al. have studied the heat transfer characteristics of water-based CuO nanofluid PCM (NFPCM) during solidification for cool thermal energy storage system. The experiments show that the NFPCMs exhibit a noteworthy decrease in solidification time due to augmented heat transfer properties. The enhanced heat transfer rate eliminates the problem of subcooling, saves energy, and is beneficial for many cool thermal energy storage systems [7].

On a different study, Chandrasekaran et al. have reported the thermal performance of NFPCM using multiwalled carbon nanotube (MWCNT) nanoparticles in deionized water (DI water) employing pseudomonas as a nucleating agent for cool thermal energy storage. The use of nanofluid in the PCM enhances the heat transfer properties, eliminates undesired subcooling, and accelerates charging. Thus, using nanofluid in PCM will enable energy efficiency design for the cool thermal energy storage systems [8] (Fig. 7).

Cingarapu et al. have presented the thermal and rheological properties of nanofluid consisting of core/shell silica-encapsulated tin (Sn/SiO_2) nanoparticles dispersed in a synthetic HTF Therminol66 (TH66) at loadings up to 5 vol.% as a potential heat transfer fluid in concentrating solar power systems. The incorporation of nanofluid in the HTF has increased the thermal conductivity and improved the total heat adsorption through the latent heat of tin core melting. The addition of the

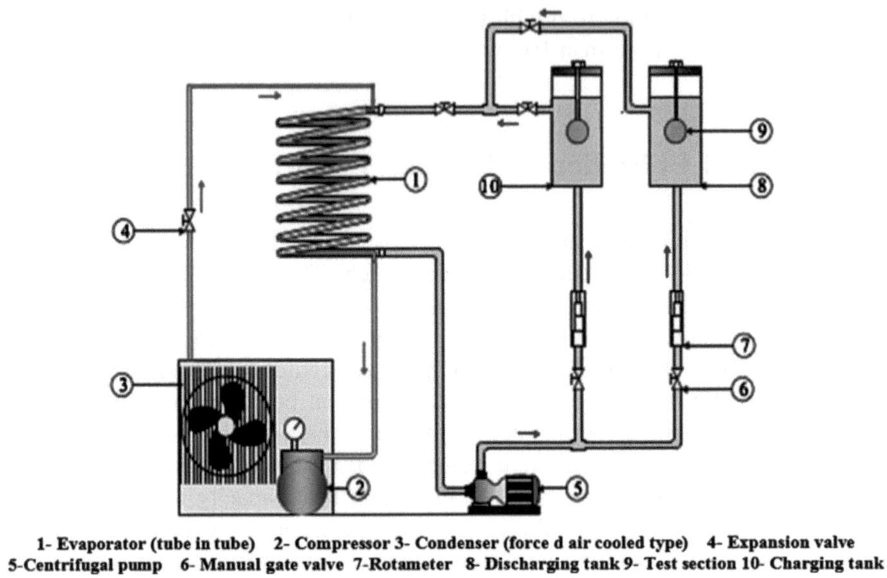

Fig. 6 Schematic diagram of the experimental test rig used in [3]

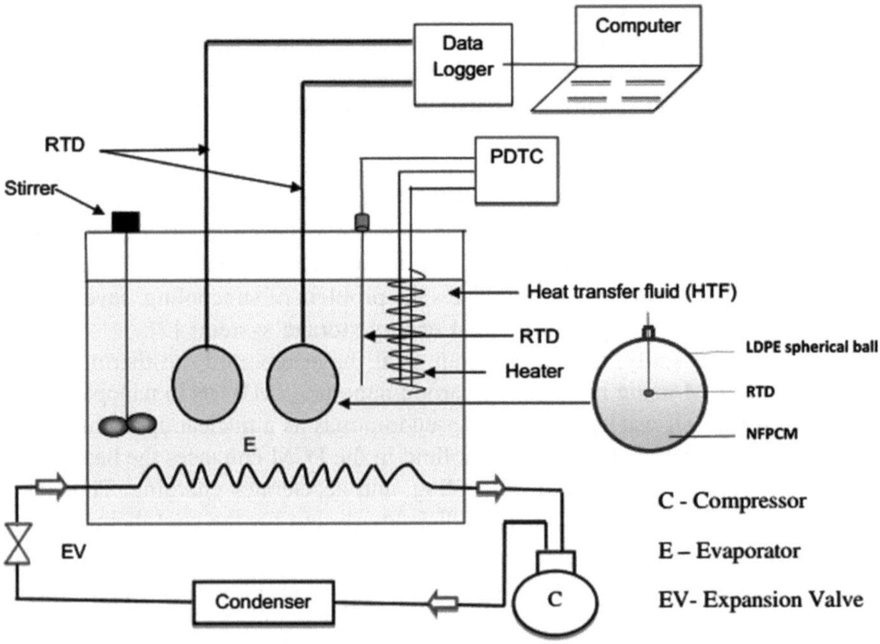

Fig. 7 Experimental setup used in Chandrasekaran et al. [7, 8]

ceramics-encapsulated phase change nanoparticles has enhanced the heat transfer and thermal storage properties [12].

Harikrishnan et al. have investigated CuO–oleic acid nanofluids as a phase change material for the cool thermal energy storage system. The dispersion of nanoparticles in base fluid has significantly enhanced heat transfer rate and thermal conductivity of the PCMs, and the nanofluid showed good thermal stability. As compared to the base fluid, the solidification and melting time of the NFPCM have been reduced by incorporating the nanoparticles. The authors have suggested CuO–oleic acid nanofluids as a potential candidate in cool thermal energy storage due to its thermal properties, thermal and chemical reliability, and augmented heat transfer rate [30] (Fig. 8).

On a different study, Harikrishnan et al. have investigated the phase change behavior of stearic acid–TiO_2 nanofluids as composite phase change materials (PCMs) in solar heating system. The thermal conductivity of the NFPCM was found higher as compared to pure stearic acid. Although the viscosity of the NFPCM was enhanced, it does not have any remarkable negative effect on the solidification and melting process. The charging and discharging time for the PCM was also reduced [31].

He et al. have presented the thermal and rheological characteristics of a new nanofluid-based PCM where TiO_2 nanoparticles were suspended in saturated $BaCl_2$ aqueous solution. Due to the incorporation of the nanoparticles, the thermal conductivity and viscosity of the nanofluid increase. However, the increase of viscosity

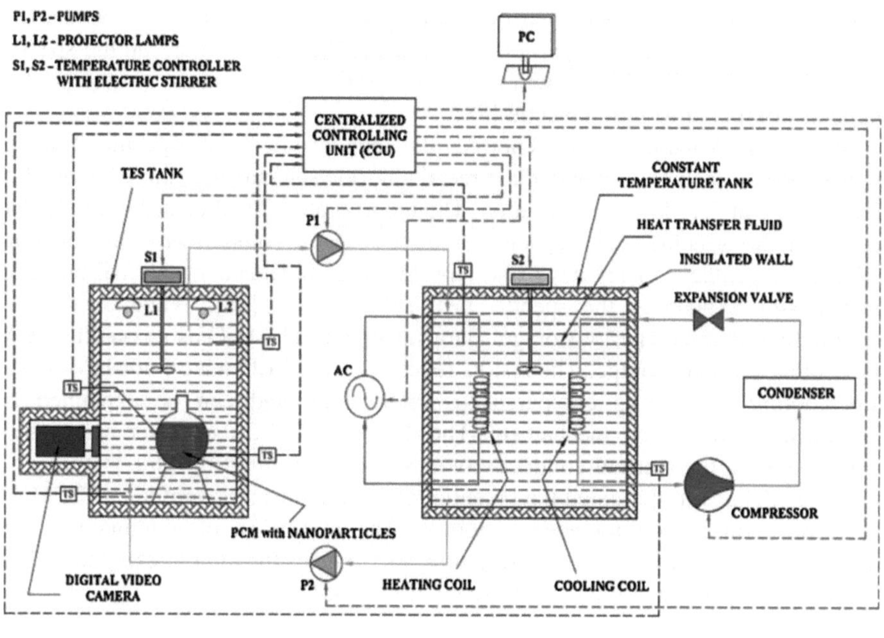

Fig. 8 Experimental setup used by Harikrishnan et al. [30, 31]

does not affect the cool storage system. The latent heat and specific heat slightly decrease with the suspending nanoparticles. The supercooling degree of TiO_2–$BaCl_2$–H_2O nanofluids is enhanced greatly. The authors have suggested this NFPCM as a suitable candidate for the low-temperature energy storage in the industries [32].

Kumarasen et al. have studied the solidification behavior of multiwalled carbon nanotubes (MWCNTs) dispersed in deionized water as a phase change material. A maximum reduction of 14% and 20.1% solidification times was observed using this NFPCM at a surrounding bath temperature of −9 °C and −12°C, respectively. The use of MWCNT-based PCM containing minimum concentration of nanoparticles reduces the undesired subcooling. 6–9% energy saving is possible in cool thermal energy storage by incorporating this NFPCM [42].

Kwak et al. have explored the applicability of SiO_2-molten salt and SiO_2–Therminol VP-1 (conventional heat transfer fluid) in solar thermal energy storage applications especially in concentrating solar power systems. The sensible heat capacity of SiO_2 dispersed in Therminol VP-1 has increased by ~6% as compared to Therminol VP1, whereas silica nanoparticles in alkali metal salt carbonate eutectic have increased the sensible heat capacity by ~17% [43].

Nouranu et al. have investigated the thermal behavior of nanoalumina-dispersed paraffin (using sodium as surfactant) as a potential thermal energy material. The results show that the melting rate of paraffin with alumina is 27% higher as compared to paraffin. The prepared NFPCM showed enhanced thermal conductivity, increased heat transfer rate, and excellent thermal reliability [67]. Schuller et al. have studied the effect of adding alumina nanoparticle to nitrate salt on the specific heat. The experiments conducted on the prepared nitrate–alumina show that the introduction of alumina nanoparticles has a parabolic enhancement on the specific heat values. The NFPCMs show a maximum 30.6% enhancement in specific heat value as compared to pure eutectic [88].

Shin and Banarjee have reported the enhancement of specific heat of high-temperature nanofluid and alkali metal chloride salt eutectics mixed with silica nanoparticle at very low concentration. The incorporation of silica nanoparticles has increased the specific heat capacity of by 14.5% in comparison with the pure chloride salt eutectic. They have also reported three independent thermal transport mechanisms explaining the specific heat enhancement [90]. Wang et al. have explored the thermal energy storage characteristics of Cu–H_2O as a potential PCM for cooling thermal storage. The results reveal that Cu–H_2O nanofluids have lower supercooling degrees as compared to water alone and can be explained by heterogenous nucleation mechanism. The dispersion of Cu in water can significantly increase the thermal conductivity and reduce the freezing time [102].

Wu et al. have investigated the potential of alumina–water nanofluid as PCM for the application in cooling energy storage. The dispersion of alumina in water advances the beginning of the freezing time and reduces the total freezing time. The freezing rate of the nanofluid is also enhanced [105].

7 Application of Nanofluid in Heat Exchangers

Numerous scientific researches have been performed with nanofluid employed in different types of heat exchangers for optimum thermal performance and to reduce the energy consumption for the fabrication of compact heat exchangers in industrial applications [1, 33, 56, 71, 73]. With careful designing, heat exchangers employing nanofluids may be used in geothermal power plants for enhanced heat transfer. Duangthongsuk and Wongwises performed experimental investigation with TiO_2–water nanofluid in a double-tube counterflow heat exchanger to study the improvement in heat transfer coefficient by 6–11% compared to water under turbulent flow condition [18]. Farajollahi et al. conducted experiment with γAl_2O_3–water and TiO_2–water nanofluids employed in a shell and tube heat exchanger and investigated the effects of Peclet number, particle type, and volume fraction on the enhancement of heat transfer [24]. Haghshenas et al. carried out numerical and experimental investigation with concentric tube and plate heat exchanger to explore the effect of ZnO–water nanofluid on the enhancement of thermal efficiency of heat exchanger in terms of convective heat transfer coefficient [28]. Leong et al. numerically investigated the thermal and energy performance of shell and tube heat recovery exchanger using nanofluid-based coolant and observed a maximum 4.53% heat transfer augmentation for laminar flow condition [46].

Gunnasegaran et al. described a numerical model of compact heat exchanger using Cu, diamond, and SiO_2 nanoparticles dispersed in ethylene glycol base fluid and reported the enhancement of thermal performance for designing more efficient and reliable heat exchangers [26]. Pantzali et al. experimented by employing CuO–water nanofluid with 4% volume fraction in a commercial herringbone-type plate heat exchanger, and the efficacy of nanofluid as a coolant has been investigated [72]. The finding suggested that along with the thermophysical properties of nanofluid, the type of flow inside the heat exchanging equipment also affects the performance of nanofluid as a coolant. Zamzamian et al. performed experimental investigation on Al_2O_3–EG and CuO–EG nanofluid double pipe and plate heat exchangers to observe the enhancement of forced convective heat transfer up to 50% under turbulent flow condition [112]. Kannadasan et al. reported an augmentation of convective heat transfer coefficients in horizontal and vertical helically coiled heat exchanger with CuO–water nanofluid with 0.1 and 0.2% volume concentration. The results showed that horizontal and vertical position of heat exchanger had little or no effect on the change of heat transfer and friction factor [34]. Ehsan et al. have reported that heat transfer fluid with 2% volume of nanoparticles in a typical heat exchanger having a heat transfer coefficient of 20,000 W m^{-2} K^{-1} has yielded around 3% heat transfer area reduction and a power advantage of 25–36% [21].

Pandey and Nema reported an experimental analysis of heat transfer and friction factor with Al_2O_3–water nanofluid as coolant with 1–4% volume concentration employed in a counterflow-corrugated plate heat exchanger and finally with the experimental data, proposed a correlation for Nusselt number and friction factor [70]. Elias et al. studied the effect of different nanoparticle shapes (cylindrical, bricks, blades, platelets, and spherical) on the thermal performance of a shell and tube heat exchanger using nanofluid as a mixture of boehmite alumina nanoparticles and mixture of water–ethylene glycol. Among the five different nanoparticle shapes, cylindrical shape showed better performance in terms of heart transfer characteristics [22]. Sonawane et al. showed an improvement in average heat transfer rate with Al_2O_3–water nanofluid as coolant in a concentric tube heat exchanger. Heat transfer characteristics of nanofluid improved with an increase of Reynolds number and volume fraction [91].

8 Application of Nanofluid in Refrigeration and Air-Conditioning System

Mahbubul et al. proposed a numerical model to enhance the thermal performance in refrigeration system by employing Al_2O_3-R141b nanorefrigerant which is one kind of nanofluids with optimum volume concentration [54]. Peng et al. experimented with Cu-R113 nanorefrigerant to explore the effect of nanoparticles on heat transfer inside a horizontal smooth tube, and the results showed a maximum 29.7% improvement in heat transfer coefficient with refrigerant-based nanofluid compare to that of pure refrigerant [75]. The experiment was conducted to investigate the effect of surfactant additives on nucleate pool boiling heat transfer of refrigerant-based nanofluid using surfactants including sodium dodecyl sulfate (SDS), cetyltrimethylammonium bromide (CTAB), and sorbitan monooleate (Span-80) [77]. Saidur et al. reported a review on the performance of nanoparticles suspended with refrigerants and lubricating oils in refrigeration systems, and heat transfer performance of different nanorefrigerants with varying volume concentrations was evaluated [86].

Bi et al. performed experiment to investigate the reliability and reduced energy consumption of a domestic refrigerator with HFC134a-mineral oil with TiO_2 nanoparticles [6]. Liu et al. observed the significant augmentation of system performance on a water chiller system by the dispersion of Cu, CuO, and carbon nanotube (CNT) dispersed in base fluid such as ethylene glycol, water, and synthetic engine oil [48]. Ali et al. examined the effect of Cu-ultrafine particles on heat and mass transfer between air and falling solution film in a cross-flow configuration, and it was reported that at low air Reynolds number, dehumidification and cooling processes were enhanced by the addition of Cu-volume fraction [2]. Peng et al. employed horizontal smooth tube to investigate the effect of nanoparticle on the frictional pressure drop with refrigerant-based nanofluid (CuO-R113 refrigerant) and proposed an empirical correlation [76]. TiO_2 nanoparticles were dispersed in R11 refrigerant for the enhancement of heat pipe efficiency, and efficiency was raised up to 1.4 times than that with pure refrigerant [62].

Pool boiling heat transfer characteristics were examined with TiO_2 nanoparticles dispersed in refrigerant HCFC 141b with 0.01, 0.02, and 0.05% volume concentration. For designing effective and compact NH_3/H_2O absorption refrigeration system, Al_2O_3 and CNT were combined to make binary nanofluids mixed in binary mixture of NH_3/H_2O and significant improvement in heat transfer rate and absorption rate was reported [45]. Absorption performance was found to be increased up to 5.32 times by the addition of both surfactants (2-ethyl-1-hexanol, n-octanol, and 2-octanol) and nanoparticles (Cu, CuO and Al_2O_3) into NH_3/H_2O binary solution during the ammonia bubble absorption process [38]. Yang et al. performed a comprehensive investigation to observe the dispersion stability of suspension, using 20 types of nanoparticles and 10 types of dispersants added in ammonia water. Results revealed that nanoparticles with better dispersive characteristics in dry condition could be easily dispersed in ammonia water [110].

9 Application of Nanofluids in Electronics Cooling

The generation of heat in the microprocessor is a great constrain to the processing rate. A huge amount of heat is generated in the processing unit so removal of heat is necessary, otherwise the chip will burn. The power density of integrated circuit and microprocessor in the recent decade has increased dramatically. Recently, the international technology road map for semiconductors projected that the integrated circuits will contain more than 9.8 billion transistors in a chip area of 280 mm^2, by 2018. So future computers and processors have to dissipate higher power in the range of 100–300 W/cm^2. Conventional air cooling technology has reached to its saturation aptitude, and liquid cooling technologies are being and have been developed for replacing it. Nanofluid, single-phase liquid, and two-phase liquid all are good candidates over air cooling system due to their increased heat transfer capabilities.

Heat pipes are widely used heat spreader used in latest electronic devices such as laptop, desktop, pc. To compare the heat transfer capabilities between deionized water and nanofluid (water-based) in heat pipe, Tsai et al. conducted an experiment [97]. The result showed a significant amount of reduction in thermal resistance of heat pipe with nanofluid compared to water. To investigate heat transport capability, Ma et el. conducted similar type of experiment, and in that case, he used oscillating-type heat pipe [51, 52]. According to his research, for input power of 80 W, 1% volume fraction of nanoparticle-based nanofluid can reduce the temperature difference between evaporator and condenser from 40.9 to 24.3 °C. These results are very positive indeed, and to increase the processing capability of any processor, a good cooling system is required first and nanofluid-based cooling system will be a very good candidate for that one.

Lin et al. investigated nanofluids in pulsating heat pipes by using silver nanoparticles and discovered encouraging results. The silver nanofluid improved heat transfer characteristics of the heat pipes [47].

A new study by Tasciuchas shows that liquids embedded with nanoparticles show superior performance and stability when exposed to electric fields, which could lead to new types of miniature camera lenses, cell phone displays, and other microscale fluidic devices. This study reported that this type of study may open up a new vista for using nanofluids in microscale and nanoscale actuator device applications [96].

The manipulation of small volumes of liquid is critical for fluidic digital display devices, optical devices, and micro-electromechanical system (MEMS) such as lab-on-chip analysis systems. Most research work of such systems has been conducted with regular liquids, but not with nanofluids. Tasciuc's team replaced the conventional fluid with droplets of water-based solutions containing bismuth telluride nanoparticles onto a teflon-coated silicon wafer. They found significant augmentation of the performance of nanofluid in place of base fluid.

Nguyen et al. experimentally inspected the behavior and heat transfer augmentation of an Al_2O_3/water nanofluid flowing inside a closed system that is used for cooling of microelectronic components. Results showed that the insertion of nanoparticles into distilled water produced a substantial augmentation of the cooling block convective heat transfer coefficient. According to his research, for a 6.8 vol.% concentration of nanoparticle, the heat transfer coefficient has been found to increase as much as 40% compared to that of the single base fluid. Experimental results also showed that a nanofluid with 36-nm particle size provides higher convective heat transfer coefficients than the ones given by a nanofluid with 47-nm particles. These positive results are encouraging the nonstop research and development of nanofluids for such applications [63].

Authors also explored microchannel cooling using Al_2O_3/water nanofluids and observed that higher heat transfer coefficients were achieved mostly in the entrance region of microchannel. However, the enhancement was weaker in the fully developed region, proving that nanoparticles have an appreciable effect on thermal boundary layer development. Higher concentrations also produced greater sensitivity to heat flux. Despite this enhancement, the overall cooling efficacy of nanoparticles was quite microscopic because of the large axial temperature rise associated with the decreased specific heat for the nanofluid compared to the base fluid. For two-phase cooling, nanoparticles cause calamitous failure by depositing into large clusters near the channel exit due to localized evaporation once boiling commenced. These and other practical disadvantages bring into question the overall merit of using nanofluids in microchannel heat sinks [100].

10 Application of Nanofluid in Transportation

Mixture of water and ethylene glycol, the universally used automotive coolant, has relatively poor heat transfer characteristic compare to water alone. Another heat transfer medium, engine oil, produces even worse scenario. Addition of nanoparticles with a suitable engine coolant has a great potential to improve cooling rate. Such improvement can be used to take out heat with a reduced cooling system.

Smaller coolant system results in smaller and lighter radiators which lead to increased fuel economy.

A prominent engine coolant is pure ethylene glycol with nanoparticles. If the resulting heat transfer rate of nanofluid approaches the (50% water +50% ethylene glycol) heat transfer rate, there are some other distinct features which make nanofluid more suitable. The most protruding among the advantages is the low-pressure operation of the nanofluid compared to (50% water +50% ethylene glycol) mixture. The nanofluid also features high boiling point which is desirable for maintaining single-phase coolant flow.

Choi et al. revealed that nanofluids have the potential of being recognized as a new generation of coolants for vehicle thermal management due to their ominously higher thermal conductivities than the base fluids. Due to the necessity of more power generation, the heat rejection requirements of automobiles and trucks are frequently increasing [11].

Ollivier et al. numerically inspected the possible application of nanofluids as a jacket water coolant in a gas spark ignition engine. For that purpose, authors accomplished numerical simulations of unsteady heat transfer through the cylinder. Authors reported that because of higher thermal diffusivity of nanofluids, the thermal signal variations for knock detection increased by 15% over that forecast using water alone [68].

The cooling of automatic transmission is also a great challenge for some advanced vehicle where high local temperatures occurred at high rotating speed. A research has been conducted by Tzeng et al. (3) with dispersed CuO and Al_2O_3 nanoparticles into transmission oil. The temperature distribution was measured at four engine operating speeds (400, 800, 1200, and 1600 rpm). The investigation result showed that CuO nanofluid produced the lowest transmission temperatures both at high and at low rotating speed. So, from the thermal advantage view point, the nanofluid has a clear advantage in automatic transmission [99].

The inclination toward higher engine power inevitably leads to larger radiators and increased frontal areas, resulting in supplementary aerodynamic drag and increased fuel consumption. According to Choi, cooling is one of the top technical challenges facing the truck industry [99].

Choi reported the limitations of existing technologies as follows:

- Liquid-side: Traditional coolants and oils have inherently poor heat transfer properties which are not enough for fulfilling our current requirement.
- Air-side: Radiator designs for increasing air-side heat transfer by extending the surface technology have reached to its saturation capacity.
- Therefore, there is a progressively increasing need for new concepts and technology for improving cooling system performance [10].

Choi reported that in USA, a project was initiated to target fuel savings for the HV industry through the development of energy-efficient nanofluids and smaller and lighter radiators. A major goal of the nanofluid scheme is to lessen the size and weight of the HV cooling systems by >10%, thereby increasing fuel efficiency

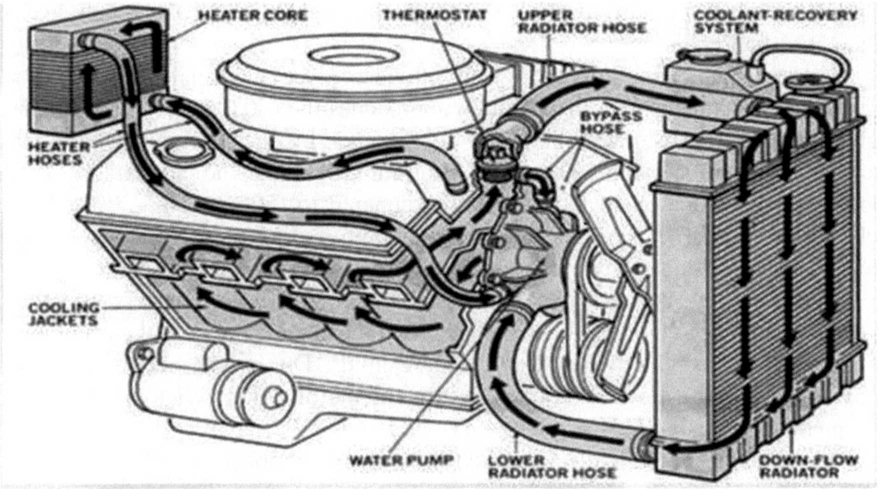

Fig. 9 Radiator section of an engine. Adapted from [10]

by >5%, regardless of the cooling demands of higher power engines. Nanofluids enable the potential to allow higher temperature coolants and higher heat rejection to reduce the radiator size by perhaps 30%. This translates into reduced aerodynamic drag and fluid pumping and fan requirements, leading to perhaps a 10% fuel savings [10] (Fig. 9).

A research has been conducted in automotive lubrication application by Zhang and Que where surface-modified nanoparticles with mineral oil have been used which is reported to be effective in reducing wear and enhancing load-carrying capacity. Such results are encouraging for improving heat transfer rate in transportation system through the use of nanofluid. However, consideration must be given to such factors as particle settling, particle agglomeration, and composition of nanofluid [113].

11 Concluding Remarks

In this chapter, the utilization of nanofluid, colloidal mixture of nanoparticles in base fluid, has been reviewed in various clean energy and energy efficiency applications. Recent developments in this field demonstrate the potential of nanofluid utilization in such applications. Nanofluid is being used in solar collectors as heat transfer fluid for its enhanced thermal and optical properties to obtain improved performance. Nanofluid-based phase change materials are promising in thermal energy systems as energy storage materials. With nanofluid, efficient, reliable, and compact heat exchanger can be designed for applications in geothermal energy harvesting. Nanorefrigerants are used in refrigeration and air-conditioning systems to enhance the thermal performance and reduce energy consumption. Electronics cooling and transportation are some other potential application area where the enhanced thermophysical properties of nanofluids can be utilized. However, challenges associated with nanofluid are present which are needed to overcome for optimal performance of this working fluid.

References

1. Abu-Nada, E., Masoud, Z., & Hijazi, A. (2008). Natural convection heat transfer enhancement in horizontal concentric annuli using nanofluids. *International Communications in Heat and Mass Transfer, 35*(5), 657–665.
2. Ali, A., Vafai, K., & Khaled, A. R. (2004). Analysis of heat and mass transfer between air and falling film in a cross flow configuration. *International Journal of Heat and Mass Transfer, 47*(4), 743–755.
3. Altohamy, A. A., Rabbo, M. A., Sakr, R. Y., & Attia, A. A. (2015). Effect of water based Al_2O_3 nanoparticle PCM on cool storage performance. *Applied Thermal Engineering, 84,* 331–338.
4. Arthur, O., & Karim, M. A. (2016). An investigation into the thermophysical and rheological properties of nanofluids for solar thermal applications. *Renewable and Sustainable Energy Reviews, 31*(55), 739–755.
5. Batchelor, G. K. (1977). The effect of Brownian motion on the bulk stress in a suspension of spherical particles. *Journal of Fluid Mechanics, 83*(01), 97–117.
6. Bi, S. S., Shi, L., & Zhang, L. L. (2008). Application of nanoparticles in domestic refrigerators. *Applied Thermal Engineering, 28*(14), 1834–1843.
7. Chandrasekaran, P., Cheralathan, M., Kumaresan, V., & Velraj, R. (2014). Enhanced heat transfer characteristics of water based copper oxide nanofluid PCM (phase change material) in a spherical capsule during solidification for energy efficient cool thermal storage system. *Energy, 72,* 636–642.
8. Chandrasekaran, P., Cheralathan, M., Kumaresan, V., & Velraj, R. (2014). Solidification behavior of water based nanofluid phase change material with a nucleating agent for cool thermal storage system. *International Journal of Refrigeration, 41,* 157–163.
9. Chen, M., He, Y., Zhu, J., & Kim, D. R. (2016). Enhancement of photo-thermal conversion using gold nanofluids with different particle sizes. *Energy Conversion and Management, 112,* 21–30.
10. Choi, S. (2006, April 18–20). Nano fluids for improved efficiency in cooling systems. In: *Heavy vehicle systems review*. Argonne National Laboratory.
11. Choi, C., Yoo, H. S., & Oh, J. M. (2008). Preparation and heat transfer properties of nanoparticle-in-transformer oil dispersions as advanced energy efficient coolants. *CurrApplPhys, 8,* 710–712.
12. Cingarapu, S., Singh, D., Timofeeva, E. V., & Moravek, M. R. (2014). Nanofluids with encapsulated tin nanoparticles for advanced heat transfer and thermal energy storage. *International Journal of Energy Research, 38*(1), 51–59.
13. Colangelo, G., Favale, E., Miglietta, P., Milanese, M., & de Risi, A. (2016). Thermal conductivity, viscosity and stability of Al_2O_3-diathermic oil nanofluids for solar energy systems. *Energy, 95,* 124–136.
14. Corcione, M. (2010). Heat transfer features of buoyancy-driven nanofluids inside rectangular enclosures differentially heated at the sidewalls. *International Journal of Thermal Sciences, 49*(9), 1536–1546.

15. Cui, Y., & Zhu, Q. (2012). Study of photovoltaic/thermal systems with MgO-water nanofluids flowing over silicon solar cells. In *Power and Energy Engineering Conference (APPEEC), 2012 Asia-Pacific Year*, pp. 1–4.
16. Das, S. K., Putra, N., Thiesen, P., & Roetzel, W. (2003). Temperature dependence of thermal conductivity enhancement for nanofluids. *Journal of Heat Transfer, 125*(4), 567–574.
17. Delfani, S., Karami, M., & Akhavan-Behabadi, M. A. (2016). Performance characteristics of a residential-type direct absorption solar collector using MWCNT nanofluid. *Renewable Energy, 87,* 754–764.
18. Duangthongsuk, W., & Wongwises, S. (2009). Heat transfer enhancement and pressure drop characteristics of TiO2–water nanofluid in a double-tube counter flow heat exchanger. *International Journal of Heat and Mass Transfer, 52*(7), 2059–2067.
19. Eastman, J. A., Choi, S. U. S., Li, S., Yu, W., & Thompson, L. J. (2001). Anomalously increased effective thermal conductivities of ethylene glycol-based nanofluids containing copper nanoparticles. *Applied Physics Letters, 78*(6), 718–720.
20. Ebrahimnia-Bajestan, E., Moghadam, M. C., Niazmand, H., Daungthongsuk, W., & Wongwises, S. (2016). Experimental and numerical investigation of nanofluids heat transfer characteristics for application in solar heat exchangers. *International Journal of Heat and Mass Transfer, 31*(92), 1041–1052.
21. Ehsan, M. M., Noor, S., Salehin, S., Sadrul Islam, A. K. M. (2015). Application of nanofluid in heat exchangers for energy savings. In M. Masud Khan, Nur Md. Sayeed Hassan (Eds.), *Thermofluids modeling for energy efficiency applications* (pp. 73–101). Academic Press Inc., Elsevier.
22. Elias, M. M., Miqdad, M., Mahbubul, I. M., Saidur, R., Kamalisarvestani, M., Sohel, M. R., et al. (2013). Effect of nanoparticle shape on the heat transfer and thermodynamic performance of a shell and tube heat exchanger. *International Communications in Heat and Mass Transfer, 44,* 93–99.
23. Faizal, M., Saidur, R., Mekhilef, S., & Alim, M. A. (2013). Energy, economic and environmental analysis of metal oxides nanofluid for flat-plate solar collector. *Energy Conversion and Management, 76,* 162–168.
24. Farajollahi, B., Etemad, S. G., & Hojjat, M. (2010). Heat transfer of nanofluids in a shell and tube heat exchanger. *International Journal of Heat and Mass Transfer, 53*(1), 12–17.
25. Gorji, T. B., Ranjbar, A. A., & Mirzababaei, S. N. (2015). Optical properties of carboxyl functionalized carbon nanotube aqueous nanofluids as direct solar thermal energy absorbers. *Solar Energy, 119,* 332–342.
26. Gunnasegaran, P., Shuaib, N. H., Abdul Jalal, M. F., & Sandhita, E. (2012). *Numerical study of fluid dynamic and heat transfer in a compact heat exchanger using nanofluids*. ISRN Mechanical Engineering.
27. Gupta, H. K., Agrawal, G. D., & Mathur, J. (2015). An experimental investigation of a low temperature Al_2O_3-H_2O nanofluid based direct absorption solar collector. *Solar Energy, 118,* 390–396.
28. Haghshenas, F. M., Talaie, M. R., & Nasr, S. (2011). Numerical and experimental investigation of heat transfer of ZnO/water nanofluid in the concentric tube and plate heat exchangers. *Thermal Science, 15*(1), 183–194.
29. Hamilton, R. L., & Crosser, O. K. (1962). Thermal conductivity of heterogeneous two-component systems. *Industrial and Engineering Chemistry Fundamentals, 1*(3), 187–191.
30. Harikrishnan, S., & Kalaiselvam, S. (2012). Preparation and thermal characteristics of CuO–oleic acid nanofluids as a phase change material. *Thermochimica Acta, 533,* 46–55.
31. Harikrishnan, S., Magesh, S., & Kalaiselvam, S. (2013). Preparation and thermal energy storage behaviour of stearic acid–TiO_2 nanofluids as a phase change material for solar heating systems. *Thermochimica Acta, 565,* 137–145.
32. He, Q., Wang, S., Tong, M., & Liu, Y. (2012). Experimental study on thermophysical properties of nanofluids as phase-change material (PCM) in low temperature cool storage. *Energy Conversion and Management, 64,* 199–205.

33. Jung, J. Y., Oh, H. S., & Kwak, H. Y. (2009). Forced convective heat transfer of nanofluids in microchannels. *International Journal of Heat and Mass Transfer, 52*(1), 466–472.
34. Kannadasan, N., Ramanathan, K., & Suresh, S. (2012). Comparison of heat transfer and pressure drop in horizontal and vertical helically coiled heat exchanger with CuO/water based nano fluids. *Experimental Thermal and Fluid Science, 42,* 64–70.
35. Karami, M., Akhavan-Bahabadi, M. A., Delfani, S., & Raisee, M. (2015). Experimental investigation of CuO nanofluid-based direct absorption solar collector for residential applications. *Renewable and Sustainable Energy Reviews, 52,* 793–801.
36. Karami, M., Akhavan-Behabadi, M. A., Dehkordi, M. R., & Delfani, S. (2016). Thermo-optical properties of copper oxide nanofluids for direct absorption of solar radiation. *Solar Energy Materials and Solar Cells, 144,* 136–142.
37. Khullar, V., Tyagi, H., Hordy, N., Otanicar, T. P., Hewakuruppu, Y., Modi, P., et al. (2014). Harvesting solar thermal energy through nanofluid-based volumetric absorption systems. *International Journal of Heat and Mass Transfer, 77,* 377–384.
38. Kim, J. K., Jung, J. Y., & Kang, Y. T. (2007). Absorption performance enhancement by nano-particles and chemical surfactants in binary nanofluids. *International Journal of Refrigeration, 30*(1), 50–57.
39. Kim, H., Ham, J., Park, C., & Cho, H. (2016). Theoretical investigation of the efficiency of a U-tube solar collector using various nanofluids. *Energy, 94,* 497–507.
40. Koo, J., & Kleinstreuer, C. (2004). A new thermal conductivity model for nanofluids. *Journal of Nanoparticle Research, 6*(6), 577–588.
41. Kulkarni, D. P., Das, D. K., & Chukwu, G. A. (2006). Temperature dependent rheological property of copper oxide nanoparticles suspension (nanofluid). *Journal of Nanoscience and Nanotechnology, 6*(4), 1150–1154.
42. Kumaresan, V., Chandrasekaran, P., Nanda, M., Maini, A. K., & Velraj, R. (2013). Role of PCM based nanofluids for energy efficient cool thermal storage system. *International Journal of Refrigeration, 36*(6), 1641–1647.
43. Kwak, H. E., Shin, D., & Banerjee, D. (2010, January). Enhanced sensible heat capacity of molten salt and conventional heat transfer fluid based nanofluid for solar thermal energy storage application. In *ASME 2010 4th International Conference on Energy Sustainability* (pp. 735–739). American Society of Mechanical Engineers.
44. Lee, S., Choi, S. S., Li, S. A., & Eastman, J. A. (1999). Measuring thermal conductivity of fluids containing oxide nanoparticles. *Journal of Heat Transfer, 121*(2), 280–289.
45. Lee, J. K., Koo, J., Hong, H., & Kang, Y. T. (2010). The effects of nanoparticles on absorption heat and mass transfer performance in NH_3/H_2O binary nanofluids. *International Journal of Refrigeration, 33*(2), 269–275.
46. Leong, K. Y., Saidur, R., Mahlia, T., & Yau, Y. (2012). Modeling of shell and tube heat recovery exchanger operated with nanofluid based coolants. *International Journal of Heat and Mass Transfer, 55*(4), 808–816.
47. Lin, Y.-H., Kang, S.-W., & Chen, H.-L. (2008). Effect of silver nano-fluid on pulsating heat pipe thermal performance. *Applied Thermal Engineering, 28*(11–12), 1312–1317.
48. Liu, M., Lin, M. C., & Wang, C. (2011). Enhancements of thermal conductivities with Cu, CuO, and carbon nanotube nanofluids and application of MWNT/water nanofluid on a water chiller system. *Nanoscale Research Letters, 6*(1), 1–13.
49. Liu, Z. H., Hu, R. L., Lu, L., Zhao, F., & Xiao, H. S. (2013). Thermal performance of an open thermosyphon using nanofluid for evacuated tubular high temperature air solar collector. *Energy Conversion and Management, 73,* 135–143.
50. Lu, L., Liu, Z. H., & Xiao, H. S. (2011). Thermal performance of an open thermosyphon using nanofluids for high-temperature evacuated tubular solar collectors: Part 1: Indoor experiment. *Solar Energy, 85*(2), 379–387.
51. Ma, H. B., Wilson, C., Borgmeyer, B., et al. (2006). Effect of nano fluid on the heat transport capability in an oscillating heat pipe. *Applied Physics Letters, 88*(14), Article ID 143116, 3.

52. Ma, H. B., Wilson, C., Yu, Q., Park, K., Choi, U. S., & Tirumala, M. (2006). An experimental investigation of heat transport capability in a nano fluid oscillating heat pipe. *Journal of Heat Transfer, 128*(11), 1213–1216.
53. Mahbubul, I. M., Saidur, R., & Amalina, M. A. (2012). Latest developments on the viscosity of nanofluids. *International Journal of Heat and Mass Transfer, 55*(4), 874–885.
54. Mahbubul, I. M., Saidur, R., & Amalina, M. A. (2013). Thermal conductivity, viscosity and density of R141b refrigerant based nanofluid. *Procedia Engineering, 56,* 310–315.
55. Maïga, S. E. B., Nguyen, C. T., Galanis, N., & Roy, G. (2004). Heat transfer behaviours of nanofluids in a uniformly heated tube. *Superlattices and Microstructures, 35*(3), 543–557.
56. Mapa, L. B., &Mazhar, S. (2005, April). Heat transfer in mini heat exchanger using nanofluids. In *Conference of American Society for Engineering Education, DeKalb, IL* (pp. 1–2).
57. Maxwell, J. C. (1881). *A treatise on electricity and magnetism* (Vol. 1). Clarendon Press.
58. Meibodi, S. S., Kianifar, A., Niazmand, H., Mahian, O., & Wongwises, S. (2015). Experimental investigation on the thermal efficiency and performance characteristics of a flat plate solar collector using SiO_2/EG–water nanofluids. *International Communications in Heat and Mass Transfer, 65,* 71–75.
59. Menbari, A., & Alemrajabi, A. A. (2016). Analytical modeling and experimental investigation on optical properties of new class of nanofluids (Al_2O_3–CuO binary nanofluids) for direct absorption solar thermal energy. *Optical Materials, 52,* 116–125.
60. Michael, J. J., & Iniyan, S. (2015). Performance of copper oxide/water nanofluid in a flat plate solar water heater under natural and forced circulations. *Energy Conversion and Management, 95,* 160–169.
61. Mooney, M. (1951). The viscosity of a concentrated suspension of spherical particles. *Journal of Colloid Science, 6*(2), 162–170.
62. Naphon, P., Thongkum, D., & Assadamongkol, P. (2009). Heat pipe efficiency enhancement with refrigerant–nanoparticles mixtures. *Energy Conversion and Management, 50*(3), 772–776.
63. Nguyen, C. T., Roy, G., Gauthier, C., & Galanis, N. (2007). Heat transfer enhancement using aluminum oxide–water Nano fluid for an electronic liquid cooling system. *Applied Thermal Engineering, 27*(8–9), 1501–1506.
64. Nielsen, L. E. (1970). Generalized equation for the elastic moduli of composite materials. *Journal of Applied Physics, 41*(11), 4626–4627.
65. Noor, S., Ehsan, M. M., Mayeed, M. S., & Islam, A. S. (2014, November). Study of convective heat transfer for turbulent flow of nanofluids through corrugated channels. In *ASME 2014 International Mechanical Engineering Congress and Exposition, American Society of Mechanical Engineers.*
66. Noor, S., Ehsan, M. M., Salehin, S., & Islam, Sadrul, A. K. M. (2014). Heat transfer and pumping power using nanofluid in a corrugated tube. In *Proceedings of AFMC 2014 Conference*. Heat Transfer, 8, 11.
67. Nourani, M., Hamdami, N., Keramat, J., Moheb, A., & Shahedi, M. (2016). Thermal behavior of paraffin-nano-Al_2O_3 stabilized by sodium stearoyllactylate as a stable phase change material with high thermal conductivity. *Renewable Energy, 88,* 474–482.
68. Ollivier, E., Bellettre, J., Tazerout, M., & Roy, G. C. (2006). Detection of knock occurrence in a gas SI engine from a heat transfer analysis. *Energy Conversion and Management, 47*(7–8), 879–893.
69. Otanicar, T. P., Phelan, P. E., Prasher, R. S., Rosengarten, G., & Taylor, R. A. (2010). Nanofluid-based direct absorption solar collector. *Journal of Renewable and Sustainable Energy, 2*(3), 033102.
70. Pandey, S. D., & Nema, V. K. (2012). Experimental analysis of heat transfer and friction factor of nanofluid as a coolant in a corrugated plate heat exchanger. *Experimental Thermal and Fluid Science, 38,* 248–256.

71. Pantzali, M. N., Kanaris, A. G., Antoniadis, K. D., Mouza, A. A., & Paras, S. V. (2009). Effect of nanofluids on the performance of a miniature plate heat exchanger with modulated surface. *International Journal of Heat and Fluid Flow, 30*(4), 691–699.
72. Pantzali, M. N., Mouza, A. A., & Paras, S. V. (2009). Investigating the efficacy of nanofluids as coolants in plate heat exchangers (PHE). *Chemical Engineering Science, 64*(14), 3290–3300.
73. Parvin, S., Nasrin, R., Alim, M. A., Hossain, N. F., & Chamkha, A. J. (2012). Thermal conductivity variation on natural convection flow of water–alumina nanofluid in an annulus. *International Journal of Heat and Mass Transfer, 55*(19), 5268–5274.
74. Paul, G., Philip, J., Raj, B., Das, P. K., & Manna, I. (2011). Synthesis, characterization, and thermal property measurement of nano-Al 95 Zn 05 dispersed nanofluid prepared by a two-step process. *International Journal of Heat and Mass Transfer, 54*(15), 3783–3788.
75. Peng, H., Ding, G., Jiang, W., Hu, H., & Gao, Y. (2009). Heat transfer characteristics of refrigerant-based nanofluid flow boiling inside a horizontal smooth tube. *International Journal of Refrigeration, 32*(6), 1259–1270.
76. Peng, H., Ding, G., Jiang, W., Hu, H., & Gao, Y. (2009). Measurement and correlation of frictional pressure drop of refrigerant-based nanofluid flow boiling inside a horizontal smooth tube. *International Journal of Refrigeration, 32*(7), 1756–1764.
77. Peng, H., Ding, G., & Hu, H. (2011). Effect of surfactant additives on nucleate pool boiling heat transfer of refrigerant-based nanofluid. *Experimental Thermal and Fluid Science, 35*(6), 960–970.
78. Polvongsri, S., & Kiatsiriroat, T. (2014). Performance analysis of flat-plate solar collector having silver nanofluid as a working fluid. *Heat Transfer Engineering, 35*(13), 1183–1191.
79. Prabhat, N. (2010). *Critical evaluation of anomalous thermal conductivity and convective heat transfer enhancement in nanofluids*. Doctoral dissertation, Massachusetts Institute of Technology.
80. Prasher, R., Bhattacharya, P., & Phelan, P. E. (2005). Thermal conductivity of nanoscale colloidal solutions (nanofluids). *Physical Review Letters, 94*(2), 025901.
81. Rejeb, O., Sardarabadi, M., Ménézo, C., Passandideh-Fard, M., Dhaou, M. H., & Jemni, A. (2016). Numerical and model validation of uncovered nanofluid sheet and tube type photovoltaic thermal solar system. *Energy Conversion and Management, 110,* 367–377.
82. Sabiha, M. A., Saidur, R., Hassani, S., Said, Z., & Mekhilef, S. (2015). Energy performance of an evacuated tube solar collector using single walled carbon nanotubes nanofluids. *Energy Conversion and Management, 105,* 1377–1388.
83. Said, Z., Sabiha, M. A., Saidur, R., Hepbasli, A., Rahim, N. A., Mekhilef, S., et al. (2015). Performance enhancement of a flat plate solar collector using titanium dioxide nanofluid and polyethylene glycol dispersant. *Journal of Cleaner Production, 92,* 343–353.
84. Said, Z., Saidur, R., Sabiha, M. A., Hepbasli, A., & Rahim, N. A. (2016). Energy and exergy efficiency of a flat plate solar collector using pH treated Al_2O_3 nanofluid. *Journal of Cleaner Production, 112,* 3915–3926.
85. Saidur, R., Kazi, S. N., Hossain, M. S., Rahman, M. M., & Mohammed, H. A. (2011). A review on the performance of nanoparticles suspended with refrigerants and lubricating oils in refrigeration systems. *Renewable and Sustainable Energy Reviews, 15*(1), 310–323.
86. Saidur, R., Leong, K. Y., & Mohammad, H. A. (2011). A review on applications and challenges of nanofluids. *Renewable and Sustainable Energy Reviews, 15*(3), 1646–1668.
87. Sattler, K. D. (Ed.). (2010). *Handbook of nanophysics: Nanoparticles and quantum dots.* CRC Press.
88. Schuller, M., Shao, Q., & Lalk, T. (2015). Experimental investigation of the specific heat of a nitrate–alumina nanofluid for solar thermal energy storage systems. *International Journal of Thermal Sciences, 91,* 142–145.
89. Shende, R., & Sundara, R. (2015). Nitrogen doped hybrid carbon based composite dispersed nanofluids as working fluid for low-temperature direct absorption solar collectors. *Solar Energy Materials and Solar Cells, 140,* 9–16.

90. Shin, D., & Banerjee, D. (2011). Enhancement of specific heat capacity of high-temperature silica-nanofluids synthesized in alkali chloride salt eutectics for solar thermal-energy storage applications. *International Journal of Heat and Mass Transfer, 54*(5), 1064–1070.
91. Sonawane, S. S., Khedkar, R. S., & Wasewar, K. L. (2013). Study on concentric tube heat exchanger heat transfer performance using Al_2O_3–water based nanofluids. *International Communications in Heat and Mass Transfer, 49,* 60–68.
92. Taylor, R. (2011). *Thermal energy conversion in nanofluids*. Doctoral dissertation, Arizona State University.
93. Taylor, R. A., Phelan, P. E., Otanicar, T. P., Walker, C. A., Nguyen, M., Trimble, S., et al. (2011). Applicability of nanofluids in high flux solar collectors. *Journal of Renewable and Sustainable Energy, 3*(2), 023104.
94. Tiwari, A. K., Ghosh, P., & Sarkar, J. (2013). Solar water heating using nanofluids—A comprehensive overview and environmental impact analysis. *International Journal of Emerging Technology and Advanced Engineering, 3*(3), 221–224.
95. Tong, Y., Kim, J., & Cho, H. (2015). Effects of thermal performance of enclosed-type evacuated U-tube solar collector with multi-walled carbon nanotube/water nanofluid. *Renewable Energy, 83,* 463–473.
96. Topnews. (2009). Retrieved Novomber 7, 2009, from http://www.topnews.in/nanofluids-be-used-make-new-types-cameras-microdevices-and-displays-221378.
97. Tsai, C. Y., Chien, H. T., Dieng, P. P., Chan, B., Luh, T. Y., & Chen, P. H. (2004). Effect of structural character of Gold Nano particles in Nano fluid on heat pipe thermal performance. *Materials Letters, 58*, 1461–1465.
98. Tyagi, H., Phelan, P., & Prasher, R. (2009). Predicted efficiency of a low-temperature nanofluid-based direct absorption solar collector. *Journal of Solar Energy Engineering, 131* (4), 041004.
99. Tzeng, S. C., Lin, C. W., & Huang, K. D. (2005). Heat transfer enhancement of nano fluids in rotary blade coupling of four-wheel-drive vehicles. *Acta Mechanica, 179*(1–2), 11–23.
100. Wang, X. Q., & Mujumdar, A. S. (2008). A review on nanofluids—Part II: Experiments and applications. *Brazilian Journal of Chemical Engineering, 25*(4), 631–648.
101. Wang, X., Xu, X., Choi, S., & U, S. (1999). Thermal conductivity of nanoparticle-fluid mixture. *Journal of Thermophysics and Heat Transfer, 13*(4), 474–480.
102. Wang, X. J., Li, X. F., Xu, Y. H., & Zhu, D. S. (2014). Thermal energy storage characteristics of Cu–H_2O nanofluids. *Energy, 78,* 212–217.
103. Wen, D., & Ding, Y. (2005). Formulation of nanofluids for natural convective heat transfer applications. *International Journal of Heat and Fluid Flow, 26*(6), 855–864.
104. Wen, D., Lin, G., Vafaei, S., & Zhang, K. (2009). Review of nanofluids for heat transfer applications. *Particuology, 7*(2), 141–150.
105. Wu, S., Zhu, D., Li, X., Li, H., & Lei, J. (2009). Thermal energy storage behavior of Al_2O_3–H_2O nanofluids. *Thermochimica Acta, 483*(1), 73–77.
106. Xie, H., Wang, J., Xi, T., Liu, Y., & Ai, F. (2002). Dependence of the thermal conductivity of nanoparticle-fluid mixture on the base fluid. *Journal of Materials Science Letters, 21*(19), 1469–1471.
107. Xie, H., Wang, J., Xi, T., Liu, Y., Ai, F., & Wu, Q. (2002). Thermal conductivity enhancement of suspensions containing nanosized alumina particles. *Journal of Applied Physics, 91*(7), 4568–4572.
108. Xuan, Y., & Li, Q. (2000). Heat transfer enhancement of nanofluids. *International Journal of Heat and Fluid Flow, 21*(1), 58–64.
109. Xue, Q. Z. (2003). Model for effective thermal conductivity of nanofluids. *Physics Letters A, 307*(5), 313–317.
110. Yang, L., Du, K., Bao, S., & Wu, Y. (2012). Investigations of selection of nanofluid applied to the ammonia absorption refrigeration system. *International Journal of Refrigeration, 35* (8), 2248–2260.

111. Yu, W., & Choi, S. U. S. (2004). The role of interfacial layers in the enhanced thermal conductivity of nanofluids: A renovated Hamilton-Crosser model. *Journal of Nanoparticle Research, 6*(4), 355–361.
112. Zamzamian, A., Oskouie, S. N., Doosthoseini, A., Joneidi, A., & Pazouki, M. (2011). Experimental investigation of forced convective heat transfer coefficient in nanofluids of Al_2O_3/EG and CuO/EG in a double pipe and plate heat exchangers under turbulent flow. *Experimental Thermal and Fluid Science, 35*(3), 495–502.
113. Zhang, Z., & Que, Q. (1997). Synthesis structure and lubricating properties of dialkyldithiophosphate-modified Mo-S compound nano clusters. *Wear, 209,* 8–12.
114. Zhu, H. T., Lin, Y. S., & Yin, Y. S. (2004). A novel one-step chemical method for preparation of copper nanofluids. *Journal of Colloid and Interface Science, 277*(1), 100–103.

Gaseous and Particle Emissions from a Compression Ignition Engine Fueled with Biodiesel–Diesel Blends

M. Mofijur, M.G. Rasul, N.M.S. Hassan, M.M.K. Khan and H.K. Rashedul

Abstract This chapter investigates the sustainability of rice bran biodiesel from environmental point of view. In this study, 5 and 20% biodiesel was tested in a naturally aspirated four-stroke multi-cylinder diesel engine at different load and speed conditions. It was found that all biodiesel blended fuel reduces the brake power (BP) and increases brake specific fuel consumption (BSFC) slightly than diesel fuel. Engine emission results indicated that blended fuel reduces the average particulate matter (PM), carbon monoxide (CO), and hydrocarbons (HC) except nitric oxides (NO) emissions than diesel fuel. Finally, it can be concluded that up to 20% rice bran biodiesel could replace diesel fuel to help in controlling the air pollution to a great extent without sacrificing engine power significantly.

1 Introduction

Diesel engines are widely used in the transportation sector as a source of power due to their higher efficiency [1]. Despite the advantages, the exhaust emission of the diesel engines has received lots of attention since the first CI engine emissions standards were introduced by the United States Environmental Protection Agency (USEPA) in 1988 [2]. The small fraction of unburnt hydrocarbon fuel in diesel engines produces a number of incomplete combustion products that directly affect human health, urban air quality, and global climate. The main particulate fraction of

M. Mofijur (✉) · M.G. Rasul · M.M.K. Khan
School of Engineering & Technology, Central Queensland University, Rockhampton, QLD 4702, Australia
e-mail: m.rahman@cqu.edu.au

H.K. Rashedul
Department of Mechanical Engineering, University Malaya, 50603 Kuala Lumpur, Malaysia

N.M.S. Hassan
School of Engineering & Technology, Central Queensland University, Cairns, QLD 4870, Australia

M.M.K. Khan et al. (eds.), *Application of Thermo-fluid Processes in Energy Systems*, Green Energy and Technology, https://doi.org/10.1007/978-981-10-0697-5_2

diesel exhaust contains fine particles, and due to their modest size, these molecules can penetrate deep into the lungs [3].

It has been reported that "vehicle emissions account for some 65% of urban air pollution" [4] which causes more than 25,000 chronic bronchitis in adults and more than 290,000 episodes of bronchitis (children); more than 0.5 million asthma attacks. Australia's air pollution death toll is higher than fatalities from road accidents. "Each year, on average, 2400 of the 140,000 Australians deaths are linked to air quality and its health related issues—much more than the 1700 people who die on Australian roads. That is an average of a death every 4 h. This number increases if long-term effects of air toxics on cancer are included" [5].

Exhaust pollution includes coarse, fine and ultra-fine particles, gaseous irritants, and polycyclic aromatic hydrocarbons (PAHs). Different emissions have different effect on human health. For example, the PM emission causes lung cancer and cardiopulmonary deaths, NOx emission irritates the lungs and cause edema, bronchitis, and pneumonia; and result in increased sensitivity to dust and pollen in asthmatics, CO emission affects fatal growth in pregnant women and tissue development of young children, HC emission has a synergistic action with other pollutants to promote morbidity in people with respiratory or circulatory problems, and PAHs cause eye irritation, coughing and sneezing, drowsiness, and symptoms akin to drunkenness. Some hydrocarbons have a close affinity for diesel particulates and may contribute to lung disease [6].

Therefore, it is imperative to find the ways to mitigate the air pollution by lowering the exhaust gasses from diesel engines. For this reason, researchers around the world have become concerned about the reduction of exhaust gasses emission. Due to the environmental impact as well as the depletion tendency, several alternative sources have been proposed as potential replacement [7]. The most promising option so far seen is the use of renewable fuel such as biodiesel. Biodiesel has several advantages as a renewable energy resource, i.e., biodiesel is safer to handle, less toxic, biodegradable, and a higher flash point than conventional fossil fuel. Biodiesel has decreased the dependency on crude oil and reduced environmental pollution [8]. It has an ideal combustion-emission profile compared to diesel, such as low emission of carbon dioxide (CO_2), particulate matter (PM), and unburned hydrocarbon (HC) [9]. Commonly biodiesel is produced from vegetable oils [10, 11]. Though the biodiesel fuels have a good number of benefits as a future bioenergy, serious concerns have been raised about the sustainability of future bioenergy development. Sustainability involves economic, environmental, and social issues. This study focuses on the environmental issues of a particular biodiesel.

In particular, the study investigates the sustainability of rice bran biodiesel from the environmental aspects. In this present investigation, 20% rice bran biodiesel was mixed with 80% diesel fuel to use in diesel engine as the researchers believe that the introduction of mandatory biofuel blends (petrol with 10% ethanol and diesel with 20% biodiesel) would reduce the negative environment and health impacts [3].

2 Literature Review

2.1 Description of Rice Bran Oil

Rice bran is the outer layer of the husked rice kernel that is obtained at the time of milling to produce polished rice. Rice bran comprises pericarp, tegmen, aleurone, and sub-aleurone. The oil contents of rice bran are 15–23% [12]. The crude rice bran oil is dark in color, and the triglyceride contents are relatively lower compared to other vegetable oils. Rice bran mostly was used as a cattle feed, and the produced oil was used for industrial purposes. Rice bran oil is a non-conventional oil, and among the non-conventional oils, it is cheaper and readily available. So the production of biodiesel from crude rice bran oil is also cheaper compared to other traditional oils [13].

2.2 Emission Study of Different Biodiesel in Diesel Engine

Labeckas and Slavinskas [14] found that 5–30% rapeseed biodiesel reduces the unburned hydrocarbon significantly at 1400 and 2200 rpm. The 5, 10, 20, and 30% Jatropha biodiesel blends lowered the HC, CO, CO_2, and smoke density significantly than diesel fuel [15]. Rajaraman et al. [16] found that (20–100%) Moringa biodiesel blend combustion in a direct injection diesel engine produces lower HC, CO, PM emission, but higher NOx emission than diesel fuel. Pure Jatropha biodiesel and preheated Jatropha biodiesel produce less engine emission except for NOx emission than diesel fuel due to the presence of higher unsaturated fatty acids of Jatropha biodiesel [17]. It has been found that 5% palm and 5% coconut oil blends in diesel reduce CO emission 7.3–21% and HC emissions 17–23%. Apart from this, 5% coconut oil blends reduce 1 and 5% palm oil blends increases 2% of NOx emission [18]. Biodiesel blended fuels (B5–B75) reduce CO, UHC, and smoke density by 24.7%, 32.5%, and 63%, respectively, but increase NOx emission than diesel fuel because of the higher oxygen contents in biodiesel [19]. Venkanna and Reddy [20] showed that *C. Inophyllum* biodiesel blends fuel (CB10–CB30) gives 11–20% reduction of smoke opacity than diesel fuel at lower load operation. Also, blends provide lower CO emission and higher HC emission than fossil diesel fuel. The addition of isobutanol with diesel–biodiesel blends lowers the PM and NOx emission by 22.4–53.1% and 32.5%, respectively. The increase of biodiesel percentages in the blends lowers 3.45% CO emission, 32.5% NOx emission, and 38.5% PM emission [21]. The Oleander oil (OOME), Kusum oil (KOME), and bitter Groundnut oil (BGOME) biodiesel lower the CO, HC emissions, and smoke opacity than diesel fuel. But it gives higher NOx emissions due to the higher oxygen contents and exhaust gas temperature [22]. Man et al. [23] studied the regulated and unregulated emission of waste cooking oil biodiesel blends (B10–B30) at different engine load and speed condition. They reported that the increase of biodiesel fuel in the blends offers the reductions of HC, CO, and particulate mass concentrations, but an increase in NOx emission than diesel fuel. The unregulated emission, acetaldehyde, and formaldehyde emissions increase with increasing the percentage of biodiesel in the blends. But a reduction in the toluene and xylene emissions was reported for biodiesel blends. From all the emission results, it is seen that the emissions are affected by the engine operating condition, biodiesel type, and biodiesel percentages.

2.3 Formation of Gaseous Emission

The emissions are mainly due to the incomplete combustion in the cylinder [24]. So it is important to know the theoretical description of how they are formed so that mitigation steps of the harmful emission to the environment can be undertaken.

The oxidation of nitrogen (from the intake air) under high-temperature conditions is responsible for the formation of nitrogen oxides (NOx). NOx collectively refers to two chemical compounds, namely: nitric oxide (NO) and nitrogen dioxide (NO_2) [25]. The NO_2/NOx ratio is highly dependent on the engine technology and after-treatment system utilized. The theory of NOx formation is best commenced by discussing the well-known extended Zeldovich mechanism, which consists of the following three chemical equations [24]:

$$O + N_2 \leftrightarrow NO + N$$
$$ON + O_2 \leftrightarrow NO + O$$
$$N + OH \leftrightarrow NO + H$$

A notable feature of these three chemical equations is their strong temperature sensitivity.

HCs are another incomplete combustion product found in CI engine exhaust [26]. The primary cause of HC emissions is due to the preparation of an over-rich or over-lean air–fuel mixture that is not able to support complete combustion [24]. Over-leaning predominates under idle and light load conditions; however, over-fueling can occur under high load operation due to decrease in the air–fuel ratio in the spray core and near the combustion chamber walls [27]. Incomplete oxidation of the carbonaceous component of hydrocarbon fuels to carbon dioxide under conditions of low exhaust gas temperatures is responsible for the formation of CO [27]. The primary parameter that governs the formation of CO in internal combustion engines is the air–fuel ratio [24]. Fuel-rich conditions are responsible for forming CO as there is insufficient oxygen available to oxidize carbon to CO_2 fully. Oxidation catalysts are very effective at removing CO emissions from CI engine exhaust, especially under high load conditions.

Diesel particulate matter (DPM) is a complex, multi-pollutant mixture of solid and liquid particles suspended in a gas [28]. DPM is a very dynamic physical and chemical system that exhibits very strong spatial and temporal dependency regarding its composition [29]. DPM is a very dynamic physical and chemical system that exhibits adamant spatial and temporal dependency regarding its composition [30], the engine operating condition (e.g., speed/load, injection timing, and strategy), the presence of after-treatment devices (such as a diesel particle filter), the maintenance status of the engine, as well as the type of fuel and lubricants used.

3 Materials and Method

3.1 Materials

In this study, rice bran oil has been used as a biodiesel source as the crude rice bran oil is a low-cost feedstock, and it is not a common source of edible oil compared to other conventional seed sources. Crude rice bran oil was collected from a colleague through personal communication. All other reagents, methanol, filter paper 150 mm were available in the chemical laboratory.

3.2 Biodiesel Fuel Production Procedure

Free fatty acid (FFA) and acid values are the primary identifiers of the production process. The acid value of crude oil was found 1.6 g/KOH, which indicates that esterification (the chemical reaction in which an alcohol and an acid form an ester as the reaction product) is not necessary to produce biodiesel from crude rice bran oil. Because, if the crude oil contains a higher acid value, then two-step processes are required considering the formation of fatty acid salts during the conversion of FFA into Fatty acid methyl ester (FAME) using an alkaline catalyst [17]. The fatty acid salt prevents the separation of FAME layer from glycerin.

In this study, a small-scale water jacketed laboratory reactor 1.8 L in size equipped with reflux condenser, thermometer, and magnetic stirrer was used to produce biodiesel from crude rice bran oil. Figure 1 shows the flow process of rice bran biodiesel, and Fig. 2 shows the production process of rice bran biodiesel. In this process, 1 L of preheated crude rice bran oil was allowed to react with

Fig. 1 Flow process of rice bran biodiesel

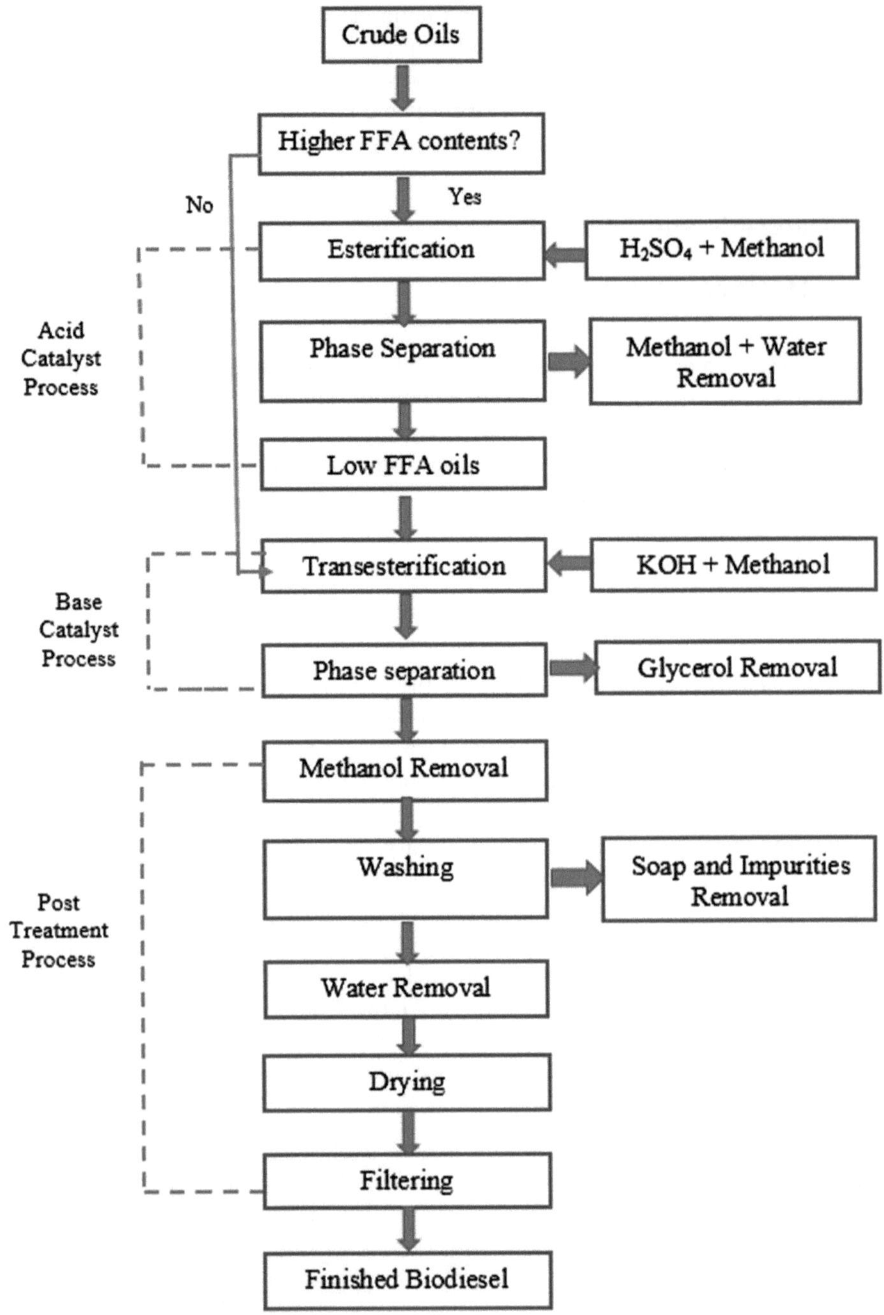

Fig. 2 Production process of rice bran biodiesel

6:1 molar ratio (methanol to oil) in the presence of 1% (w/w) of potassium hydroxide (KOH) catalyst. The reaction was maintained at 60 °C for 2 h at the speed of 800 rpm. After completion of the reaction, the mixture was poured into a separation funnel left for 12 h to be cooled, settled and glycerol to be separated from biodiesel. The upper part of the funnel contains biodiesel, and the bottom was glycerin which contains excess methanol and impurities. The biodiesel was collected, and the glycerin was drawn off. The produced biodiesel was then heated at 65 °C to remove remain methanol. Then the biodiesel was washed using warm distilled water to remove all impurities. Finally, the washed biodiesel was dried using Na_2SO_4 and filtered through a filter paper. The final product was collected and stored for characterization. The biodiesel yield was found more than 90% which was calculated using following formula:

$$\%\,Yield : \left(\frac{amount\ of\ methyl\ ester\ produced}{amount\ of\ oil\ taken}\right) \times 100.$$

3.3 *Characterization of Biodiesel Fuel*

The physical and chemical properties of the crude oil and biodiesel fuels were tested according to the ASTM D6751 standards. Table 1 shows the list of the equipment used in this study to characterize the biodiesel. Fatty acid composition was tested using gas chromatography (Agilent 6890 model, USA). Cetane number (CN), iodine value (IV), saponification value (SV), degree of unsaturation (DU),

Table 1 List of equipment used in this study

Property	Equipment	Standard method	Accuracy
Kinematic viscosity	NVB classic (Norma lab, France)	ASTM D445	±0.01 mm^2/s
Density	DM40 LiquiPhysics™ density meter (Mettler Toledo, Switzerland)	ASTM D127	±0.1 kg/m^3
Flash point	NPM 440 Pensky-martens flash point tester (Norma Lab, France)	ASTM D93	±0.1 °C
Cloud and pour point	NTE 450 cloud and pour point tester (Norma lab, France)	ASTM D2500	±0.1 °C
Heating value	6100EF semi auto bomb calorimeter (Perr, USA)	ASTM D240	±0.001 MJ/kg
Acid value	Automation titration rondo 20 (Mettler Toledo, Switzerland)	ASTMD664 and EN 14111	±0.001 mg KOH/g
Oxidation stability, 110 °C	873 Rancimat (Metrohm, Switzerland)	EN 14112	±0.01 h

and long chain saturated factor (LCSF) were determined using the following equations [31]:

$$CN = 46.3 + (5458/SV) - (0.225 \times IV) \quad (1)$$

$$SV = \Sigma(560 \times A_i)/M_{wi} \quad (2)$$

$$IV = \Sigma(254 \times A_i \times D)/M_{wi} \quad (3)$$

$$\begin{aligned} LCSF = {} & 0.1 \cdot (C16:0, wt.\%) + 0.5 \cdot (C18:0\,wt.\%) + 1 \cdot (C20:0\,wt.\%) \\ & + 1.5 \cdot (C22:0\,wt.\%) + 2.0 \cdot (C24:0\,wt.\%) \end{aligned} \quad (4)$$

$$DU = \Sigma(Mono\ Unsaturated\ Fatty\ Acid + 2.\ Poly\ Unsaturated\ Fatty\ Acids) \quad (5)$$

where *Ai* is the percentage of each component, *D* is the number of double bonds, and *Mw* is the molecular mass of each component.

3.4 Engine Test

First, the engine was run using diesel fuel for a few minutes to warm up before switching to the biodiesel blend. Furthermore, the engine was run with diesel fuel before it was shut down. Figure 3 shows the engine test bed, and Fig. 4 shows the schematic diagram of the engine test bed. Table 2 shows the specifications of the engine used in this study. A CODA 5 exhaust gas analyzer (Fig. 5a) was used to measure the NOx, HC, and CO emissions from the engine, and an MPM-4 M particulate meter (Fig. 5b) was used to measure the PM emission.

Fig. 3 Engine test bed

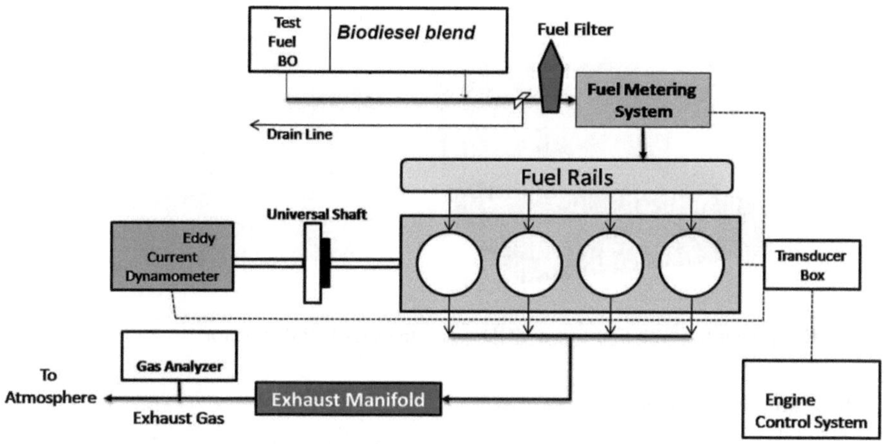

Fig. 4 Schematic diagram of engine test bed

Table 2 The specifications of the engine used in this study

Model	Kubota V3300
Type	Vertical, 4 cycle liquid cooled diesel
No. of cylinder	4
Total displacement (L)	3.318
Bore × Stroke (mm)	98 × 110
Combustion system	E-TVCS
Intake system	Natural aspired
Output	
Gross intermittent (kW/rpm)	54.5/2600
Net intermittent (Rated power output) (kW/rpm)	50.7/2600
Net continuous (kW/rpm)	44.1/2600
Rated torque (N m/rpm)	230/1400
Compression ratio	22.6
No load high idling speed (rpm)	2800
No load low idling speed (rpm)	700–750
Direction of rotation	Counter clockwise (viewed from flywheel side)
Governing	Centrifugal fly weight high-speed governor
Fuel	Diesel fuel No-2-D (ASTM D975)
Starter capacity (V-kW)	12-2.5
Alternator capacity (V-A)	12-60

A Kubota (model V3300) multi-cylinder diesel engine was used to perform the performance and emission test. In the performance and emission study, B5 and B20 fuels were used based on the suggestion from the literature [32, 33] that up to 20%

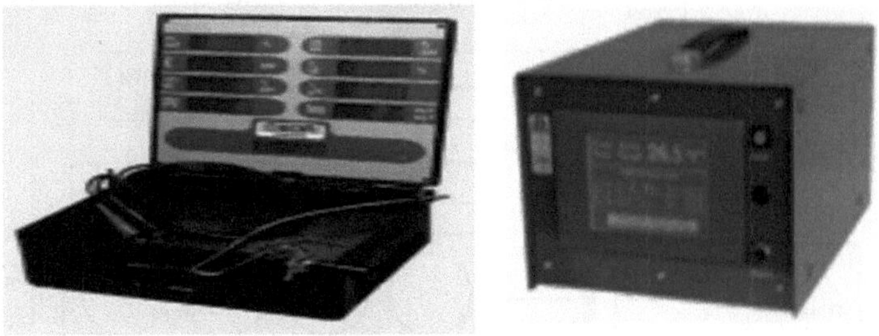

Fig. 5 Equipment for emission test **a** CODA gas analyzer **b** PM monitor

Table 3 Properties of crude rice bran oil

Properties	Rice bran oil
Kinematic viscosity (mm^2/s) at 40 °C	52
Density (kg/m^3) at 15 °C	924
Higher heating value (MJ/kg)	39.5
Acid value (mg KOH/g)	1.3
Flash point (°C)	301
Pour point (°C)	0
CFPP (°C)	16

biodiesel can be used in a diesel engine with no modifications. Engine performance data were collected at full load condition and at different speeds ranging from 1200 to 2400 rpm at an interval of 200 rpm, whereas emission data was collected at a different speed (at idle speed and the speed at which maximum torque was found) and at full load conditions. The emission data also have been investigated at intermittent speed (1800 rpm) and different load conditions (25–100%).

4 Results and Discussion

4.1 Properties of Fuel Samples

The characterization of biodiesel is very important as the suitability of biodiesel as a diesel fuel is determined by the physical and chemical properties. The physical and chemical properties results of crude rice bran oil and rice bran biodiesel are given in the following sections.

4.1.1 Properties of Crude Oils

The crude rice bran oil was characterized by viscosity, density, flash point, acid value, higher heating value. The properties of rice bran oil are presented in Table 3.

The viscosity of crude oil was found 52 mm^2/s which are 15–16 times greater than conventional diesel fuel. The flash point, pour point, and cold filter plugging point were found 301 °C, 0 °C, and 16 °C, respectively. The acid value was determined 1.3 mg KOH/g which is similar with the other conventional biodiesel feedstocks.

4.1.2 Physical Properties of Biodiesel Fuel

The fuel properties of rice bran biodiesel were analyzed and compared with diesel and ASTM D6751 standards. Table 4 shows the fuel properties of rice bran biodiesel. It was found that the kinematic viscosity (KV) of rice bran is 5.37 mm^2/s. However, all these results are within the specified limit ASTM D6751 standards (1.9–6 mm^2/s). The flash point (FP) was found 174.5 °C, which is much higher than diesel fuel (68.5 °C) that indicates rice bran biodiesel fuel is safer to handle and storage. The oxidation stability and higher heating value of rice bran biodiesel were found 1.61 h and 39.50 MJ/kg, respectively.

4.1.3 Chemical Properties of Biodiesel Fuel

Fatty acid composition is an important chemical property of biodiesel fuel. Fatty acids are categorized into saturated and unsaturated fatty acid. A fatty acid that does not contain double bond is known as saturated fatty acid, and that contains double bond is known as unsaturated fatty acid. Table 5 shows the fatty acid composition of rice bran oil. It can be seen that rice bran biodiesel has 24.20% saturated and 75.80% unsaturated fatty acids. Oleic acid (18:1) was the predominant fatty acid (43.5%) in rice bran biodiesel sample. The degree of unsaturation and long chain saturated factor was found 107.40 and 5.08, respectively.

Table 4 Properties of biodiesel and diesel fuel

Properties	Unit	RB100	Diesel	ASTMD 6751 [34]
Kinematic viscosity at 40 °C	mm^2/s	5.37	3.23	1.9–6
Density at 15 °C	kg/m^3	887	827.2	–
Higher heating value	MJ/kg	39.50	45.30	–
Oxidation stability	h	1.61	–	3 min
Flash point	°C	174.5	68.5	130 min
Pour point	°C	−3	0	–
Cloud point	°C	0	8	Report
CFPP	°C	2	5	–
Cetane number	–	51.30	48	47 min
Iodine number	–	97.52	–	–
Saponification value	–	202.55	–	–

Table 5 Fatty acid composition of rice bran biodiesel

Fatty acids	Molecular weight	Structure	Formula	Rice bran biodiesel (wt%)
Lauric	200	12:0	$C_{12}H_{24}O_2$	0.1
Myristic acid	228	14:0	$C_{14}H_{28}O_2$	0.3
Palmitic	256	16:0	$C_{16}H_{32}O_2$	20.3
Palmitoleic	254	16:1	$C_{16}H_{30}O_2$	0.2
Stearic	284	18:0	$C_{18}H_{36}O_2$	2.0
Oleic	282	18:1	$C_{18}H_{34}O_2$	43.5
Linoleic	280	18:2	$C_{18}H_{32}O_2$	31.0
Linolenic	278	18:3	$C_{18}H_{30}O_2$	0.6
Arachidic	312	20:0	$C_{20}H_{40}O_2$	0.8
Eicosenoic	310	20:1	$C_{20}H_{38}O_2$	0.5
Behenic	340	22:0	$C_{22}H_{44}O_2$	0.3
Lignoceric	368	24:0	$C_{24}H_{48}O_2$	0.4
Total saturated fatty acid				24.20
Total monounsaturated fatty acid (MUFA)				44.20
Total polyunsaturated fatty acid (PUFA)				31.60
Degree of unsaturation (DU)				107.40
Long chain saturated factor (LCSF)				5.08

4.2 *Engine Performance Study*

4.2.1 Brake Torque

Figure 6 shows the engine brake torque (BP) output of B5 (5% rice bran biodiesel and 95% diesel fuel), B20 (20% rice bran biodiesel and 80% diesel fuel), and B0 (pure diesel fuel) at different engine speeds. It is clear that engine torque for diesel fuel is higher than that for biodiesel blended fuel, which is also supported by other researchers [35, 36]. The engine torque lowered with increasing the engine speed and increasing the percentage of biodiesel fuel in the blend. In an average, diesel fuel gives the highest torque followed by the B5 and B20, which is 202 Nm, 195 Nm, and 185 Nm, respectively. The maximum torque for B0, B5, and B20 fuel samples were observed 220 Nm, 214 Nm, and 204 Nm, respectively, at 1400 rpm. At all test speeds, B5 and B20 fuels reduce the torque by 3 and 8% than diesel fuel. The reduction of torque for biodiesel–diesel blended fuel can be attributed to the heating value of biodiesel fuel [37]. The heating value of rice bran biodiesel is lower than diesel fuel (Table 4). Aydin et al. [38] also reported that the engine torque was decreased with the increased percentage of biodiesel in the blends due to the higher viscosity and lower heating value of biodiesel fuel.

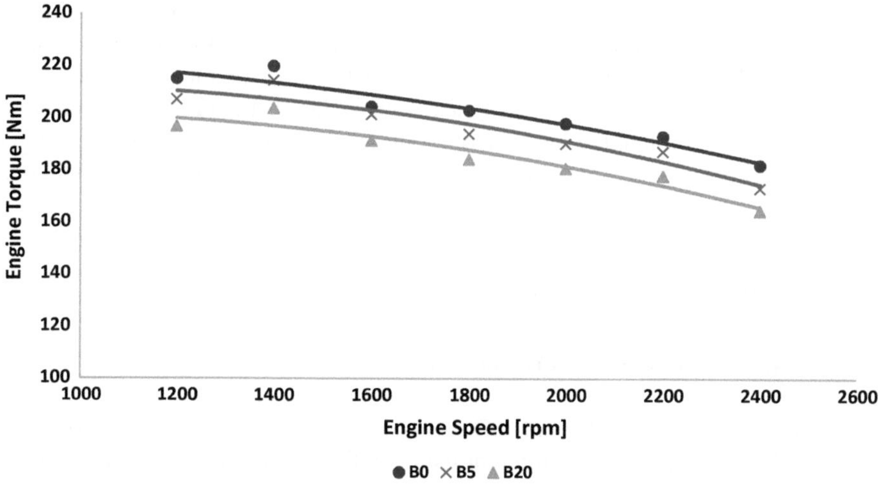

Fig. 6 Variation of brake torque on the speed

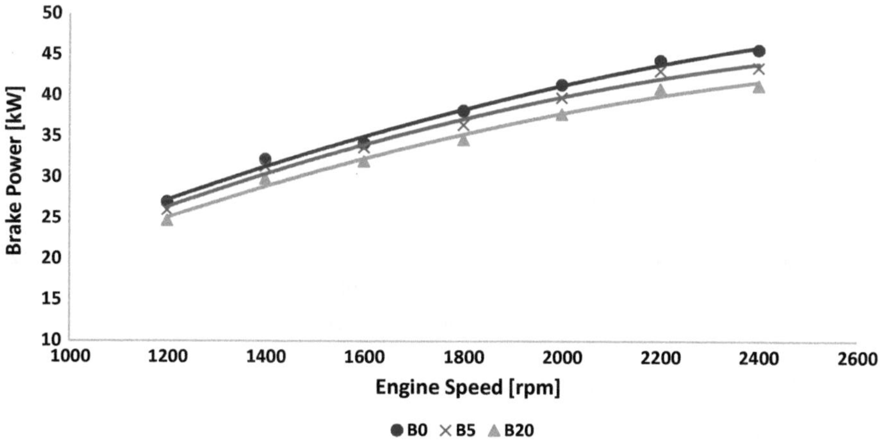

Fig. 7 Variation of brake power on the speed

4.2.2 Brake Power

The engine brake power (BP) output of B5, B20, and B0 at different engine speeds are shown in Fig. 7. It can be seen that the brake power for diesel fuel is higher than that for biodiesel blended fuel, which is also supported by other researchers [35, 36]. The brake power is lowered with increasing the proportion of biodiesel in the blends and increased with increasing the speeds. The average brake power for B0, B5, and B20 is 37.6, 36.3, and 34.5 kW, respectively. The maximum brake power was found 45.7 kW for diesel fuel at 2400 rpm. The blends, B5 and B20 fuels reduce the brake power by 3.46 and 8.29% compared to diesel fuel. The reduction of BP for biodiesel–diesel blended fuel can be attributed to the heating value of biodiesel fuel [37] which is commonly agreed by most of the researchers. The heating value of rice bran biodiesel is lower than diesel fuel (Table 4). Carraretto et al. [39] also found that the increase of biodiesel percentage in the blends resulted in a slight decrease of power over the entire speed range for different combinations.

4.3 Engine Emission Study

4.3.1 Carbon Monoxide (CO) Emission

Figure 8a shows the variation of CO emission by running the engine using B5, B20, and B0 at 1800 rpm and different load conditions. It is seen that CO emission

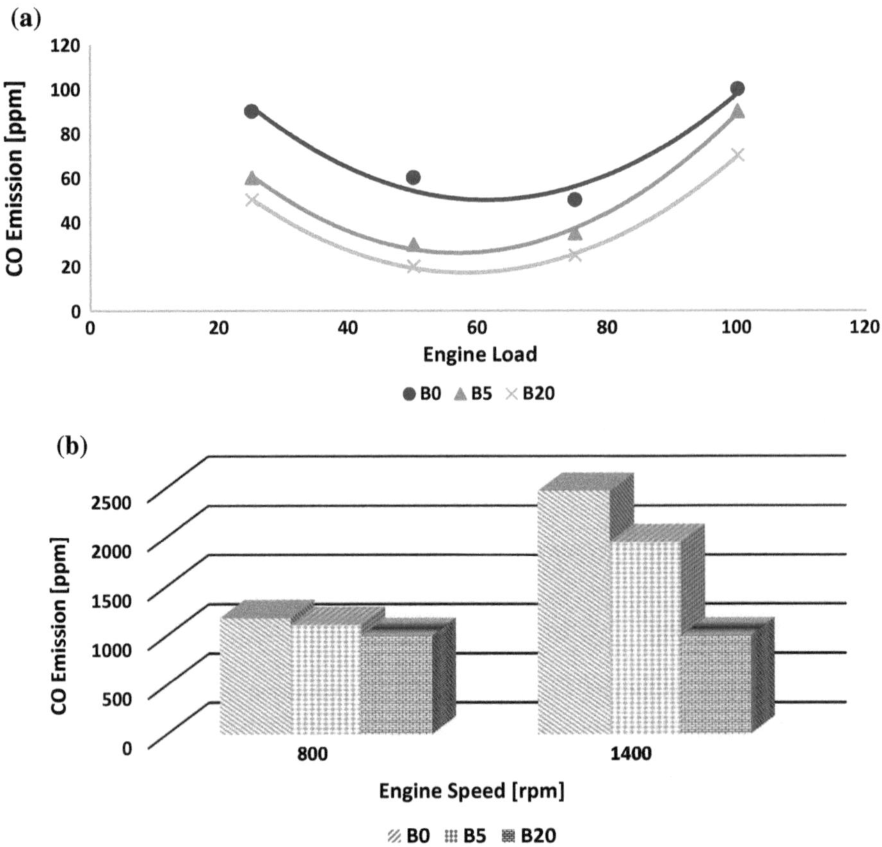

Fig. 8 Variation of CO emission **a** at 1800 rpm and different loads **b** at idle speed and 1400 rpm

increases significantly at lower load condition, decreased at moderate load, and increased slightly at full load. The air–fuel mixing process is affected by the atomization of the blend at low load due to its higher viscosity. The resulting locally rich mixtures cause more CO emission during combustion. Authors in [40] also found that CO emissions decreased as load increased, but they increased slightly at heavy load or full load.

The variation of CO emission of biodiesel blended fuel with engine speed is shown in Fig. 8b. It is seen (Fig. 8b) that over the entire range of speed, the biodiesel fuels provide lower CO emission than diesel fuel. Similar results were also found by the researchers [41]. CO emission is increased with increasing engine speed from idle speed to 1400 rpm. The increase of biodiesel proportion in the blends lowers the CO emission significantly. At 800 rpm fuel samples B0, B5, and B20 produce 1180 ppm, 1110, and 1000 ppm, respectively, while at 1400 rpm B0, B5, and B20 produce 1100 ppm, 1010, and 600 ppm, respectively. Biodiesel blended fuel reduce CO emission by (5.9–59.5%) at both speed condition. Lowering CO emission for biodiesel fuel can be explained by the higher oxygen contents and higher cetane number of biodiesel fuel. Biodiesel fuel contains 12% higher oxygen, which allows more carbon molecules to burn and combustion becomes completed [15].

4.3.2 Hydrocarbon (HC) Emission

Figures 9a, b show the variation of HC emission by running the engine using B5, B20, and B0 at a different speed and different load conditions. It is found that in all operating conditions, HC emission is decreased as the percentages of biodiesel increased in the blends. It is also evident that at lower load condition HC emission is higher, at moderate load is lower, and at a full load slightly higher. Authors [42, 43] also observed that the HC emissions of biodiesel fuel are lower at low load, but it was reported that a greater decrease occurred at intermediate load than lower and higher load condition. However, at lower speed condition, the HC emission is lower compared to the higher speed condition for all fuel samples. Similar results were also reported by the other researchers [41]. For example, Wu et al. [40] found 45–67% reduction in HC emission on average than diesel fuel.

Unlike to the CO emission, the HC emission decreased with the engine speed. At 800 rpm fuel samples B0, B5, and B20 produce 7 ppm, 5, and 2 ppm, respectively, while at 1400 rpm B0, B5, and B20 produce 6 ppm, 4, and 3 ppm, respectively. In average, biodiesel blended fuel reduce HC emission by (28–71%) at all operating conditions. The reason of lowering HC emission could be attributed to the combined effect of oxygen content and cetane number as explained for CO emission [17, 37].

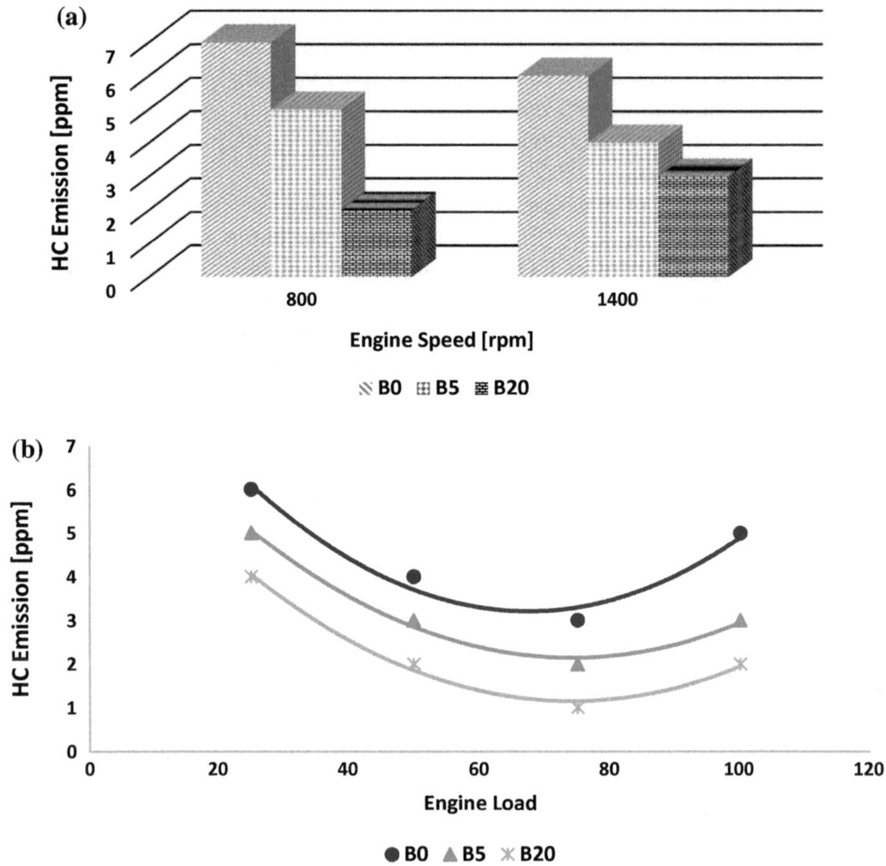

Fig. 9 Variation of HC emission **a** at idle speed and 1400 rpm **b** at 1800 rpm and different loads

4.3.3 Nitrogen Oxides (NOx) Emission

NOx emission is reported by the many authors to be increased for biodiesel blended fuel. The variation NOx emission with different engine speed and load for various blended fuel is shown in Fig. 10a, b. As shown therein, the NOx emission increased with the engine load at intermediate speed due to the higher combustion temperature and local stoichiometry of the blends. It is also found that biodiesel blended fuel gives (6.8–12.3%) higher NOx than diesel fuel and NOx emission is increased as the percentages of biodiesel increased in the mixture.

The NOx variation of fuel samples with engine speed showed a similar trend of load condition. NOx emission also increased with the engine speed. At idle speed condition, B5 gives the highest NOx emission, whereas B20 showed maximum NOx emission at 1400 rpm than diesel fuel. The higher combustion temperatures in the cylinder and presence of oxygen in the blends caused higher NOx emission [25].

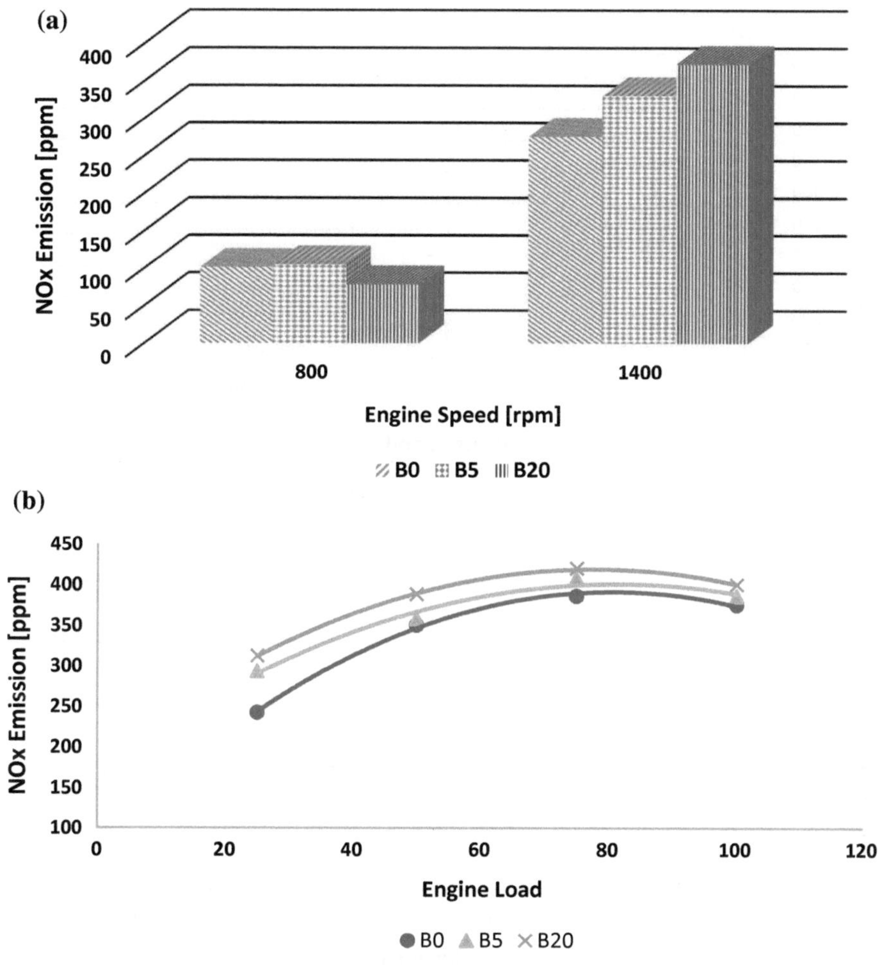

Fig. 10 Variation of NOx emission **a** at idle speed and 1400 rpm **b** at 1800 rpm and different loads

Also, higher exhaust gas temperatures with the fuels provided the higher NOx emission. It is also agreed that inherent oxygen in the biodiesel fuel is more efficient to produce higher NOx than the oxygen supplied with the air [38].

4.3.4 Particulate Matter (PM) Emission

The variation of PM emission for different fuels as a function of load and speed is presented in Fig. 11a, b. It is found that biodiesel fuel lowers 5–9.85 and 60–74% PM emission in both idle speed and 1400 rpm, respectively, than diesel fuel. Also,

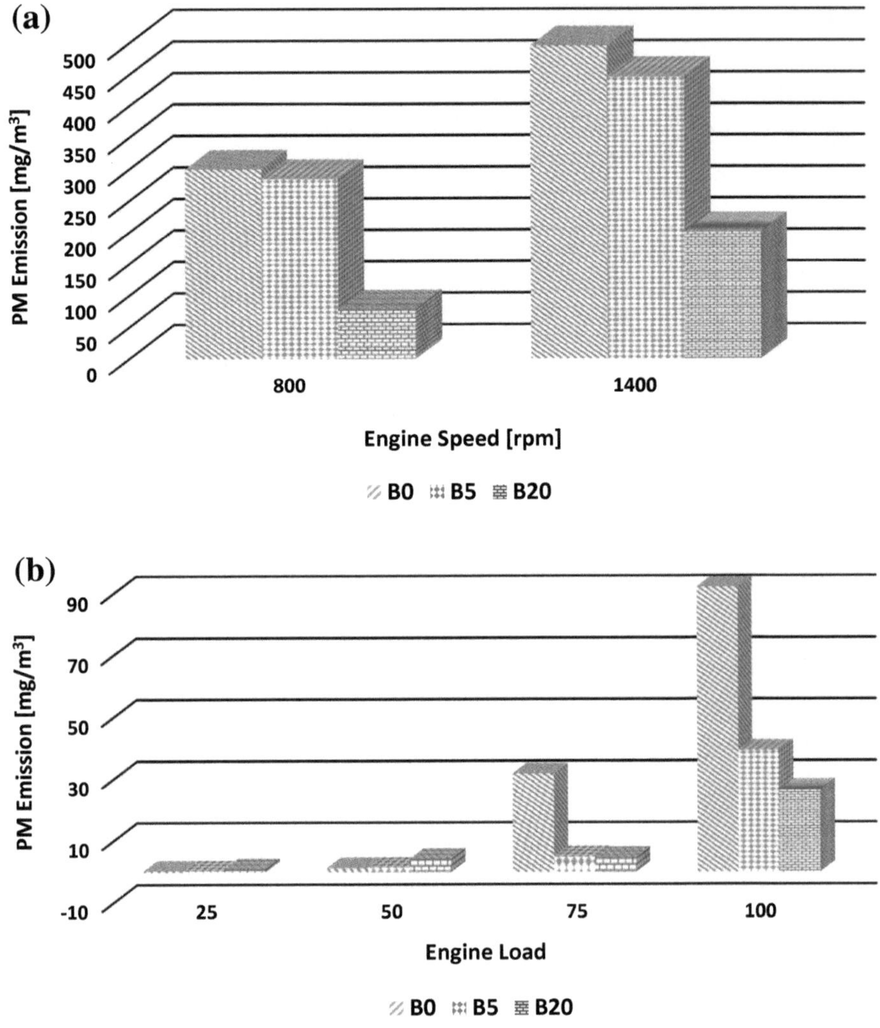

Fig. 11 Variation of PM emission **a** at idle speed and 1400 rpm **b** at 1800 rpm and different loads

PM emission increased as the biodiesel percentage increased in the blends. Similar results were also reported by other researchers [44, 45]. Hass et al. [46] found 30% more reduction in PM emission when then they increased biodiesel percentage from 20 to 100%. The reason of lowering PM emission can be attributed to the higher oxygen contents in biodiesel fuel. The oxygen causes complete combustion of biodiesel in the cylinder. Also, a lack of sulfur and aromatic compound further reduce the PM emissions [47].

The variation of PM emission with load showed a different trend that was in speed. At lower load condition, biodiesel showed a bit higher PM emission, whereas at higher load condition biodiesel fuel reduced the PM emission significantly. It happened because particles are mainly formed during the diffusion combustion, and most of the combustion is diffusive at high load, which means that the oxygen content of biodiesel is more efficient in reducing the PM emission. A significant reduction in PM emission at high load and the increase in the lower load were also found by [48] when they used soybean biodiesel.

5 Conclusions

The development of sustainable liquid transportation fuels is essential to ensure the future security of energy supply in the world. Biodiesel produced from different feedstock represents a sustainable source of energy to be used in the transportation sector. The following conclusions could be drawn from this study:

- The use of rice bran biodiesel in diesel engine favors reducing the 56.5% of CO emission compared to the diesel fuel. Such trend is attributed to the higher oxygen content and lower carbon to hydrogen ratio compared with diesel fuel.
- The HC emission is reduced significantly when rice bran biodiesel blends are used instead of diesel fuel. This reduction is mainly due to the higher oxygen contents and cetane number of rice bran biodiesel.
- The PM emission for rice bran biodiesel is reduced to a large extent compared to diesel fuel. The reduction is mainly due to the higher oxygen and lower aromatic compounds present in the biodiesel.
- An increase of NOx emission in rice bran biodiesel blends was found due to the higher combustion temperature, higher cetane number, and higher oxygen contents of biodiesel fuel.
- The use of rice bran biodiesel blends in the diesel engine lowers the engine power slightly due to the lower heating value of biodiesel fuel compared to diesel fuel.

Finally, it can be concluded that up to 20% rice bran biodiesel could replace diesel fuel to help in controlling the air pollution to a great extent without sacrificing engine power significantly. Therefore, rice bran biodiesel is more sustainable than the existing energy supplies from an environmental perspective as the rice bran biodiesel considerably reduce the harmful emission.

Acknowledgements This work was conducted under the IPRA Scholarship funded by the Central Queensland University, Australia.

References

1. Mofijur, M., Atabani, A. E., Masjuki, H. H., Kalam, M. A., & Masum, B. M. (2013). A study on the effects of promising edible and non-edible biodiesel feedstocks on engine performance and emissions production: A comparative evaluation. *Renewable and Sustainable Energy Reviews, 23,* 391–404.
2. Hesterberg, T. W., Long, C. M., Bunn, W. B., Sax, S. N., Lapin, C. A., & Valberg, P. A. (2009). Non-cancer health effects of diesel exhaust: a critical assessment of recent human and animal toxicological literature. *Critical Reviews in Toxicology, 39,* 195–227.
3. Katter, H. B. (2015). *Towards a clean energy economy: Achieving a biofuel mandate for Queensland.*
4. Kearney, R. (2006). Department of Infectious Diseases and Immunology at the University of Sydney.
5. CSIRO Medial Release, 2 March 2004, Air Pollution Death Toll Needs Solutions.
6. Mofijur, M., Rasul, M. G., Hyde, J., Azad, A. K., Mamat, R., & Bhuiya, M. M. K. (2016). Role of biofuel and their binary (diesel–biodiesel) and ternary (ethanol–biodiesel–diesel) blends on internal combustion engines emission reduction. *Renewable and Sustainable Energy Reviews, 53,* 265–278.
7. Demirbas, A. (2009). Progress and recent trends in biodiesel fuels. *Energy Conversion and Management, 50,* 14–34.
8. Jayed, M. H., Masjuki, H. H., Saidur, R., Kalam, M. A., & Jahirul, M. I. (2009). Environmental aspects and challenges of oilseed produced biodiesel in Southeast Asia. *Renewable and Sustainable Energy Reviews, 13,* 2452–2462.
9. Liaquat, A. M., Kalam, M. A., Masjuki, H. H., & Jayed, M. H. (2010). Potential emissions reduction in road transport sector using biofuel in developing countries. *Atmospheric Environment, 44,* 3869–3877.
10. Harch, C. A., Rasul, M. G., Hassan, N. M. S., & Bhuiya, M. M. K. (2014). Modelling of engine performance fuelled with second generation biodiesel. *Procedia Engineering, 90,* 459–465.
11. Habibullah, M., Masjuki, H. H., Kalam, M. A., Rahman, S. M. A., Mofijur, M., Mobarak, H. M., et al. (2015). Potential of biodiesel as a renewable energy source in Bangladesh. *Renewable and Sustainable Energy Reviews, 50,* 819–834.
12. Ju, Y.-H., & Vali, S. R. (2005). Rice bran as a potential source for biodiesel: A review. *Journal of Scientific and Industrial Research, 64,* 866–882.
13. Lin, L., Ying, D., Chaitep, S., & Vittayapadung, S. (2009). Biodiesel production from crude rice bran oil and properties as fuel. *Applied Energy, 86,* 681–688.
14. Labeckas, G., & Slavinskas, S. (2015). Combustion phenomenon, performance and emissions of a diesel engine with aviation turbine JP-8 fuel and rapeseed biodiesel blends. *Energy Conversion and Management, 105,* 216–229.
15. Chauhan, B. S., Kumar, N., & Cho, H. M. (2012). A study on the performance and emission of a diesel engine fueled with Jatropha biodiesel oil and its blends. *Energy, 37,* 616–622.
16. Rajaraman, S., Yashwanth, G. K., Rajan, T., Kumaran, R. S., & Raghu, P. (2009, November 13–19). Experimental investigations of performance and emission characteristics of Moringa oil methyl ester and its diesel blends in a single cylinder direct injection diesel engine. In *Proceedings of the ASME 2009 International Mechanical Engineering Congress and Exposition, Lake Buena Vista, Florida, USA 2009.*
17. Rao, P. V. (2011). Experimental investigations on the influence of properties of Jatropha biodiesel on performance, combustion, and emission characteristics of a DI-CI engine. *World Academy of Science, Engineering and Technology, 75,* 855–868.
18. Kalam, M. A., Masjuki, H. H., Jayed, M. H., & Liaquat, A. M. (2011). Emission and performance characteristics of an indirect ignition diesel engine fuelled with waste cooking oil. *Energy, 36,* 397–402.

19. Selvam, D. J. P., & Vadivel, K. (2012). Performance and emission analysis of DI diesel engine fuelled with methyl esters of beef tallow and diesel blends. *Procedia Engineering, 38,* 342–358.
20. Venkanna, B. K., & Venkataramana, R. C. (2013). Effect of injector opening pressure on performance, emission and combustion characteristics of DI diesel engine fueled with diesel and honne oil methyl ester. *Environmental Progress and Sustainable Energy, 32,* 148–155.
21. Yang, P.-M., Lin, K. C., Lin, Y.-C., Jhang, S.-R., & Chen, S.-C. (2016). Emission evaluation of a diesel engine generator operating with a proportion of isobutanol as a fuel additive in biodiesel blends. *Applied Thermal Engineering, 100,* 628–635.
22. Yadav, A. K., Khan, M. E., Dubey, A. M., & Pal, A. (2016). Performance and emission characteristics of a transportation diesel engine operated with non-edible vegetable oils biodiesel. *Case Studies in Thermal Engineering, 8,* 236–244.
23. Man, X. J., Cheung, C. S., Ning, Z., Wei, L., & Huang, Z. H. (2016). Influence of engine load and speed on regulated and unregulated emissions of a diesel engine fueled with diesel fuel blended with waste cooking oil biodiesel. *Fuel, 180,* 41–49.
24. Heywood, J.B. (1988). *Internal combustion engine fundamentals* McGraw hill book co.
25. Palash, S. M., Kalam, M. A., Masjuki, H. H., Masum, B. M., Rizwanul Fattah, I. M., & Mofijur, M. (2013). Impacts of biodiesel combustion on NOx emissions and their reduction approaches. *Renewable and Sustainable Energy Reviews, 23,* 473–490.
26. Mofijur, M., Masjuki, H. H., Kalam, M. A., & Atabani, A. E. (2013). Evaluation of biodiesel blending, engine performance and emissions characteristics of Jatropha curcas methyl ester: Malaysian perspective. *Energy, 55,* 879–887.
27. Majewski, W. A., & Jääskeläinen, H. (2008). *Emission formation in diesel engines*. Dieselnet Technology Guide, Ecopoint Inc. http://www.dieselnet.com/tech/diesel_emiform.html.
28. Eastwood, P. (2008). *Particulate emissions from vehicles* (pp. 1–493). Chichester: Wiley.
29. Surawski, N. C. (2012). *An investigation of gaseous and particulate emissions from compression ignition engines operated with alternative fuels, injection technologies, and combustion strategies*.
30. Zielinska, B., Samy, S., McDonald, J. D., & Seagrave, J. C. (2010). *Atmospheric transformations of diesel emissions* (pp. 5–60). Health Effects Institute.
31. Islam, M. A., Brown, R. J., Brooks, P. R., Jahirul, M. I., Bockhorn, H., & Heimann, K. (2015). Investigation of the effects of the fatty acid profile on fuel properties using a multi-criteria decision analysis. *Energy Conversion and Management, 98,* 340–347.
32. Bari, S. (2014). Performance, combustion and emission tests of a metro-bus running on biodiesel-ULSD blended (B20) fuel. *Applied Energy, 124,* 35–43.
33. Ge, J., Kim, M., Yoon, S., & Choi, N. (2015). Effects of pilot injection timing and EGR on combustion, performance and exhaust emissions in a common rail diesel engine fueled with a canola oil biodiesel-diesel blend. *Energies, 8,* 7312.
34. Atabani, A. E., & da Silva César, A. (2014). Calophyllum inophyllum L.—A prospective non-edible biodiesel feedstock. Study of biodiesel production, properties, fatty acid composition, blending and engine performance. *Renewable and Sustainable Energy Reviews, 37,* 644–655.
35. Ali, O. M., Mamat, R., Abdullah, N. R., & Abdullah, A. A. (2016). Analysis of blended fuel properties and engine performance with palm biodiesel–diesel blended fuel. *Renewable Energy, 86,* 59–67.
36. Altaie, M. A. H., Janius, R. B., Rashid, U., Taufiq-Yap, Y. H., Yunus, R., Zakaria, R., et al. (2015). Performance and exhaust emission characteristics of direct-injection diesel engine fueled with enriched biodiesel. *Energy Conversion and Management, 106,* 365–372.
37. Xue, J., Grift, T. E., & Hansen, A. C. (2011). Effect of biodiesel on engine performances and emissions. *Renewable and Sustainable Energy Reviews, 15,* 1098–1116.
38. Aydin, H., & Bayindir, H. (2010). Performance and emission analysis of cottonseed oil methyl ester in a diesel engine. *Renewable Energy, 35,* 588–592.
39. Carraretto, C., Macor, A., Mirandola, A., Stoppato, A., & Tonon, S. (2004). Biodiesel as alternative fuel: Experimental analysis and energetic evaluations. *Energy, 29,* 2195–2211.

40. Wu, F., Wang, J., Chen, W., & Shuai, S. (2009). A study on emission performance of a diesel engine fueled with five typical methyl ester biodiesels. *Atmospheric Environment, 43,* 1481–1485.
41. Pali, H. S., Kumar, N., & Alhassan, Y. (2015). Performance and emission characteristics of an agricultural diesel engine fueled with blends of Sal methyl esters and diesel. *Energy Conversion and Management, 90,* 146–153.
42. Cheung, C. S., Zhu, L., & Huang, Z. (2009). Regulated and unregulated emissions from a diesel engine fueled with biodiesel and biodiesel blended with methanol. *Atmospheric Environment, 43,* 4865–4872.
43. Lapuerta, M., Armas, O., & Rodríguez-Fernández, J. (2008). Effect of biodiesel fuels on diesel engine emissions. *Progress in Energy and Combustion Science, 34,* 198–223.
44. Zhu, L., Zhang, W., Liu, W., & Huang, Z. (2010). Experimental study on particulate and NOx emissions of a diesel engine fueled with ultra low sulfur diesel, RME-diesel blends and PME-diesel blends. *Science of the Total Environment, 408,* 1050–1058.
45. Kim, H., & Choi, B. (2010). The effect of biodiesel and bioethanol blended diesel fuel on nanoparticles and exhaust emissions from CRDI diesel engine. *Renewable Energy, 35,* 157–163.
46. Haas, M. J., Scott, K. M., Alleman, T. L., & McCormick, R. L. (2001). Engine performance of biodiesel fuel prepared from soybean soapstock: A high quality renewable fuel produced from a waste feedstock. *Energy and Fuels, 15,* 1207–1212.
47. Karavalakis, G., Stournas, S., & Bakeas, E. (2009). Light vehicle regulated and unregulated emissions from different biodiesels. *Science of the Total Environment, 407,* 3338–3346.
48. Durbin, T. D., & Norbeck, J. M. (2002). Effects of biodiesel blends and arco EC-diesel on emissions from light heavy-duty diesel vehicles. *Environmental Science and Technology, 36,* 1686–1691.

Correlation Between Physicochemical Properties and Quality of Biodiesel

M.I. Jahirul, R.J. Brown and W. Senadeera

Abstract Biodiesel produced from renewable feedstocks represents a sustainable source of energy and will therefore play a significant role in providing the energy requirements for transportation in the near future. Biodiesel offers many benefits over conventional petroleum fuels, including the wide regional distribution of biomass feedstocks, high greenhouse gas reduction potential, biodegradability and a significant contribution to sustainability. Chemically, all biodiesels are fatty acid methyl esters (FAME), produced from raw vegetable oil and animal fat. However, clear differences in chemical structure are apparent when comparing one feedstock to the next in terms of chain length, degree of unsaturation and number of double bonds—all of which determine the fuel properties and quality of biodiesel as a diesel engine fuel. In this chapter, biodiesel feedstocks, production processes, chemical compositions, standards, physicochemical properties and in-use performance are discussed. A correlation study between the properties of biodiesel and its chemical composition is analysed using principal component analysis (PCA). The necessary data regarding the chemical composition and fuel properties of biodiesel were obtained from more than 100 papers published in recognised international journals. The PCA indicated that individual biodiesel properties have a complex correlation with the parameters of chemical composition. The average chain length and average number of double bonds are the most influential parameters that affect all biodiesel properties. The results of this analysis are presented graphically and discussed in this chapter. Therefore, this chapter will provide the reader a clearer understanding of the physicochemical properties of biodiesel.

Keywords Biodiesel · Chemical composition · Fuel properties · PCA

M.I. Jahirul (✉) · R.J. Brown · W. Senadeera
Biofuel Engine Research Facility (BERF), Science and Engineering Faculty,
Queensland University of Technology (QUT), Brisbane 4000, Australia
e-mail: md_jahirul@yahoo.com; m1.islam@qut.edu.au

M.M.K. Khan et al. (eds.), *Application of Thermo-fluid Processes in Energy Systems*,
Green Energy and Technology, https://doi.org/10.1007/978-981-10-0697-5_3

1 Introduction

Globally, the transport sector occupies third place when total energy consumption and greenhouse gas emissions are considered (following the trade and building sectors). It is also the fastest growing sector. By 2030, the energy consumption and, therefore, the CO_2 emissions of this sector are predicted to be eight percent higher than current levels [86]. In addition, the energy supply depends heavily on non-renewable petroleum fuel (for production of gasoline and diesel) and currently consumes 30% of the world's petroleum oil, increasing to 60% by 2030 [85]. Furthermore, the supply of petroleum oil is geographically restricted, and the era of low-cost and secure oil is almost over. These facts have forced automobile researchers to look for alternative carbon-neutral transport fuels that promise a harmonious correlation with sustainable development, energy conversion, energy potency and environmental preservation [59]. However, such an alternative fuel for the transportation sector is yet to be developed. Moreover, cars with no greenhouse gas emissions (electric, solar, hydrogen, etc.) are far away from changing into mainstream vehicles. Therefore, development of a sustainable long-run alternative fuel has become essential, and biodiesel is receiving significant attention and is coming to the forefront as a sustainable alternative to standard fossil fuels [85].

Biodiesel is liquid fuel created from various oilseed crops and animal fat. Biodiesels offer many socio-economic advantages over petroleum-based fuels in automobile engine applications, in particular the fact that they are renewable, biodegradable, non-toxic and eco-friendly [61]. However, the majority of current vehicle engines are not optimised for the utilisation of biodiesel. Therefore, these engines are unlikely to be efficient when using biodiesel and show several technical issues such as carbon deposition, corrosion, high lubricating oil contamination, poor low-temperature performance and heavy gum and wax formation when compared to petroleum diesel [50]. The distinctions between petroleum diesel and biodiesel could be attributed to the variation of physical properties and chemical compositions. Petroleum diesel consists of hundreds of compounds boiling at completely different temperatures (determined by the petroleum refining method and crude oil raw material), whereas biodiesel contains compounds that are primarily eight to twenty-four carbon chain length alkyl esters (determined entirely by the feedstock) [57, 78]. Besides the main fatty ester components, the minor constituents of biodiesel embrace intermediary glycerides and free fatty acids resulting from the transesterification reaction, methanol, free fatty acids, etc. As engines are manufactured for petroleum diesel, original equipment makers (OEMs) and industry associations have shown a cautious response in their acceptance of biodiesel, especially those from new sources and the concept of using biodiesel blends as fuel [78].

2 Biodiesel

Fatty acid methyl or ethyl esters, commonly referred to as "biodiesel", are a liquid fuel alternative to diesel. They are made of agricultural products, forest organic matter and animal fat feedstocks. Biodiesel is the only currently available alternative transport fuel made from oilseed crops and animal fat which can be used directly in conventional, unmodified diesel engines. Biodiesel is safer to handle, store and transport compared to petroleum diesel because it is biodegradable, non-toxic and has a higher flashpoint than diesel [61]. One of the major advantages of biodiesel is that it has a potential to reduce dependency on imported petroleum through the use of domestic feedstocks for production [51].

In fuel property terms, biodiesel has a higher cetane rating than petroleum diesel, which improves engine performance. Also, it has better lubricant properties than petroleum diesel, which can extend engine life [50]. The use of biodiesel reduces particulate emissions by up to 75% when compared with conventional diesel fuel. Biodiesel also substantially reduces unburned hydrocarbons, carbon monoxides and particulate matters, including an elimination of sulphur dioxide in exhaust emissions. The exhaust emissions of particulate matter from biodiesel have been found to be 30% lower than overall particulate matter emissions from fossil diesel. The exhaust emissions of total hydrocarbons are up to 93% lower for biodiesel than for diesel fuel [105].

As a fuel, there are currently several disadvantages to using biodiesel in diesel engine applications. These mainly result from the differences in chemical composition between petroleum diesel and biodiesel. These major disadvantages are lower energy density, higher viscosity, higher copper strip corrosion and issues with the degradation of fuel in storage for prolonged periods. Biodiesel also has a higher cold-filter plugging point temperature than fossil diesel, which means it will crystallise into a gel at lower temperatures when used in its pure form. Biodiesel can also cause dilution of engine lubricant oil, requiring more frequent oil changes than when using petroleum diesel fuels in conventional diesel engines. This increase in dilution and polymerisation of engine sump oil is due to the higher viscosity at lower temperatures of biodiesel when compared to petroleum diesel.

3 Biodiesel Feedstock

Feedstocks for biodiesel production can be classified into four groups. These are (1) virgin vegetable oil feedstocks such as rapeseed, soya bean, sunflower and palm oil; (2) waste vegetable oils; (3) animal fats including beef tallow, lard and yellow grease; and (4) non-edible oils such as jatropha, neem oil and castor oil. The prevalence of these feedstocks varies around the world (Fig. 1). The regional availability of feedstocks for biodiesel production depends greatly on climate, soil conditions and options for alternate land use [83]. Consequently, different regions

Fig. 1 Biodiesel feedstocks around the world [55]

are focussing their efforts on different feedstocks. As an example, the widespread use of soya beans in the USA as a food product has led to the emergence of soya bean biodiesel in that country. In Europe, rapeseed is the most common source of biodiesel production. In India and south-east Asia, the jatropha tree is used in biodiesel production, and in Malaysia and Indonesia, palm oil is used as a significant biodiesel source.

4 Biodiesel Production

More than 100 years ago, Rudolf Diesel (1858–1913) demonstrated the operation of a diesel engine using vegetable oil as a fuel, so the potential of using these feedstocks has been long recognised. However, vegetable oils are extremely viscous, with viscosity ranging from 10 to 17 times higher than that of petroleum diesel [57]. This makes vegetable oil unsuitable to use as a direct fuel in the modern diesel engine. As a consequence, researchers and scientists have developed various methods to reduce the viscosity of bio-oils to make them suitable for diesel engine use. Some of these methods include dilution with other fuels, esterification, micro-emulsification, pyrolysis and catalytic cracking. Of these techniques, esterification is the most promising and widely used solution due to its high conversion efficiency, simplicity, low conversion cost and the fuel qualities of the product.

Transesterification of bio-oils with alcohols to produce esters is a widely used technique for commercial biodiesel production [83].

Transesterification is a chemical reaction in which oils (triglycerides) are converted into esters as shown in Fig. 2. Triglycerides react with alcohols (e.g. methanol and ethanol) under acid- or base-catalysed conditions, producing fatty acid alkyl esters and glycerol. A catalyst is used to improve the reaction rate and yield. Because the transesterification reaction is reversible, excess alcohol is used to shift the equilibrium to favour production of the ester. The yield of biodiesel in transesterification is affected by several process parameters. These include the reaction temperature, the molar ratio of alcohol to oil, the type and concentration of catalyst and the reaction time [58]. After the reaction is complete, glycerol is removed as a by-product. The biodiesel produced may be denominated by the feedstock used and the ester formed including fatty acid methyl ester (FAME), fatty acid ethyl ester (FAEE), soya bean methyl ester (SME) and rapeseed methyl ester (RME). The total ester content in biodiesel is the measure of the completeness of the transesterification reaction [106].

Alkali-catalysed transesterification cannot be directly used to produce high-quality biodiesel from feedstocks containing high levels of free fatty acids (FFAs). This is because FFAs react with the catalyst to form soap (Fig. 3), resulting in emulsification and separation problems. To overcome this problem, a pre-esterification process may be used to reduce the content of FFAs in the feedstock. A typical pre-esterification process uses homogeneous acid catalysts, such as sulphuric acid, phosphorous acid combined with sulphonic acid, or heterogenous "solid-acid" catalysts, to esterify the free fatty acids as shown in Fig. 4.

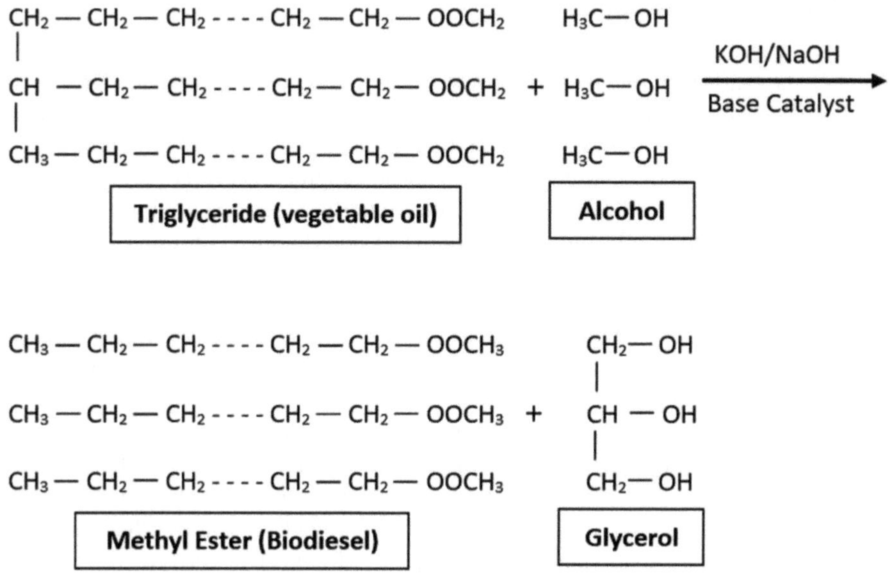

Fig. 2 Transesterification reaction [55]

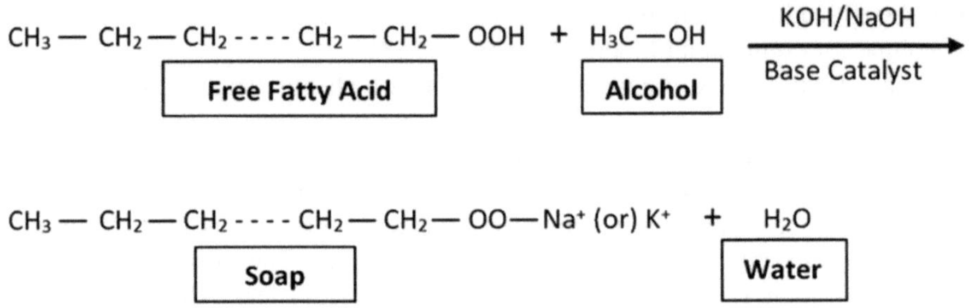

Fig. 3 Soap formation during biodiesel production [55]

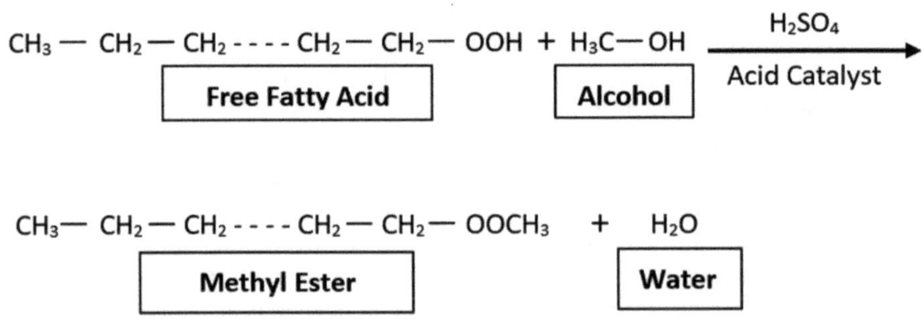

Fig. 4 Acid pre-esterification [55]

5 Biodiesel Standards

Quality standards are crucial for the commercial use of any fuel product. They serve as guidelines for production, assure customers that they are buying high-quality fuels and provide authorities with approved tools for a common approach to transport, storage and handling. Modern diesel engines using common rail fuel injection systems are more sensitive to fuel quality. Therefore, engine and automotive manufacturers rely on fuel standards in determining consumer warranties. However, the chemical compositions of biodiesel and petroleum diesel are very different, and these differences result in varying physicochemical properties. To improve the viability of biodiesel for use as a commercial fuel for direct replacement of petroleum diesel, the properties of biodiesel need to reflect a functional equivalence with diesel.

Biodiesel can be used as a pure fuel (B100) or blended with petroleum diesel in varying concentrations. For B100, the most internationally recognised standards are EN14214 (Europe) and ASTM D-6751 (USA). Both standards are similar in content, with only minor differences in some parameters. Many other countries have defined their own standards, which are frequently derived from either EN14214 or ASTM D-6751 [51]. As a part of the Fuel Quality Standards Act 2000,

Table 1 International standards of biodiesel [18, 126]

Properties	Units	USA ASTM D-6751	Europe EN 14214	Australia
Viscosity, 40 °C	mm^2/s	1.9–6.0	3.5–5.0	3.5–5.0
Density	gm/m^3	n/a	0.86–0.90	0.86–0.90
Cetane number	–	47 min	51 min	51 min
Flashpoint	°C	130 min	120 min	120 min
Cloud point	°C	Report	Report	Report
Acid number	mg KOH/g	0.80 max	0.5 max	0.8 max
Free glycerine	wt%	0.02 max	0.02 max	0.02 max
Total glycerine	wt%	0.24 max	0.25 max	0.25 max
Iodine number	–	–	120 max	n/a
Oxidation stability	h	–	6 min	n/a
Monoglyceride	Mass (%)	–	0.8 max	n/a
Diglyceride	Mass (%)	–	0.2 max	n/a
Triglyceride	Mass (%)	–	0.2 max	n/a

the Australian government released a biodiesel fuel standard, "Fuel Standard (Biodiesel) Determination 2003". A summary of the major fuel quality parameters in these standards is detailed in Table 1.

6 Data Collection

Data were collected from more than 150 papers, mostly published in the last decade, and which contain experimental results of the chemical composition of biodiesel along with corresponding fuel properties. During data collection, special care was taken to ensure the quality of the data and to eliminate duplication. Data have been taken only from literature in which the experiments were conducted following recognised international standards. Some extreme data have been excluded from the database due to unexpected results contained therein. Data were also eliminated from the database if it was found to differ greatly to fuel properties in the primary data collection results. Furthermore, the experimental results for density and kinematic viscosity of biodiesel are highly dependent on temperature. Although 15 and 40 °C temperatures are recommended for density and kinematic viscosity respectively, some researchers did not mention the test temperature. Therefore, those data have also been excluded from the database. Since the properties of particular biodiesels can be varied depending on the type of alcohol (methyl, ethyl, etc.) used in the production process, this study only considers the methyl esters for inclusion in the database. The list of papers including feedstock use and the country of the authors is given in Table 2.

Table 2 Biodiesel datasets investigated in this study

Feedstock	References
Algae	[37, 54]
Almond	[10, 46]
Babassu	[12, 89, 94, 114, 118]
Beauty leaf	[56]
Camelina	[27, 44, 91, 118, 129, 137]
Canola	[2, 21–23, 31, 37, 40, 49, 52, 69, 73, 88, 118]
Coconut	[6, 21, 40, 43, 75, 92, 114, 118, 131]
Coffee	[32, 133]
Corn	[29, 30, 81, 114, 123]
Cottonseeds	[2, 35, 89, 111, 115, 132]
Fish oil	[13, 82, 113]
Grape	[29, 109]
Hazelnut	[35, 73, 90]
Hepar	[118]
Jatropha	[8, 22–24, 26, 60, 76, 118, 120, 125, 133, 135]
Lard	[15, 36, 69, 80, 138]
Linseed	[41, 48, 79, 101, 103, 114, 117]
Mahua	[34, 45, 47, 66]
Mustard	[11, 62]
Neem	[8, 103, 104, 119, 128]
Olive	[21, 29, 38, 65, 109]
Palm	[12, 14, 21, 28, 29, 37, 63, 64, 67, 74, 84, 88, 93, 97, 99, 109, 122, 134]
Peanut	[12, 29, 31, 68, 81, 90, 99, 109, 130]
Poppyseed	[35]
Rapeseed	[21, 35, 44, 67, 79, 81, 97, 99, 109, 110, 116, 122, 133, 136]
Rice bran	[127]
Rubber seed	[53, 107]
Safflower	[110]
Sesame	[1, 16]
Soya bean	[2–5, 9, 12, 17, 19–21, 31, 40, 43, 48, 69, 81, 87–89, 97–99, 102, 109, 114, 120, 121, 124, 132, 136, 138]
Soap nut	[22, 23]
Sunflower	[7, 12, 29, 35, 41, 65, 81, 88, 99, 109, 112, 115, 120, 123]
Tallow	[3, 7, 12, 29, 35, 65, 79, 88, 95, 99, 108, 112]
Terebinth	[96]
Terminalia	[39]
Turnip	[124]
Walnut	[90]
Waste cooking oil	[3, 21, 25, 33, 42, 48, 73, 77, 81, 82, 100]
Yellow grease	[19, 69]
Pure methyl ester	[70–72, 89, 114]

7 Chemical Composition of Biodiesel

Petroleum diesel fuels are saturated straight-chain hydrocarbons with carbon chain lengths of 12–18, whereas vegetable oils and animal fats consist of 90–98% triglycerides, small amounts of monoglycerides and free fatty acids. The fatty acid compositions of triglycerides differ in chain length, the degree of unsaturation and the presence of other functional groups. The fatty acid compositions are feedstock-dependent and are affected by factors such as climatic conditions, soil type, plant health and plant maturity upon harvest. Using the carboxyl reference system, fatty acids are designated by two numbers: the first number denotes the total number of carbon atoms in the fatty acid, and the second is the number of double bonds indicating the degree of unsaturation. For example, 18:1 designates oleic acid which has 18 carbon atoms and one C=C double bond. The most common fatty acids found in biodiesels and their structures are listed in Table 3.

The biodiesels are mainly comprised of the methyl esters of various fatty acids. The most common components found in biodiesel samples are mono-unsaturated oleic acid (C18:1) and di-unsaturated linoleic acid (C18:2) methyl esters. These two fatty acids (C18:1 and C18:2) were found in almost every biodiesel sample with an average weight percentage of 34.9 and 24.7, respectively, as shown in Fig. 5. Also, a significant amount of unsaturated erucic acid (22:1) and oleic acid (18:1) methyl ester were found in biodiesels. Furthermore, the oleic (C18:1) and linoleic (C18:2) are not only most commonly found in the biodiesel samples, but also showed highest in average weight percentage in the biodiesel samples, at approximately 40 and 32%, respectively. By contrast, an average of 7.5 and 6.5% of linolenic acid (C18:3) and stearic (18:0) acids methyl esters were present in the samples. Apart from fatty acid methyl esters, other chemical compositions usually found in the biodiesel are mainly unreacted monoglycerides and free fatty acids represented as the acid value.

Table 3 Chemical structure of common fatty acids in biodiesels

Fatty acid		Chemical structure
1.	Caprylic (8:0)	$CH_3(CH_2)_6COOH$
2.	Capric (10:0)	$CH_3(CH_2)_8COOH$
3.	Lauric (12:0)	$CH_3(CH_2)_{10}COOH$
4.	Myristic (14:0)	$CH_3(CH_2)_{12}COOH$
5.	Palmitic (16:0)	$CH_3(CH_2)_{14}COOH$
6.	Palmitoileic (16:1)	$CH_3(CH_2)_6$ CH=CH $(CH_2)_6$ COOH
7	Stearic (18:0)	$CH_3(CH_2)_{16}COOH$
8.	Oleic (18:1)	$CH_3(CH_2)_7$ CH=CH $(CH_2)_7$ COOH
9.	Linoleic (18:2)	$CH_3(CH_2)_4$ CH=CHCH$_2$CH=CH $(CH_2)_7$ COOH
10.	Linolenic (18:3)	$CH_3(CH_2)_2$CH=CHCH$_2$CH=CHCH$_2$CH=CH$(CH_2)_7$ COOH
11.	Gondonic (20:1)	$CH_3(CH_2)_7$ CH=CH $(CH_2)_9$ COOH
12.	Erucic (22:1)	$CH_3(CH_2)_9$ CH=CH $(CH_2)_9$ COOH

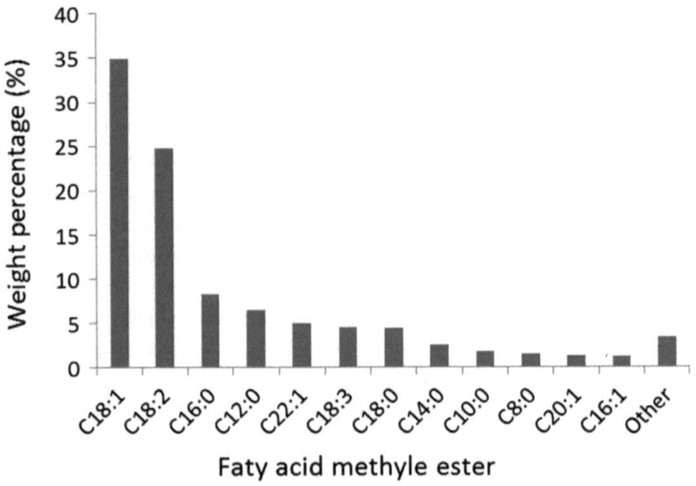

Fig. 5 Average weight in percentages of fatty acid methyl esters found in the collected data samples

8 Fuel Properties

The quality of biodiesel and its performance as an internal combustion engine application is largely determined by the fuel properties. One important liquid fuel property is kinematic viscosity (KV), which indicates the resistance or flow of liquid fuel. It plays a dominant role in the fuel spray, fuel–air mixture formation and the combustion process in diesel engine application. In a diesel engine, the liquid fuel is sprayed into compressed air and atomised into small droplets near the nozzle exit. In the engine combustion chamber, the fuel forms a cone-shaped spray at the nozzle exit which is affected by the viscosity. In addition to that viscosity also affects the fuel atomisation quality, penetration and size of the fuel droplet [18]. Higher viscosities result in higher drag in the fuel line and injection pump, higher engine deposits, higher fuel pump duties and increased wear in the fuel pump elements and injectors. Moreover, the mean diameter of the fuel droplets from the injector and their penetration increases with an increase in fuel viscosity. Higher pressure in the fuel line can cause early injection, moving the combustion of the fuel closer to top dead centre, thus increasing the maximum pressure and temperature in the combustion chamber [80]. Studies show that in a light-duty diesel engine, the CO and UHC could increase by 0.02% (by volume) and 1 ppm (by volume), respectively, with 1 cSt. increase of fuel viscosity [72]. On the other hand, low fuel viscosity is undesirable because it does not provide sufficient lubrication for the precision fit of fuel injection pumps, resulting in leakage or increased wear [50]. Therefore, all biodiesel standards define the upper limit and lower limit of viscosity. Heating value is another fuel property indicating the energy content in the fuel, along with biodiesel density. It determines the amount of energy taken by the engine in certain volume. When injecting the fuel in a diesel engine, the fuel pumps measure fuel on the basis of volume, not by mass. Therefore, the change in biodiesel density directly affects the engine output power and hence engine performance. This property is also correlated with engine exhaust emissions, particularly particulate matter (PM), nitrogen oxides (NOX) and carbon mono-oxides (CO) [105]. This is because the higher density of biodiesel increases the diameter of the fuel droplets in the combustion chamber, which consequently affects the fuel atomisation, combustion process and exhaust emission formation. Another significant fuel property that directly affects engine output power is the higher heating value (HHV). It determines the suitability of biodiesel as an engine fuel, as it indicates the energy content in the fuel. In general, biodiesels are approximately 10% less energy dense as compared with petroleum diesel, depending on the oxygen content in the hydrocarbon molecules. Furthermore, of great concern when considering biodiesels for engine fuel is oxidation stability (OS), which reflects resistance to oxidation during long-term storage. Biodiesels from any sources usually show very poor oxidation stability when compared with mineral diesel due to their chemical composition. During the oxidation process, the quality of fuel declines due to gum formation which remains in the biodiesel. This gum does not combust completely, resulting in poor combustion, carbon deposits in the combustion chamber and lubrication oil thickening. Therefore, high oxidation stability is desirable for a good quality biodiesel.

The average fuel properties reported in the collected data were found to be within the limits of European (EU), American (US) and Australian (AU) biodiesel standards, except for oxidation stability (OS). The average OS was found to be 4.73 h, which is much lower than the minimum OS requirement (6 h minimum) of EU and AU biodiesel standards. These results indicate that a vast number of the investigated biodiesels were unlikely to fulfil EU biodiesel standards, and this may be one of the major issues that restrict the widespread use of biodiesel in conventional diesel engines. The biodiesels that showed poor OS and which were rich in unsaturated FAME include soya bean, sunflower, safflower, corn, cottonseeds, linseeds, jatropha and camelina. European and Australian biodiesel standards also impose tight restrictions on kinematic viscosity (KV), limiting it to a minimum of 3.5 and a maximum of 5 cSt. However, the KV of biodiesels in the secondary data ranged from 0.99 to 7.21 cSt, which means that many of them would be unlikely to meet EU and AU biodiesel standards regarding KV. US biodiesel standards are laxer regarding KV (1.9–6 cSt). However, they place tighter restrictions on other biodiesel properties, which means that many biodiesels identified in the secondary dataset would still be unlikely to meet US standards. Overall, most of the maximum and minimum values for fuel properties were outside the range of biodiesel standards, demonstrating the significant level of variation in the data. This was not unexpected because the data were collected from a large number of different biodiesels with a wide variety of chemical structures. Therefore, the collected secondary data were useful for conducting an in-depth correlation study between fuel properties and the chemical composition of biodiesel (Table 4).

Table 4 Summary of data for biodiesel properties

Properties	Biodiesel standard			Max.[a]	Min.[a]	Ave.[a]
	ASTM D7651	EN 4214	Australian			
Kinematic viscosity (cSt.)	1.9–6	3.5–5	3.5–5	6	2.15	4.42
Density (kg/l)	n/a	0.86–0.9	0.86–0.90	0.924	0.829	0.876
Higher heating value (Mj/kg)	n/a	n/a	n/a	41.6	35.86	39.91
Oxidation stability (h)	3 min	6 min	6 min	11.4	0.2	4.73

[a]Max. = maximum value; Min. = minimum value; Ave. = average value

9 Correlation of Chemical Composition and Fuel Properties

The fuel properties of biodiesel are generally controlled by its chemical composition. Due to variations in the chemical structure in fatty acid methyl esters, the fuel properties of biodiesel significantly differ from one another. Figure 6 shows the effect of average chain length (ACL) on kinematic viscosity (KV), density, higher heating value (HHV) and oxidation stability (OS). The ACL was correlated with all the fuel properties investigated in this study. There was a very strong positive correlation with KV, as shown in Fig. 4a. This is mainly due to the increase in carbon content, as well as random intermolecular interaction in the FAME, which consequently increased the KV. For the same reason, ACL was also found to have a strong positive correlation with density and HHV. It is also interesting to found that biodiesels with an ACL less than 14 were unlikely to meet the lower limit of both US and EU standards. On the other hand, biodiesel with a very high ACL (over 19) is more likely to exceed the upper limit of biodiesel standards regarding KV.

The average number of double bonds (ANDB) in the biodiesel (which indicates the concentration of unsaturated fatty acid methyl esters) was found to be another influential factor affecting most of the biodiesel properties investigated in this study. Figure 7 shows the effect of ANDB on the kinematic viscosity (KV), density, higher heating value (HHV) and oxidation stability (OS). Oxidation stability (OS) may have a slight negative correlation with ACL, as shown in Fig. 4d. However, this property has a very strong negative correlation with ANDB. Figure 5d indicates that the OS of biodiesel decreased rapidly with an increase in ANDB. This is because a higher number of double bonds in the fatty acid chain of biodiesel make it much more susceptible to oxidation. ANDB also has a moderate

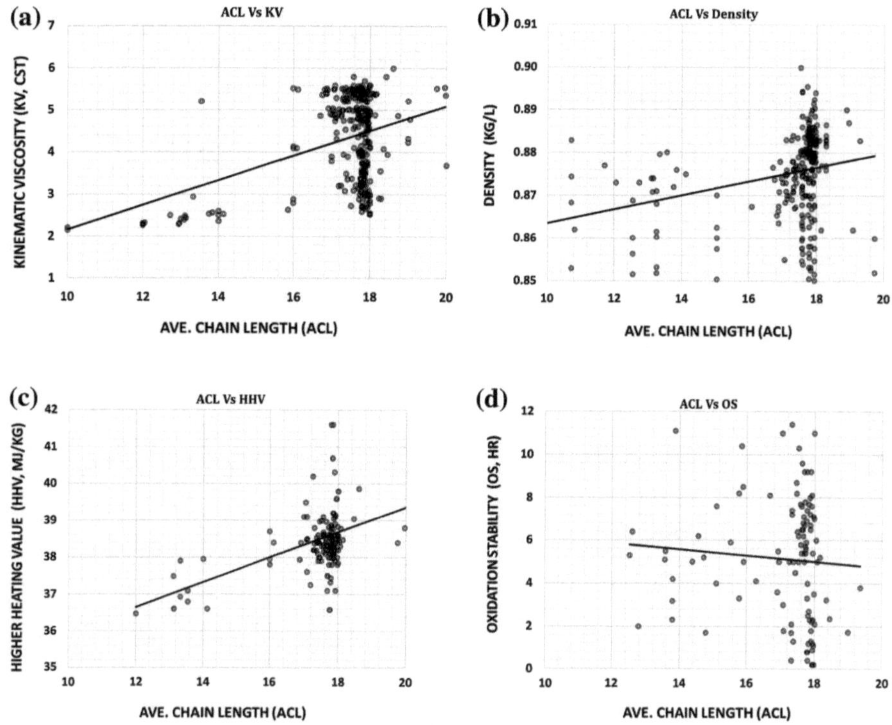

Fig. 6 Effect of ACL on biodiesel **a** kinematic viscosity (KV), **b** density, **c** higher heating value (HHV) and **d** oxidation stability (OS)

positive correlation with density and HHV, but no correlation was found between ANDB and KV, as shown in Fig. 5a.

The result of this analysis indicated that ACL and ANDB have certain effects on the four biodiesel properties investigated in this study. However, with a close look at Figs. 4 and 5, it can be seen that ACL and ANDB are the only parameters that control all fuel properties of biodiesel, rather other parameters and their combined effect may involve in determining the properties. Therefore, a multicriteria data analysis is required to investigate the correlation of biodiesel's properties with its chemical composition, and this is investigated in the next section.

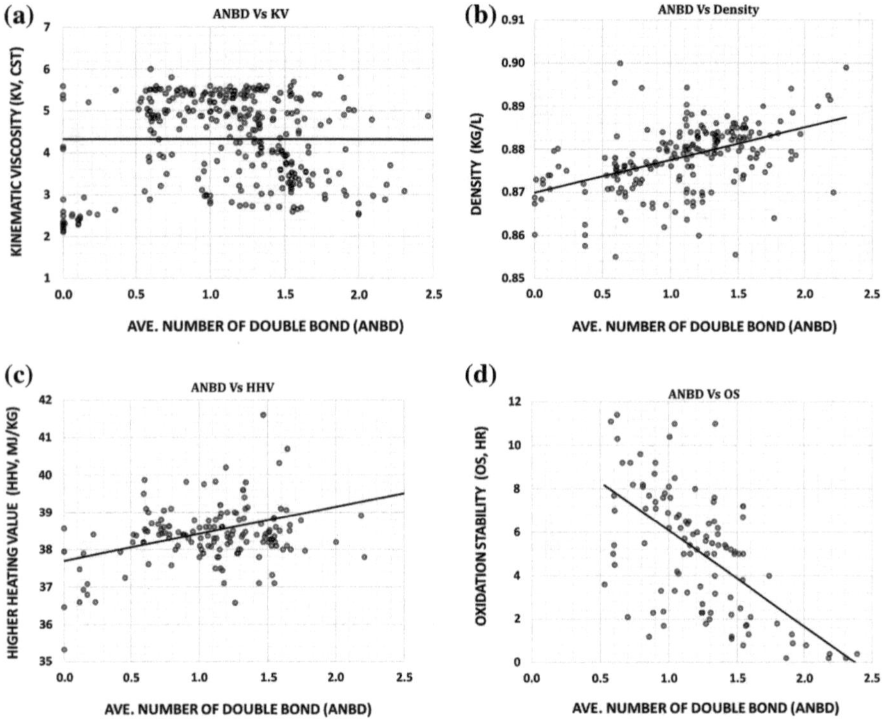

Fig. 7 Effect of ANDB on biodiesel **a** kinematic viscosity (KV), **b** density, **c** higher heating value (HHV) and **d** oxidation stability (OS)

10 Principal Component Analysis

The findings of the correlation study reported in the previous section indicate a complex relationship between biodiesel quality and its chemical composition. A particular fuel property does not depend on a single chemical parameter; rather it is influenced by multiple parameters and factors. Therefore, multivariate data analysis is required to gain a detailed understanding of this relationship. Principal component analysis (PCA) is one of the popular multivariate data analysis techniques used by almost all scientific disciplines. PCA is used to analyse datasets with highly intercorrelated dependent variables. It reduces the complexity and dimensionality of the problem, thereby extracting the most important information and analysing the structure of the observations and variables. PCA changes the input variables into principal components (PCs) that are an independent and linear combination of input variables. PCA also represents patterns in the observations and variables by displaying them as points on a diagram. In this study, PCA was conducted using Microsoft XLSTAT software to observe the influence of chemical composition on individual fatty acid compositions. The variables used for the principal components were individual fatty acid methyl esters chain length ranging from 8 to 22, while the interaction terms included average chain length (ACL), average number of double bonds (ANDB) and weight percentages of oxygen (O_2), hydrogen (H), carbon (C), saturation (percentage of saturated fatty acid), mono-unsaturated fraction (MUFA) and poly-unsaturated fraction (PUFA). The variables also included the most commonly found impurities in biodiesel, namely monoglyceride and free fatty acid contents regarding acid number (AN). In general, variables which lie close to (±45°) an observation are correlated, those lying in opposite directions (135°–225°) are anti-correlated, and those lying in an orthogonal direction have less or no influence. The direction and length of the variables are indicative of their influence on the observation, with a short length indicative of little influence. The results of four fuel properties are graphically shown in Figs. 8, 9, 10 and 11.

As discussed earlier, the results of principal component analysis as shown in Fig. 8 also indicate a strong positive correlation between KV and ACL. Therefore, fatty acid methyl esters (FAME) with a carbon chain length 18 or above increase the KV, and the inverse is true for short chain fatty acids (<C16). Moreover, the presence of unsaturated FAME also had a moderate positive influence on KV. Among the unsaturated FAME, MUFA has more influence compared to PUFA on KV. The presence of other parameters that increase the KV of biodiesel are impurities and carbon content. On the other hand, an increase in the oxygen percentage reduces the KV of biodiesel. The influence of free fatty acid and hydrogen

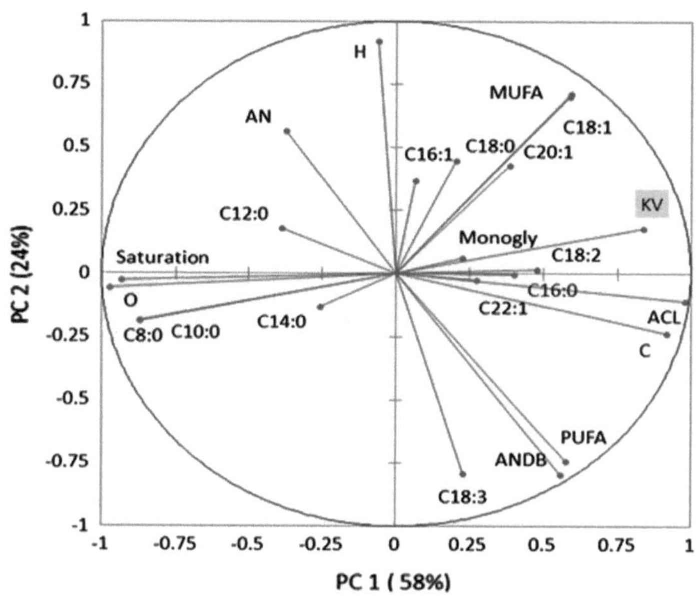

Fig. 8 Principal component analysis and correlation of kinematic viscosity with chemical composition of biodiesel

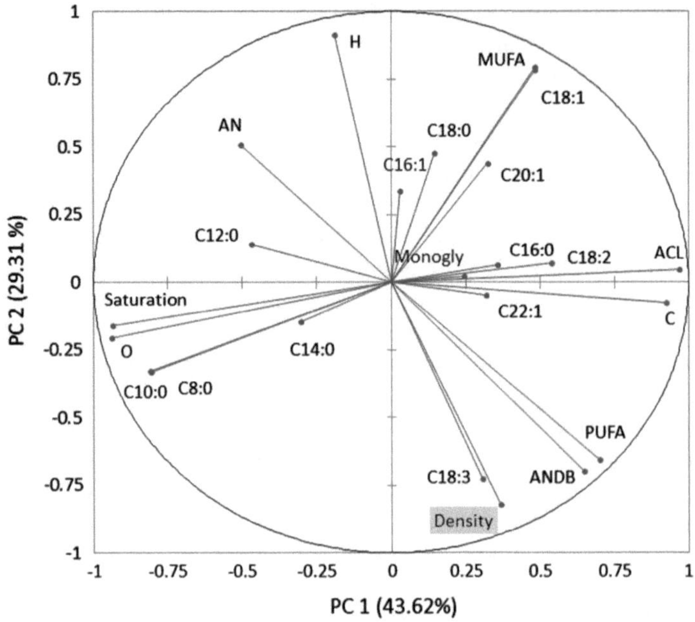

Fig. 9 Principal component analysis and correlation of density with chemical composition of biodiesel

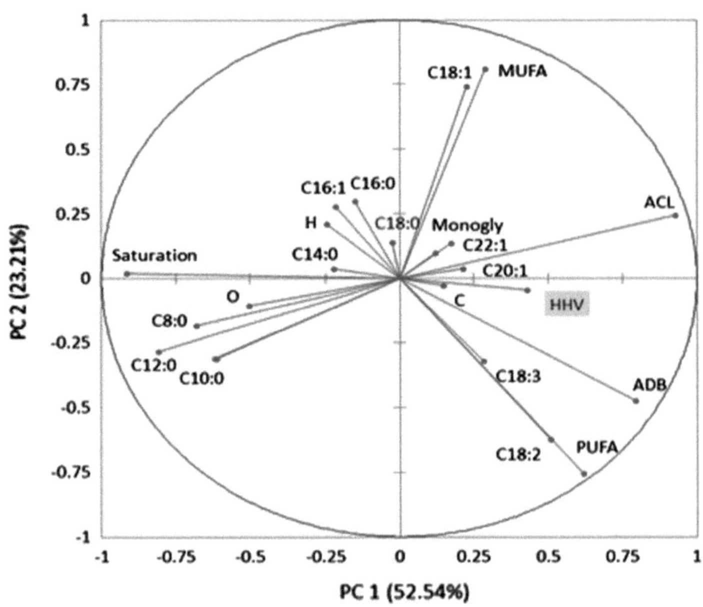

Fig. 10 Principal component analysis and correlation of higher heating value with chemical composition of biodiesel

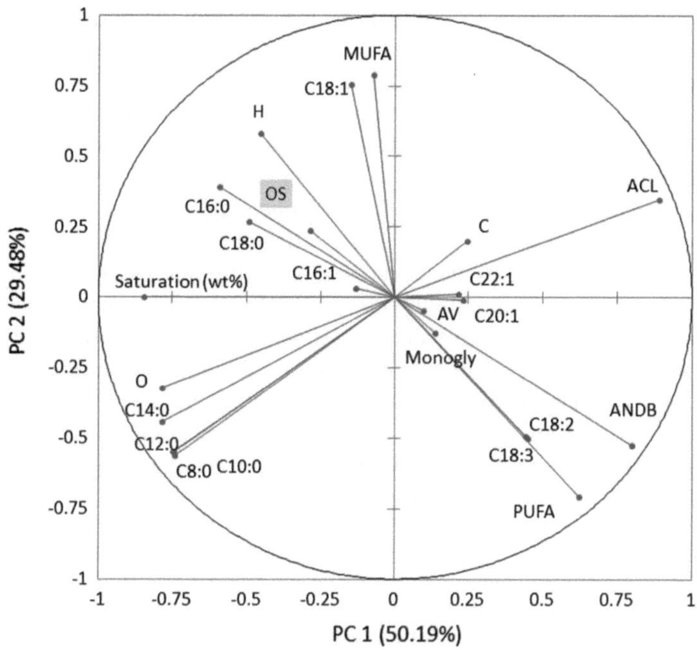

Fig. 11 Principal component analysis and correlation of oxidation stability with chemical composition of biodiesel

content on KV is insignificant. In contrast, the density of biodiesels is significantly affected by ANDB, and PUFA is shown in Fig. 9. In the PCA plan, the line for linolenic (C18:3) FAME lies very close to the density line which indicates a strong positive relationship between them. The other parameter which strongly influences the density of biodiesel is the percentage of hydrogen and AN with opposite correlations. The presence of saturated FAME significantly reduces the HHV of biodiesel as shown in Fig. 10. This figure also shows that ACL and ANDB have a positive correlation with HHV. Therefore, biodiesel with a high content of long-chain unsaturated FAME always shows a higher HHV. But the opposite correlation is evident in Fig. 11, where long-chain unsaturated FAME significantly reduces the OS of the biodiesel. It can be argued that biodiesel with a higher HHV may have less OS and vice versa.

11 Conclusion

In this chapter, a detailed investigation of key physicochemical properties of biodiesel has been carried out. During the last few decades, numerous types of biodiesel have been investigated. Formulated from a wide variety of sources, biodiesels differ from one to another as regards fuel properties and chemical composition. The aim of this chapter was to investigate the correlation between the fuel properties of biodiesel and its chemical composition, based on data collected from published literature. The fuel properties considered in this study were kinematic viscosity, density, higher heating value and oxidation stability. An investigation was conducted using principal component analysis (PCA) data analysis tool. A complex relationship was found between chemical composition and biodiesel properties. PCA indicated that the fuel properties of biodiesels are determined by a number of parameters and by the combination of different chemical compositions. An average number of double bonds (ANDB) and an average chain length (ACL) may well be the most influential parameters affecting most of the properties of biodiesels. Parameters relating to biodiesel production and purification, such as free fatty acids and glycerol content, also influence certain biodiesel properties found in this study. Thus, the future challenge is to develop an accurate model for estimating biodiesel properties and to find an optimum combination of the chemical composition of biodiesel for enhanced performance in automotive applications.

Acknowledgements The authors wish to express their thanks for the provision of the QUTPRA scholarship, and to QUT for providing funds for conducting this research. The authors would also like to thank the Biofuel Engine Research Facility (BERF) for providing facilities to complete this study. The authors would also like to thank Australian Research Council's Linkage Projects funding scheme (project number LP110200158) for providing support for this research.

References

1. Ahmad, M., Ullah, K., Khan, M., Ali, S., Zafar, M., & Sultana, S. (2011). Quantitative and qualitative analysis of sesame oil biodiesel. *Energy Sources, Part A: Recovery, Utilization, and Environmental Effects, 33,* 1239–1249.
2. Albuquerque, M., Machado, Y., Torres, A., Azevedo, D., Cavalcante JR, C., Firmiano, L. & Parente JR, E. (2009). Properties of biodiesel oils formulated using different biomass sources and their blends. *Renewable Energy, 34*, 857–859.
3. Alcantara, R., Amores, J., Canoira, L. T., Fidalgo, E., Franco, M., & Navarro, A. (2000). Catalytic production of biodiesel from soy-bean oil, used frying oil and tallow. *Biomass and Bioenergy, 18,* 515–527.
4. Ali, Y., Hanna, M., & Cuppett, S. (1995). Fuel properties of tallow and soybean oil esters. *Journal of the American Oil Chemists' Society, 72,* 1557–1564.
5. Ali, Y., Hanna, M. A., & Cuppett, S. L. (1995). Fuel properties of tallow and soybean oil esters. *Journal of the American Oil Chemists' Society, 72,* 1557–1564.
6. Alleman, T., & Mccormick, R. L. (2006). *Analysis of coconut-derived biodiesel and conventional diesel fuel samples from the Philippines. US DOE, NREL/MP-540-38643.*
7. Antolın, G., Tinaut, F., Briceno, Y., Castano, V., Perez, C., & Ramırez, A. (2002). Optimisation of biodiesel production by sunflower oil transesterification. *Bioresource Technology, 83*, 111–114.
8. Aransiola, E., Ojumu, T., Oyekola, O., & Ikhuomoregbe, D. (2012). A study of biodiesel production from non-edible oil seeds: A comparative study. *The Open Conference Proceedings Journal, M*1.

9. Armas, O., Yehliu, K., & Boehman, A. L. (2010). Effect of alternative fuels on exhaust emissions during diesel engine operation with matched combustion phasing. *Fuel, 89,* 438–456.
10. Atapour, M., & Kariminia, H.-R. (2011). Characterization and transesterification of Iranian bitter almond oil for biodiesel production. *Applied Energy, 88,* 2377–2381.
11. Bannikov, M. (2011). Combustion and emission characteristics of Mustard biodiesel. In *6th International Advanced Technologies Symposium (IATS'11), Turkey* (pp.1–5).
12. Barnwal, B., & Sharma, M. (2005). Prospects of biodiesel production from vegetable oils in India. *Renewable and Sustainable Energy Reviews, 9,* 363–378.
13. Behçet, R. (2011). Performance and emission study of waste anchovy fish biodiesel in a diesel engine. *Fuel Processing Technology, 92,* 1187–1194.
14. Benjumea, P., Agudelo, J., & Agudelo, A. (2008). Basic properties of palm oil biodiesel–diesel blends. *Fuel, 87,* 2069–2075.
15. Berrios, M., Gutiérrez, M., Martín, M., & Martin, A. (2009). Application of the factorial design of experiments to biodiesel production from lard. *Fuel Processing Technology, 90,* 1447–1451.
16. Betiku, E., & Adepoju, T. F. (2013). Methanolysis optimization of sesame (Sesamum indicum) oil to biodiesel and fuel quality characterization. *International Journal of Energy and Environmental Engineering, 4,* 1–8.
17. Canakci, M. (2005). Performance and emissions characteristics of biodiesel from soybean oil. *Proceedings of the Institution of Mechanical Engineers, Part D: Journal of Automobile Engineering, 219,* 915–922.
18. Canakci, M., & Sanli, H. (2008). Biodiesel production from various feedstocks and their effects on the fuel properties. *Journal of Industrial Microbiology and Biotechnology, 35,* 431–441.
19. Canakci, M., & van Gerpen, J. H. (2003). Comparison of engine performance and emissions for petroleum diesel fuel, yellow grease biodiesel, and soybean oil biodiesel. *Transactions of the ASAE, 46,* 937–944.
20. Candeia, R., Silva, M., Carvalho Filho, J., Brasilino, M., Bicudo, T., Santos, I., & Souza, A. 2009. Influence of soybean biodiesel content on basic properties of biodiesel–diesel blends. *Fuel, 88*, 738–743.
21. Cecrle, E., Depcik, C., Duncan, A., Guo, J., Mangus, M., Peltier, E., et al. (2012). Investigation of the effects of biodiesel feedstock on the performance and emissions of a single-cylinder diesel engine. *Energy & Fuels, 26,* 2331–2341.
22. Chhetri, A., & Watts, K. (2012). Densities of canola, jatropha and soapnut biodiesel at elevated temperatures and pressures. *Fuel, 99,* 210–216.
23. Chhetri, A., & Watts, K. (2012). Viscosities of canola, jatropha and soapnut biodiesel at elevated temperatures and pressures. *Fuel, 102,* 789–794.
24. Chhetri, A. B., Tango, M. S., Budge, S. M., Watts, K. C., & Islam, M. R. (2008). Non-edible plant oils as new sources for biodiesel production. *International Journal of Molecular Sciences, 9,* 169–180.
25. Chhetri, A. B., Watts, K. C., & Islam, M. R. (2008). Waste cooking oil as an alternate feedstock for biodiesel production. *Energies, 1,* 3–18.
26. Choudhury, S., & Bose, P. (2008). *Jatropha derived Biodiesel–Its suitability as CI engine fuel*. SAE Technical Paper.
27. Chung, K.-H. (2010). Transesterification of *Camellia japonica* and *Vernicia fordii* seed oils on alkali catalysts for biodiesel production. *Journal of Industrial and Engineering Chemistry, 16,* 506–509.
28. Crabbe, E., Nolasco-Hipolito, C., Kobayashi, G., Sonomoto, K., & Ishizaki, A. (2001). Biodiesel production from crude palm oil and evaluation of butanol extraction and fuel properties. *Process Biochemistry, 37,* 65–71.
29. Cursaru, D., Neagu, M., & Bogatu, L. (2013). Investigations on the oxidation stability of biodiesel synthesized from different vegetable oils. *Revista de Chimie, 64,* 438–441.

30. Dantas, M., Albuquerque, A., Barros, A., Rodrigues Filho, M., Antoniosi Filho, N., Sinfrônio, F., et al. (2011). Evaluation of the oxidative stability of corn biodiesel. *Fuel, 90,* 773–778.
31. Davis, J., Geller, D., Faircloth, W., & Sanders, T. (2009). Comparisons of biodiesel produced from unrefined oils of different peanut cultivars. *Journal of the American Oil Chemists' Society, 86,* 353–361.
32. Deligiannis, A., Papazafeiropoulou, A., Anastopoulos, G., & Zannikos, F. (2011) Waste coffee grounds as an energy feedstock. In *Proceedings of the 3rd International CEMEPE & SECOTOX Conference Skiathos*, ISBN, (pp. 978–960).
33. Demirbas, A. (2009). Biodiesel from waste cooking oil via base-catalytic and supercritical methanol transesterification. *Energy Conversion and Management, 50,* 923–927.
34. Demirbas, A. (2009). Potential resources of non-edible oils for biodiesel. *Energy Sources, Part B: Economics, Planning and Policy, 4,* 310–314.
35. Demirbaş, A. (2002). Biodiesel from vegetable oils via transesterification in supercritical methanol. *Energy Conversion and Management, 43,* 2349–2356.
36. Dias, J. M., Alvim-Ferraz, M., & Almeida, M. F. (2009). Production of biodiesel from acid waste lard. *Bioresource Technology, 100,* 6355–6361.
37. Do, L. D., Singh, V., Chen, L., Kibbey, T. C., Gollahalli, S. R., & Sabatini, D. A. (2011). Algae, Canola, or Palm oils—Diesel microemulsion fuels: Phase behaviors, viscosity, and combustion properties. *International Journal of Green Energy, 8,* 748–767.
38. Dorado, M., Ballesteros, E., Arnal, J., Gomez, J., & Lopez Gimenez, F. (2003). Testing waste olive oil methyl ester as a fuel in a diesel engine. *Energy & Fuels, 17,* 1560–1565.
39. Dos Santos, I., De Carvalho, S., Solleti, J., Ferreira de La Salles, W., Teixeira da Silva de La Salles, K., & Meneghetti, S. (2008). Studies of *Terminalia catappa L.* oil: Characterization and biodiesel production. *Bioresource Technology, 99,* 6545–6549.
40. Duncan, A. M., Ahosseini, A., McHenry, R., Depcik, C. D., Stagg-Williams, S. M., & Scurto, A. M. (2010). High-pressure viscosity of biodiesel from soybean, canola, and coconut oils. *Energy & Fuels, 24,* 5708–5716.
41. el Diwani, G., & el Rafie, S. (2008). Modification of thermal and oxidative properties of biodiesel produced from vegetable oils. *International Journal of Environmental Science and Technology, 5,* 391–400.
42. Encinar, J. M., Gonzalez, J. F., & Rodríguez-Reinares, A. (2005). Biodiesel from used frying oil. Variables affecting the yields and characteristics of the biodiesel. *Industrial and Engineering Chemistry Research, 44,* 5491–5499.
43. Feitosa, F. X., Rodrigues, M. D. L., Veloso, C. B., Cavalcante Jr, C. L., Albuquerque, M. C., & De Sant'ana, H. B. (2010). Viscosities and densities of binary mixtures of coconut colza and coconut + soybean biodiesel at various temperatures. *Journal of Chemical & Engineering Data,* 55, 3909–3914.
44. Fröhlich, A., & Rice, B. (2005). Evaluation of *Camelina sativa* oil as a feedstock for biodiesel production. *Industrial Crops and Products, 21,* 25–31.
45. Ghadge, S. V., & Raheman, H. (2005). Biodiesel production from mahua (*Madhuca indica*) oil having high free fatty acids. *Biomass and Bioenergy, 28,* 601–605.
46. Giwa, S., & Ogunbona, C. (2014). Sweet almond (Prunus amygdalus" dulcis") seeds as a potential feedstock for Nigerian Biodiesel Automotive Project. *Revista Ambiente & Água, 9,* 37–45.
47. Godiganur, S., Suryanarayana Murthy, C., & Reddy, R. P. (2009). 6BTA 5.9 G2-1 Cummins engine performance and emission tests using methyl ester mahua (*Madhuca indica*) oil/diesel blends. *Renewable Energy,* 34, 2172-2177.
48. Guzatto, R., de Martini, T. L., & Samios, D. (2011). The use of a modified TDSP for biodiesel production from soybean, linseed and waste cooking oil. *Fuel Processing Technology, 92,* 2083–2088.
49. Haagenson, D. M., Brudvik, R. L., Lin, H., & Wiesenborn, D. P. (2010). Implementing an in situ alkaline transesterification method for canola biodiesel quality screening. *Journal of the American Oil Chemists' Society, 87,* 1351–1358.

50. Haseeb, A., Fazal, M., Jahirul, M., & Masjuki, H. (2011). Compatibility of automotive materials in biodiesel: A review. *Fuel, 90,* 922–931.
51. Hoekman, S. K., Broch, A., Robbins, C., Ceniceros, E., & Natarajan, M. (2012). Review of biodiesel composition, properties, and specifications. *Renewable and Sustainable Energy Reviews, 16,* 143–169.
52. Hu, J., Du, Z., Li, C., & Min, E. (2005). Study on the lubrication properties of biodiesel as fuel lubricity enhancers. *Fuel, 84,* 1601–1606.
53. Ikwuagwu, O., Ononogbu, I., & Njoku, O. (2000). Production of biodiesel using rubber [*Hevea brasiliensis* (Kunth. Muell.)] seed oil. *Industrial Crops and Products, 12,* 57–62.
54. Islam, M. A., Brown, R. J., Brooks, P. R., Jahirul, M. I., Bockhorn, H., & Heimann, K. (2015). Investigation of the effects of the fatty acid profile on fuel properties using a multi-criteria decision analysis. *Energy Conversion and Management, 98,* 340–347.
55. Jahirul, M., Brown, R., Senadeera, W., O'hara, I., & Ristovski, Z. (2013a). The use of artificial neural networks for identifying sustainable biodiesel feedstocks. *Energies, 6,* 3764.
56. Jahirul, M. I., Brown, J. R., Senadeera, W., Ashwath, N., Laing, C., Leski-Taylor, J., et al. (2013). Optimisation of bio-oil extraction process from beauty leaf (calophyllum inophyllum) oil seed as a second generation biodiesel source. *Procedia Engineering, 56,* 619–624.
57. Jahirul, M. I., Brown, R. J., Senadeera, W., O'Hara, I. M., & Ristovski, Z. D. (2013). The use of artificial neural networks for identifying sustainable biodiesel feedstocks. *Energies, 6,* 3764–3806.
58. Jahirul, M. I., Koh, W., Brown, R. J., Senadeera, W., O'Hara, I., & Moghaddam, L. (2014). Biodiesel production from non-edible beauty leaf (Calophyllum inophyllum) oil: Process optimization using response surface methodology (RSM). *Energies, 7,* 5317–5331.
59. Jahirul, M. I., Masjuki, H., Saidur, R., Kalam, M., Jayed, M., & Wazed, M. (2010). Comparative engine performance and emission analysis of CNG and gasoline in a retrofitted car engine. *Applied Thermal Engineering, 30,* 2219–2226.
60. Jain, S., & Sharma, M. (2012). Application of thermogravimetric analysis for thermal stability of *Jatropha Curcas* biodiesel. *Fuel, 93,* 252–257.
61. Jayed, M., Masjuki, H., Saidur, R., Kalam, M., & Jahirul, M. I. (2009). Environmental aspects and challenges of oilseed produced biodiesel in Southeast Asia. *Renewable and Sustainable Energy Reviews, 13,* 2452–2462.
62. Jham, G. N., Moser, B. R., Shah, S. N., Holser, R. A., Dhingra, O. D., Vaughn, S. F., et al. (2009). Wild Brazilian mustard (Brassica juncea L.) seed oil methyl esters as biodiesel fuel. *Journal of the American Oil Chemists' Society, 86,* 917–926.
63. Kalam, M., & Masjuki, H. (2002). Biodiesel from palmoil—an analysis of its properties and potential. *Biomass and Bioenergy, 23,* 471–479.
64. Kalam, M. A., & Masjuki, H. H. (2002). Biodiesel from palmoil—An analysis of its properties and potential. *Biomass and Bioenergy, 23,* 471–479.
65. Kalligeros, S., Zannikos, F., Stournas, S., Lois, E., Anastopoulos, G., Teas, C., et al. (2003). An investigation of using biodiesel/marine diesel blends on the performance of a stationary diesel engine. *Biomass and Bioenergy, 24,* 141–149.
66. Kapilan, N., & Reddy, R. (2008). Evaluation of methyl esters of mahua oil (Madhuca indica) as diesel fuel. *Journal of the American Oil Chemists' Society, 85,* 185–188.
67. Karavalakis, G., Stournas, S., & Bakeas, E. (2009). Light vehicle regulated and unregulated emissions from different biodiesels. *Science of the Total Environment, 407,* 3338–3346.
68. Kaya, C., Hamamci, C., Baysal, A., Akba, O., Erdogan, S., & Saydut, A. (2009). Methyl ester of peanut (*Arachis hypogea* L.) seed oil as a potential feedstock for biodiesel production. *Renewable Energy, 34,* 1257–1260.
69. Kinast, J. (2003). Production of *Biodiesels from Multiple Feedstocks and Properties of Biodiesels and Biodiesel/Diesel Blends*. Final Report, Report 1 in a Series of 6. National Renewable Energy Lab., Golden, CO, US.
70. Knothe, G. (2005). Dependence of biodiesel fuel properties on the structure of fatty acid alkyl esters. *Fuel Processing Technology, 86,* 1059–1070.

71. Knothe, G. (2008). "Designer" biodiesel: Optimizing fatty ester composition to improve fuel properties. *Energy & Fuels, 22,* 1358–1364.
72. Knothe, G., & Steidley, K. R. (2005). Kinematic viscosity of biodiesel fuel components and related compounds. Influence of compound structure and comparison to petrodiesel fuel components. *Fuel, 84,* 1059–1065.
73. Koçak, M. S., Ileri, E., & Utlu, Z. (2007). Experimental study of emission parameters of biodiesel fuels obtained from canola, hazelnut, and waste cooking oils. *Energy & Fuels, 21,* 3622–3626.
74. Kousoulidou, M., Fontaras, G., Ntziachristos, L., & Samaras, Z. (2010). Biodiesel blend effects on common-rail diesel combustion and emissions. *Fuel, 89,* 3442–3449.
75. Kumar, G., Kumar, D., Singh, S., Kothari, S., Bhatt, S., & Singh, C. P. (2010). Continuous low cost transesterification process for the production of coconut biodiesel. *Energies, 3,* 43–56.
76. Kumar Tiwari, A., Kumar, A., & Raheman, H. (2007). Biodiesel production from jatropha oil (*Jatropha curcas*) with high free fatty acids: An optimized process. *Biomass and Bioenergy, 31,* 569–575.
77. Lapuerta, M., Herreros, J. M., Lyons, L. L., García-Contreras, R., & Briceño, Y. (2008). Effect of the alcohol type used in the production of waste cooking oil biodiesel on diesel performance and emissions. *Fuel, 87,* 3161–3169.
78. Lebedevas, S., & Vaicekauskas, A. (2006). Research into the application of biodiesel in the transport sector of Lithuania. *Transport, 21,* 80–87.
79. Lebedevas, S., Vaicekauskas, A., Lebedeva, G., Makareviciene, V., Janulis, P., & Kazancev, K. (2006). Use of waste fats of animal and vegetable origin for the production of biodiesel fuel: quality, motor properties, and emissions of harmful components. *Energy & Fuels, 20,* 2274–2280.
80. Lee, K.-T., Foglia, T. A., & Chang, K.-S. (2002). Production of alkyl ester as biodiesel from fractionated lard and restaurant grease. *Journal of the American Oil Chemists' Society, 79,* 191–195.
81. Lin, B.-F., Huang, J.-H., & Huang, D.-Y. (2009). Experimental study of the effects of vegetable oil methyl ester on DI diesel engine performance characteristics and pollutant emissions. *Fuel, 88,* 1779–1785.
82. Lin, C.-Y., & Li, R.-J. (2009). Engine performance and emission characteristics of marine fish-oil biodiesel produced from the discarded parts of marine fish. *Fuel Processing Technology, 90,* 883–888.
83. Lin, L., Cunshan, Z., Vittayapadung, S., Xiangqian, S., & Mingdong, D. (2011). Opportunities and challenges for biodiesel fuel. *Applied Energy, 88,* 1020–1031.
84. Loh, S.-K., Chew, S.-M., & Choo, Y.-M. (2006). Oxidative stability and storage behavior of fatty acid methyl esters derived from used palm oil. *Journal of the American Oil Chemists' Society, 83,* 947–952.
85. Luque, R., Herrero-Davila, L., Campelo, J. M., Clark, J. H., Hidalgo, J. M., Luna, D., et al. (2008). Biofuels: A technological perspective. *Energy & Environmental Science, 1,* 542–564.
86. Miller, R., Schmidt, G., & Shindell, D. (2006). Forced annular variations in the 20th century intergovernmental panel on climate change fourth assessment report models. *Journal of Geophysical Research: Atmospheres,* 111.
87. Moraes, M. S. A., Krause, L. C., da Cunha, M. E., Faccini, C. S., de Menezes, E. W., Veses, R. C., et al. (2008). Tallow biodiesel: Properties evaluation and consumption tests in a diesel engine. *Energy & Fuels, 22,* 1949–1954.
88. Moser, B. R. (2008). Influence of blending canola, palm, soybean, and sunflower oil methyl esters on fuel properties of biodiesel. *Energy & Fuels, 22,* 4301–4306.
89. Moser, B. R. (2011). Biodiesel production, properties, and feedstocks. *Biofuels.* Springer.
90. Moser, B. R. (2012). Preparation of fatty acid methyl esters from hazelnut, high-oleic peanut and walnut oils and evaluation as biodiesel. *Fuel, 92,* 231–238.

91. Moser, B. R., & Vaughn, S. F. (2010). Evaluation of alkyl esters from *Camelina sativa* oil as biodiesel and as blend components in ultra low-sulfur diesel fuel. *Bioresource Technology, 101,* 646–653.
92. Nakpong, P., & Wootthikanokkhan, S. (2010). High free fatty acid coconut oil as a potential feedstock for biodiesel production in Thailand. *Renewable Energy, 35,* 1682–1687.
93. Ng, H. K., & Gan, S. (2010). Combustion performance and exhaust emissions from the non-pressurised combustion of palm oil biodiesel blends. *Applied Thermal Engineering, 30,* 2476–2484.
94. Nogueira Jr, C. A., Feitosa, F. X., Fernandes, F. A., Santiago, R. L. S., & De Sant'ana, H. B. (2010). Densities and viscosities of binary mixtures of babassu biodiesel + cotton seed or soybean biodiesel at different temperatures. *Journal of Chemical & Engineering Data, 55,* 5305–5310.
95. Öner, C., & Altun, Ş. (2009). Biodiesel production from inedible animal tallow and an experimental investigation of its use as alternative fuel in a direct injection diesel engine. *Applied Energy, 86,* 2114–2120.
96. Özcanlı, M., Keskin, A., & Aydın, K. (2011). Biodiesel production from terebinth (Pistacia terebinthus) oil and its usage in diesel engine. *International Journal of Green Energy, 8,* 518–528.
97. Park, J.-Y., Kim, D.-K., Lee, J.-P., Park, S.-C., Kim, Y.-J., & Lee, J.-S. (2008). Blending effects of biodiesels on oxidation stability and low temperature flow properties. *Bioresource Technology, 99,* 1196–1203.
98. Pereira, R. G., Oliveira, C. D., Oliveira, J. L., Oliveira, P. C. P., Fellows, C. E., & Piamba, O. E. (2007). Exhaust emissions and electric energy generation in a stationary engine using blends of diesel and soybean biodiesel. *Renewable Energy, 32,* 2453–2460.
99. Pérez, Á., Casas, A., Fernández, C. M., Ramos, M. J., & Rodríguez, L. (2010). Winterization of peanut biodiesel to improve the cold flow properties. *Bioresource Technology, 101,* 7375–7381.
100. Phan, A. N., & Phan, T. M. (2008). Biodiesel production from waste cooking oils. *Fuel, 87,* 3490–3496.
101. Puhan, S., Jegan, R., Balasubbramanian, K., & Nagarajan, G. (2009). Effect of injection pressure on performance, emission and combustion characteristics of high linolenic linseed oil methyl ester in a DI diesel engine. *Renewable Energy, 34,* 1227–1233.
102. Qi, D., Geng, L., Chen, H., Bian, Y., Liu, J., & Ren, X. (2009). Combustion and performance evaluation of a diesel engine fueled with biodiesel produced from soybean crude oil. *Renewable Energy, 34,* 2706–2713.
103. Radha, K., & Manikandan, G. (2011). Novel production of biofuels from neem oil. In *World Renewable Energy Congress–Sweden, Bioenergy Technology, Linkoping, Sweden* (pp. 8–13).
104. Ragit, S., Mohapatra, S., Kundu, K., & Gill, P. (2011). Optimization of neem methyl ester from transesterification process and fuel characterization as a diesel substitute. *Biomass and Bioenergy, 35,* 1138–1144.
105. Rahman, M., Pourkhesalian, A., Jahirul, M., Stevanovic, S., Pham, P., Wang, H., et al. (2014). Particle emissions from biodiesels with different physical properties and chemical composition. *Fuel, 134,* 201–208.
106. Rajendra, M., Jena, P. C., & Raheman, H. (2009). Prediction of optimized pretreatment process parameters for biodiesel production using ANN and GA. *Fuel, 88,* 868–875.
107. Ramadhas, A. S., Jayaraj, S., & Muraleedharan, C. (2005). Biodiesel production from high FFA rubber seed oil. *Fuel,* 84, 335–340.
108. Ramalho, E., Carvalho Filho, J., Albuquerque, A., De Oliveira, S., Cavalcanti, E., Stragevitch, L., et al. (2012). Low temperature behavior of poultry fat biodiesel: diesel blends. *Fuel,* 93, 601–605.
109. Ramos, M. J., Fernández, C. M., Casas, A., Rodríguez, L., & Pérez, Á. (2009). Influence of fatty acid composition of raw materials on biodiesel properties. *Bioresource Technology, 100,* 261–268.

110. Rashid, U., & Anwar, F. (2008). Production of biodiesel through base-catalyzed transesterification of safflower oil using an optimized protocol. *Energy & Fuels, 22,* 1306–1312.
111. Rashid, U., Anwar, F., & Knothe, G. (2009). Evaluation of biodiesel obtained from cottonseed oil. *Fuel Processing Technology, 90,* 1157–1163.
112. Rashid, U., Anwar, F., Moser, B. R., & Ashraf, S. (2008). Production of sunflower oil methyl esters by optimized alkali-catalyzed methanolysis. *Biomass and Bioenergy, 32,* 1202–1205.
113. Reyes, J., & Sepulveda, M. (2006). PM-10 emissions and power of a diesel engine fueled with crude and refined biodiesel from salmon oil. *Fuel, 85,* 1714–1719.
114. Rodrigues Jr, J. D. A., Cardoso, F. D. P., Lachter, E. R., Estevão, L. R., Lima, E., & Nascimento, R. S. (2006). Correlating chemical structure and physical properties of vegetable oil esters. *Journal of the American Oil Chemists' Society,* 83, 353–357.
115. Royon, D., Daz, M., Ellenrieder, G., & Locatelli, S. (2007). Enzymatic production of biodiesel from cotton seed oil using *t*-butanol as a solvent. *Bioresource Technology, 98,* 648–653.
116. Sahoo, P., Das, L., Babu, M., & Naik, S. (2007). Biodiesel development from high acid value polanga seed oil and performance evaluation in a CI engine. *Fuel, 86,* 448–454.
117. Samios, D., Pedrotti, F., Nicolau, A., Reiznautt, Q., Martini, D., & Dalcin, F. (2009). A transesterification double step process—TDSP for biodiesel preparation from fatty acids triglycerides. *Fuel Processing Technology, 90,* 599–605.
118. Sanford, S. D., White, J. M., Shah, P. S., Wee, C., Valverde, M. A., & Meier, G. R. (2009). Feedstock and biodiesel characteristics report. *Renewable Energy Group, 416.*
119. Sardar, N., Ahmad, M., Khan, M. A., Ali, S., Zafar, M., Khalid, N., et al. (2011). Prospects and potential of non edible neem oil biodiesel based on physicochemical characterization. *Energy Sources, Part A: Recovery, Utilization, and Environmental Effects, 33,* 1422–1430.
120. Sarin, R., Sharma, M., Sinharay, S., & Malhotra, R. (2007). Jatropha–palm biodiesel blends: an optimum mix for Asia. *Fuel, 86,* 1365–1371.
121. Schwab, A., Bagby, M., & Freedman, B. (1987). Preparation and properties of diesel fuels from vegetable oils. *Fuel, 66,* 1372–1378.
122. Senatore, A., Cardone, M., Rocco, V., & Prati, M. V. (2000). *A comparative analysis of combustion process in DI diesel engine fueled with biodiesel and diesel fuel.* SAE Technical Paper.
123. Serdari, A., Fragioudakis, K., Teas, C., Sakellaropoulos, F., Zannikos, F., Stournas, S., et al. (1998). Adding biodiesel corn oil and sunflower oil to diesel fuel: the impact on the performance of conventional road vehicles. *Journal of the Institute of Energy, 71,* 126–136.
124. Shah, S. N., Iha, O. K., Alves, F. C., Sharma, B. K., Erhan, S. Z., & Suarez, P. A. (2013). Potential application of turnip oil (raphanus sativus l.) for biodiesel production: Physical-chemical properties of neat oil, biofuels and their blends with ultra-low sulphur diesel (ULSD). *BioEnergy Research, 6,* 841–850.
125. Singh, R., & Padhi, S. K. (2009). Characterization of Jatropha oil for the preparation of biodiesel. *Natural Product Radiance, 8,* 127–132.
126. Singh, S., & Singh, D. (2010). Biodiesel production through the use of different sources and characterization of oils and their esters as the substitute of diesel: A review. *Renewable and Sustainable Energy Reviews, 14,* 200–216.
127. Sinha, S., Agarwal, A. K., & Garg, S. (2008). Biodiesel development from rice bran oil: Transesterification process optimization and fuel characterization. *Energy Conversion and Management, 49,* 1248–1257.
128. Sivalakshmi, S., & Balusamy, T. (2012). Influence of ethanol addition on a diesel engine fuelled with neem oil methyl ester. *International Journal of Green Energy, 9,* 218–228.
129. Soriano Jr, N. U., & Narani, A. (2012). Evaluation of biodiesel derived from Camelina sativa oil. *Journal of the American Oil Chemists' Society, 89,* 917–923.

130. Sun, Y.-Q., Chen, B.-S., Zeng, G., Wang, J., Fang, J.-H., Xiong, W., et al. (2008). Research of low temperature properties of peanut oil and peanut biodiesel. *Internal Combustion Engines, 3,* 018.
131. Tan, R. R., Culaba, A. B., & Purvis, M. R. (2004). Carbon balance implications of coconut biodiesel utilization in the Philippine automotive transport sector. *Biomass and Bioenergy, 26,* 579–585.
132. Tang, H., Salley, S. O., & Simon Ng, K. (2008). Fuel properties and precipitate formation at low temperature in soy-, cottonseed-, and poultry fat-based biodiesel blends. *Fuel, 87,* 3006–3017.
133. Todaka, M., Kowhakul, W., Masamoto, H., Shigematsu, M., & Onwona-Agyeman, S. (2013). Thermal decomposition of biodiesel fuels produced from rapeseed, jatropha, and coffee oils with different alcohols. *Journal of Thermal Analysis and Calorimetry, 113,* 1355–1361.
134. Vedaraman, N., Puhan, S., Nagarajan, G., & Velappan, K. (2011). Preparation of palm oil biodiesel and effect of various additives on NOx emission reduction in B20: An experimental study. *International Journal of Green Energy, 8,* 383–397.
135. Wang, L.-B., Yu, H.-Y., He, X.-H., & Liu, R.-Y. (2012). Influence of fatty acid composition of woody biodiesel plants on the fuel properties. *Journal of Fuel Chemistry and Technology, 40,* 397–404.
136. Wu, F., Wang, J., Chen, W., & Shuai, S. (2008). *Effects of different biodiesels and their blends with oxygenated additives on emissions from a diesel engine*. SAE Technical Paper, 008-01-1812.
137. Wu, X., & Leung, D. Y. (2011). Optimization of biodiesel production from camelina oil using orthogonal experiment. *Applied Energy, 88,* 3615–3624.
138. Wyatt, V. T., Hess, M. A., Dunn, R. O., Foglia, T. A., Haas, M. J., & Marmer, W. N. (2005). Fuel properties and nitrogen oxide emission levels of biodiesel produced from animal fats. *Journal of the American Oil Chemists' Society, 82,* 585–591.

A Review of Microalgal Biofuels, Challenges and Future Directions

Saleh M.A. Mobin and Firoz Alam

Abstract Fossil fuels play an absolute dominant role in the global energy mix, followed by biofuels and other energy sources. Fossil fuel dependency is unsustainable due to its finite nature. Moreover, a large amount of greenhouse gas emission is generated by the use of fossil fuels. Biofuels from microalgae have the potential to provide a sustainable and carbon-neutral energy source, complementing the shortfall of fossil fuels and enhancing the mitigation of global warming. One of the notable advantages of biofuels from microalgae is that it does not pose any threat to human or animal food chain and its production can be achieved using barren land, salt water, wastewater and CO_2 emitted by thermal power plants. Despite having enormous potential, current production, harvesting and processing techniques of microalgal biomass remain not cost-effective or widely used. Hence, further optimisation of microalgal mass culture, harvesting and processing techniques, and efficient utilisation of by-products are needed to make this carbon-neutral energy source economically viable and sustainable. Coupling microalgae cultivation with wastewater and CO_2 from power plants is considered a promising route for the production of bioenergy and bio-based by-products. Significant challenges remain to be addressed to utilise the full potential of third-generation biofuel derived mainly from microalgal biomasses. This chapter presents a brief view of the current progress on microalgal biofuel production, its future directions and challenges.

Keywords Biofuel · Microalgae · Biodiesel · Bioethanol · Renewable energy

S.M.A. Mobin (✉) · F. Alam
School of Aerospace, Mechanical and Manufacturing Engineering,
RMIT University, Plenty Road, Bundoora, Melbourne, VIC 3083, Australia
e-mail: salehmobin2013@gmail.com

M.M.K. Khan et al. (eds.), *Application of Thermo-fluid Processes in Energy Systems*,
Green Energy and Technology, https://doi.org/10.1007/978-981-10-0697-5_4

1 Introduction

Continued use of petroleum as an energy source has long been considered unsustainable because fossil fuels are finite in amount and their combustion produces carbon dioxide (CO_2) and other greenhouse gases (GHG) that contribute to global warming, resulting in a significant climate change [1]. Carbon dioxide emissions from fossil fuel combustion showed an increase of 58% in 2012 above the 1990 emissions level and predicted to be 65% in 2015 [2, 3]. To confront energy shortages, global climate change, rising crude oil prices, and land and water degradation, biofuels along with other alternative energy sources such as wind and solar power have received increasing attention from researchers, industry and governments for sustainable development [4–6]. Renewable and carbon-neutral fuels are necessary for environmental and economic sustainability.

Biofuels generally refer to solid, liquid or gaseous fuels derived from biomass, which is organic material that stores solar energy in the form of chemical energy through physical and chemical processes, and these fuels can be used as a substitute for petroleum fuels [7–9]. Biofuels are generally classified into two groups, namely primary biofuel and secondary biofuel. The secondary biofuel is sub-classified into three groups: (a) first-generation biofuel, (b) second-generation biofuel and (c) third-generation biofuel. A detailed classification of biofuels with production sources is shown in Fig. 1.

Biofuel from oil crops is a potential renewable and carbon-neutral alternative to petroleum fuels. However, the biomass of currently produced biofuel is human foodstock, which is believed to cause shortages of food and worldwide dissatisfaction, especially in developing nations [4, 6].

Microalgae have become popular oil sources used to make biofuel due to their high lipid content, rapid growth rate, ease of cultivation and the volume at which

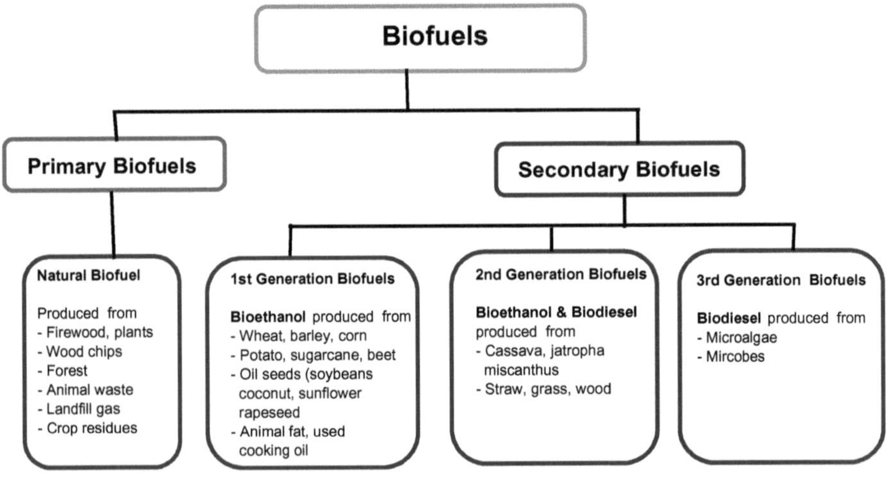

Fig. 1 Biofuel production sources (biomasses). Adapted from [4, 5, 6, 10]

they can be grown [11]. Microalgae do not compete with human and animal food crops and can be grown on non-arable land and in saline water.

The feedstock that attained economic levels of commercial production for first-generation biofuels includes rapeseed oil and palm oil, sugarcane, sugar beet, wheat, barley and corn [12]. First-generation biofuel feedstocks are limited by sustainability due to the scale and also due to required agricultural inputs (land, water, fertiliser, pesticides and herbicides) [13]. Despite having potential for producing carbon-neutral biofuels, first-generation biofuels generate notable economic, environmental and political concerns. The most alarming issue associated with such fuels is that, with the increase of production capacity, more arable agricultural land is needed for the production of first-generation biofuel feedstock. The reduced land available for human and animal food production will increase direct competition for other agricultural inputs [4, 6]. It is also important to consider that the production of a biofuel crop has a carbon footprint, derived from organic carbon released from the soil, and is an offender for GHG emission during production [14]. In addition, large-scale cultivation of biofuel crops can be detrimental to ecosystems and biodiversity [15].

Increased pressure on arable land currently used for food production leads to severe food shortages. With a growing world population, the demand for food is increasing while arable land is decreasing. Intensive use of agricultural inputs such as fertiliser, pesticides and freshwater on limited farming land can not only reduce the food production capacity of land but also cause significant environmental damage [16]. Hence, enthusiasm for first-generation biofuels has declined over the time.

As first-generation biofuels are not viable, researchers focused on second-generation biofuels. The primary intention here is to produce biofuels using non-food-based products and agricultural wastes, most commonly lignocellulosic biomass, which is plant cell wall material available in the woody part of the plant. Lignocellulosic biomass and lignocellulosic residues, such as by-products of agricultural processes including cereal straw, sugarcane bagasse, forest harvesting residues, wood processing residues (e.g. saw dust), wastes (organic components of municipal solid wastes, trimmed branches, leaves, wood chips, etc.), as well as dedicated energy crops (purpose-grown vegetative grasses and short rotation forests), do not compete with the human food chain directly and make them more sustainable than first-generation feedstock [10, 17]. Dedicated energy crops, produced on marginal land, represent a second biomass source for use as a second-generation biofuel feedstock. Jatropha, a land-based plant rich in triacylglycerides (TAGs), is a leading candidate in this respect due to its drought and pest resistance [18].

A number of technologies are available for the conversion of second-generation feedstocks into usable fuel molecules. Pyrolysis is reportedly one of the few economically viable methods among the biochemical and thermochemical processes [19]. Pyrolysis or gasification is the thermochemical breakdown of lignocellulosic feedstocks in the absence of oxygen and results in the production of chemical building blocks: hydrogen, carbon monoxide, CO_2 and other gases, which can be assembled into final products 'alcohols' (methanol and ethanol). However, converting the lignocellulosic feedstocks into fermentable sugars through biological conversion requires separation of cellulose and hemicelluloses from lignin, due to its inhibitory nature, followed by the hydrolysis of polysaccharides [20]. Therefore, use of sophisticated, more complex and energetically expensive technologies for the pretreatment of second-generation feedstock with special enzymes makes second-generation biofuels economically unprofitable for commercial production [10, 21, 22]. In addition, terrestrial bioenergy production systems are now facing issues related to (1) indirect emission and carbon debt from land clearance, (2) consequences of using farmed or marginal land, (3) competition for using agricultural inputs such as fertiliser, pesticide, water, (4) consequences of land-use change and changes in the carbon stocks of modified ecosystems and (5) potential threats to biodiversity or to water and nutrient cycling; hence, they are becoming a sustainability hurdle for further expansion [1, 23, 24, 25, 26]. Therefore, a more sustainable feedstock had to be evolved to overcome these limitations.

The focus of research is currently drawn to third-generation biofuels. The main component of third-generation biofuels is microalgae, as shown in Fig. 1. It is currently considered to be a feasible alternative renewable energy resource for biofuel production, overcoming the disadvantages of first- and second-generation biofuels [9, 11, 10, 27]. On an area basis, biofuel production from microalgae can be 15−300 times greater compared to traditional crops [10]. In addition, microalgae have a very short harvesting cycle (1−10 days, depending on the process) compared to conventional crop plants that are generally harvested once or twice a year. This short-cycle harvesting allows significantly increased yields [16, 10]. Furthermore, the microalgae generally have higher productivity than land-based plants, as some species have doubling times of a few hours (h) and accumulate very large amounts of triacylglycerides (TAGs). Most importantly, high-quality agricultural land is not required for microalgae biomass production [28]. Different types of renewable biofuels can be extracted from microalgae. Depending on microalgal species, these include many different kinds of lipids, hydrocarbons and other complex oils. Production of microalgae for biodiesel extraction will not compromise the production of foodstock for human consumption, fodder and other products and by-products from crops [11].

The objective of this chapter is to present a brief overview of current knowledge on biofuel production from algae, challenges that remain and the future direction of this emerging industry.

2 Microalgae for Biofuel Production

Microalgae are unicellular photosynthetic microorganisms that are found naturally in freshwater and marine environments. There are about 200,000–800,000 algae species, of which around 50,000 species have been described [29]. The sizes of microalgae depend on their species and vary from micrometres to millimetres [30]. Their position is at the bottom of food chains. Microalgae are considered to be one of the oldest living organisms on our planet. While the mechanism of photosynthesis in these microorganisms is similar to that of higher plants, microalgae are generally more efficient converters of solar energy. They can convert CO_2 and H_2O to biomass using sunlight [31]. The average photosynthetic efficiency of microalgae is 6–8% [32], which is much higher than that of terrestrial biomass (1.8–2.2%). They may be grown in shallow lagoons, raceway ponds, closed ponds, photobioreactors and sea-based systems [33] and are very efficient in utilising the nutrients from wastewater, including nitrogen and phosphorus [12]. Due to its rapid growth rate, the nutrients can be recycled back to the soil by fertilising the waste by-products.

In addition, because the cells grow in aqueous suspension, they have more efficient access to water, CO_2 and other nutrients [11, 10]. Generally, microalgae are classified in accordance with their colours. The current systems of classification of microalgae are based on (a) kinds of pigments, (b) chemical nature of storage products and (c) cell wall constituents [10].

Typical biofuel yields from various biomasses are shown in Table 1 [34]. The table clearly shows the huge potential of microalgae compared to other biomasses. The production rate (L/ha) of oil from microalgae (91%) is much higher than other feedstocks such as oil palm (3%), coconut (1.5%), jatropha (1.2%), avocado (1.4%) and rapeseed/canola (1%) [35]. Some major groups of microalgae are shown in Table 2.

Most microalgae species produce lipids, carbohydrates, fatty acids, peptides, carotenoids, antioxidants, enzyme polymers, toxins and sterols [33, 35]. Figure 2 shows potential products and by-products from algae [36]. The lipid content in relation to dry weight and lipid productivity of various microalgae are shown in

Table 1 Typical oil yields from the various biomasses. Adapted from [34]

S.N.	Crop	Oil yield (L/ha)
1	Rubber seed	80−120
2	Corn	172
3	Soybean	446
4	Safflower	779
5	Chinese tallow	907
6	Camelina	915
7	Sunflower	952
8	Peanut	1,059
9	Canola	1,190
10	Rapeseed	1,190
11	Castor	1,413
12	Jatropha	1,892
13	Karanj	2,590
14	Coconut	2,689
15	Oil palm	5,950
16	Microalgae (30% oil by wt)	58,700
17	Microalgae (70% oil by wt)	136,900

Table 2 Major microalgae groups based on their colours. Adapted from [4]

	Colour	Group
1	Yellow-green algae	Xanthophyceae
2	Red algae	Rhodophyceae
3	Golden algae	Chrysophyceae
4	Green algae	Chlorophyceae
5	Brown algae	Phaeophyceae
6	Cyanobacteria	Cyanophyceae

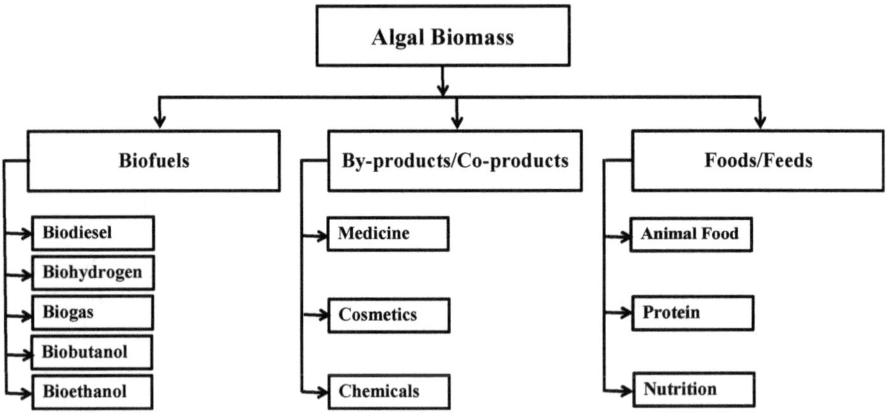

Fig. 2 Potential products from algae. Adapted from [36]

Table 3. Microalgae are unique biomasses that grow extremely rapidly, unlike other oil crops, and Table 3 shows that many of microalgae are rich in oil. The lipid content of microalgae has been found to be very high (lipid content >60% of dry weight (dwt) biomass) in a number of marine species such as *Dunaliella tertiolecta*, *Nannochloropsis* sp., *Neochloris oleoabundans* and *Porphyridium cruentum*, and in some freshwater species such as *Botryococcus braunii*, *Chlorella emersonii* [4, 6, 11, 16, 10, 27, 37, 38, 39]. A moderate lipid content (40-60% of dw) has been found in diverse range of marine microalgal species such as *Crypthecodinium cohnii*, *Dunaliella salina*, *Isochrysis galbana*, *Nannochloris* sp., *Nannochloropsis oculata* NCTU-3, *Nitzschia* sp., *Phaeodactylum tricornutum*, *Schizochytrium* sp., *Skeletonema costatum* and in a number of freshwater microalgae such as *Chlorella minutissima*, *Chlorella protothecoides*, *Chlorella vulgaris*, *Chlorella* sp., *Scenedesmus dimorphus* and *Scenedesmus obliquus* [4, 6, 11, 16, 10, 27, 37, 38, 39]. Therefore, careful selection of algal strain is very important for algal biofuel production. When comparing different microalgal types, Table 3 clearly shows that the lipid production rate (mg/L/day) in green microalgae is much higher than that in red or blood-red microalgae. As mentioned earlier, some microalgae can double their biomass within 24 h and the shortest doubling time during their growth is around 3.5 h, which makes microalgae an ideal renewable source for biofuel production [11].

Table 3 Lipid content and lipid productivity of microalgae. Adapted from [35]

	Microalgae species	Alga type	Lipid content (% dry weight biomass)/oil content	Lipid productivity (mg/L/day)
1	*Arthrospira maxima*	Blue green	20.3	–
2	*Ankistrodesmus* sp.	Green	24.0–31.0	–
3	*Botryococcus braunii*	Green	25.0–75.0	–
4	*Chaetoceros muelleris*	Diatom	33.6	21.8
5	*Chaetoceros calcitrans*	Diatom	14.6–16.4/39.8	17.6
6	*Chlorella emersonii*	Green	25.0–63.0	10.3–50.0
7	*Chlorella minutissima*	Green	57.0	–
8	*Chlorella protothecoides*	Green	14.6–57.8	1214
9	*Chlorella sorokiniana*	Green	19.0–22.0	44.7
10	*Chlorella vulgaris CCAP 211/11b*	Green	19.2	170.0
11	*Chlorella vulgaris*	Green	5.0–58.0	11.2–40.0
12	*Chlorella* sp.	Green	10.0–48.0	42.1
13	*Chlorella pyrenoidosa*	Green	11.0–26.0	–
14	*Chlorella* sp.	Green	18.0–57.0	18.7
15	*Chloroococcum* Sp. *UMACC 112*	Green	19.3	53.7
16	*Cylindrotheca* sp.	Diatom	16.0–37.0	–
17	*Crypthecodinium cohnii*	Red	20.0–51.1	–
18	*Dunaliella salina*	Green	16.0–44.0	46.0
19	*Dunaliella primolecta*	Green	23.0	–
20	*Dunaliella tertiolecta*	Green	16.7–71.0	–
21	*Ellipsoidion sp.*	*Eustigmatophytes*	27.4	47.3
22	*Euglena gracilis*	Green	14.0–20.0	–
23	*Haematococcus pluvialis*	Blood-red	15.6–33.0	–
24	*Isochrysis galbana*	*Prymnesiophytes*	7.0–40.0	–
25	*Isochrysis* sp.	*Prymnesiophytes*	7.1–33.0	37.8
26	*Monodus subterraneus* UTEX 151	*Eustigmatophytes*	16.1	30.4
27	*Monallanthus salina*	Green	20.0–22.0	–
28	*Nannochloris* sp.	Green	20.0–56.0	60.9–76.5

(continued)

Table 3 (continued)

	Microalgae species	Alga type	Lipid content (% dry weight biomass)/oil content	Lipid productivity (mg/L/day)
29	*Nannochloropsis oculata* NCTU-4	Green	30.8–50.4	142.0
30	*Nannochloropsis oculata*	Green	22.7–29.7	84.0–142.0
31	*Nannochloropsis* sp.	*Eustigmatophytes*	12.0–68.0	37.6–90.0
32	*Neochloris oleoabundans*	Green	29.0–65.0	90.0–134.0
33	*Nitzschia* sp.	Diatom	16.0–47.0	–
34	*Oocystis pusilla*	Green	10.5	–
35	*Pavlova salina*	*Prymnesiophytes*	30.9	49.4
36	*Pavlova lutheri*	*Prymnesiophytes*	35.5	40.2
37	*Phaeodactylum tricornutum*	Diatom	18.0–57.0	44.8
38	*Porphyridium cruentum*	Red	9.0–18.8/60.7	34.8
39	*Scenedesmus dimorphus*	Green	6–7/16-40	–
40	*Scenedesmus obliquus*	Green	11–22/35–55	–
41	*Scenedesmus quadricauda*	Green	1.9–18.4	35.1
42	*Schizochytrium* sp.	Yellow-green	50–57	–
43	*Scenedesmus* sp. DM	Green	19.6–21.1	40.8–53.9
44	*Skeletonema* sp.	Diatom	13.3–31.8	27.3
45	*Skeletonema costatum*	Diatom	13.5–51.3	17.4
46	*Spirulina platensis*	Green	10.3 ± 0.10	–
47	*Spirulina maxima*	Green	4.0–9.0	–
48	*Thalassiosira pseudonana*	Diatom	20.6	17.4
49	*Tetraelmis suecica*	Green	8.5–23.0	27.0–36.4
50	*Tetraelmis* sp.	Green	12.6–14.7	43.4

3 Present Technological Option for Biomass Cultivation and Harvesting of Microalgae

Microalgae production industry is still small, and presently around 6000 tonnes (dry weight) biomass is produced autotrophically per year worldwide [35]. Photosynthetic growth of third-generation biofuel feedstock requires light, CO_2, water and nutrients. Essential nutrient elements include carbon (C), nitrogen (N), phosphorus (P), iron and in some cases silicon, calcium, magnesium, potassium, iron, manganese, sulphur, zinc, copper and cobalt [6, 11]. Temperatures must remain between 20 °C and 30 °C.

A range of methods for microalgae cultivation can be used. However, efficient and cost-effective large-scale cultivation is needed for a successful renewable energy candidate. Many designs for mass algal cultivation have been devised which are generally grouped into (a) suspended cultures, including open ponds and closed photobioreactors, and (b) immobilised cultures, including matrix-immobilised systems and biofilms. Large-scale production of microalgal biomass generally uses continuous suspended culture during daylight. In this method, fresh culture medium is fed in at a constant rate and the same quantity of microalgal culture medium is withdrawn continuously. The biomass is produced during daylight, as much as 25% of that biomass may be lost from the system during night-time due to failure of respiration and by some other growth-limiting factors such as light intensity, growth temperature and the temperature at night [11]. The most widely used large-scale production systems are high-rate algal ponds or raceway ponds. Raceway ponds are open air and shallow and equipped with paddlewheels that provide circulation of the algae and nutrients. Raceways are relatively inexpensive to build and operate, but often suffer low productivity [11, 40].

Open raceway ponds have been used for over 60 years [41]. The raceway pond culture is usually no more than 30 cm deep to allow for efficient penetration of sunlight [11]. Mixing and circulation are produced in raceways by a paddlewheel, and this paddlewheel operates all the time to prevent sedimentation. Although open systems are generally easy to operate and use sunlight as an energy source, they have several disadvantages. The principal drawbacks are as follows: (i) requiring large areas of land and (ii) lacking any real control over the environmental conditions encountered. For example, there is no control over temperature, which generally can vary seasonally and diurnally. There can be significant water loss due to evaporation. The distribution of light and CO_2 through the culture is much less efficient than in photobioreactor systems. In open systems, contamination by competing algal strains and by bacteria is difficult to avoid. In many cases, it is necessary to use extremophilic organisms that can grow under conditions which other strains cannot tolerate (high temperature, pressure and salinity). These disadvantages pose a severe limit on the number of different strains which can be cultivated for mass production. Due to these limitations, open pond productivity is typically fairly low.

Photobioreactors are the most efficient productive systems to improve algal cultivation yields compared to open ponds or raceway systems, because they offer more control over culture conditions such as pH, nutrient dosing and temperature control, better protection against culture contamination, better mixing, less evaporative loss and higher cell densities [4, 5, 11, 40, 42, 43, 44, 45, 46]. Photobioreactors are most suitable for selective strain-specific algal biomass production, to increase lipid productivity (mg/L/day) and biofuel productivity (per day). Other benefits of photobioreactors are better control of CO_2 transfer and protection from climate-related impacts, such as rainfall and diurnal and seasonal temperature fluctuations [11, 40, 45]. However, closed photobioreactors are more expensive to construct. The photobioreactor consists of an array of straight transparent tubes (generally 0.1 m or less in diameter) that are usually made of plastic or glass. The photoreactor system can be sub-classified as follows: (a) vertical photobioreactors, where solar collective tubes are placed parallel to each other and organised vertically above the ground, (b) flat or horizontal photoreactors, where solar collective tubes are placed parallel to each other and flat above the ground and (c) helical photoreactors, where solar collector tubes are coiled around a supporting frame to form a helical coiled structure. The helical coiled photoreactor is considered the easiest to scale up production. Biomass sedimentation in tubular photobioreactors is prevented by maintaining a highly turbulent flow, which is produced using either a mechanical pump or a gentler airlift pump [11]. Photobioreactors can produce 13-fold greater volumetric algal biomass compared with raceway ponds [11]. However, each system has relative advantages and disadvantages as shown in Table 4.

In order to maximise the algae biomass production, a large surface area-to-volume ratio is required that can capture the required amount of sunlight. To increase this ratio, most photobioreactors are constructed to form thin tubes or plates. This has major impact on the economy of scale for the outdoor cultivation of algae. As per Zemke [47], for heterotrophic production, the biomass production of a bioreactor can be doubled by doubling the volume of the bioreactor. As the surface area (material) increases with the square of the characteristic length while volume increases as a cube of the characteristic length (square–cube law), the capital cost of doubling production is roughly $2^{2/3}$ or 59% more. However, to double the productivity of a phototrophic cultivation system, double the cultivation surface area (material) is required, resulting in nearly a 100% increase in capital costs. This is a major constraint to reduce the cost of algae production. While phototrophic cultivation may have the advantage over heterotrophic cultivation of using a 'free' energy source, the capital investment required to capture this energy is substantial [47].

One of the significant challenges of using raceways and tubular photobioreactors is biomass recovery. This challenge may be mitigated to an extent by immobilised cultures or attached algal processes [5, 6, 48]. Likewise, due to their low construction costs and ease of operation, open raceway ponds will likely be the systems of choice for mass microalgae cultivation. Figure 3 shows algae culture in controlled environment and open air systems. Figure 4 shows algae culture in various types of photobioreactor systems.

The current production and harvesting techniques of microalgal biomass are still expensive to ensure a competitive production price for biofuels from algae. In order to reduce the production cost, an optimisation of mass culture, harvesting and processing techniques and wastewater utilisation are needed. Over past decades, numerous studies have been conducted on growing microalgae on different types of wastewater. Coupling microalgae cultivation with wastewater (derived from municipal, domestic, agricultural and industrial activities) treatment is considered one of the most promising routes to produce bioenergy and bio-based by-products in an economically viable and environmentally friendly way. However, there are some major concerns when wastewater is used for algae cultivation, which include

Table 4 Relative merits and demerits of major algal culture systems. Adapted from [4, 5, 6, 11, 46]

Raceway Pond		Flat Plate		Tubular		Vertical Column	
Advantages	Disadvantages	Advantages	Disadvantages	Advantages	Disadvantages	Advantages	Disadvantages
• Low energy input	• Water loss due high evaporation rate	• Low concentration of dissolved oxygen	• Scale up requires many compartments and support materials	• Relatively higher biomass productivity	• Requires large land area	• Compact and easy to operate	• Small illumination surface area
• Low cost	• Temperature, light and pH control difficulty	• Easier position control (vertical or inclined at an optimal angle towards the sun)	• Difficulty in controlling the culture temperature	• Potential of cell damage is minimised if airlift system is used	• Risk of accumulating high concentration of oxygen	• Relatively low cost	• Cell sedimentation may occur if airlift system is not used
	• Susceptible to contamination	• Lower power consumption			• Decreasing CO_2 concentration along the tube length	• Lower power consumption	
					• Mixing difficulty in longer tubes		

(a) Laboratory culture of various algae types

(b) Mass algae culture in raceways [14]

(c) Mass algae culture in tank

Fig. 3 Algae culture in laboratory and open water systems

the following: (i) high turbidity due to the presence of solid particles, which will affect light penetration significantly; (ii) high nutrient concentration, especially high ammonia concentration, which can inhibit algal growth considerably; (iii) a large portion of the carbon sources is locked in large insoluble organic compounds and is unavailable for algae to assimilate; (iv) a large quantity of freshwater is necessary to dilute the concentrated animal wastewater, unless water recycling and reuse is enabled; and (v) high-performance algae strains adapted to the adverse environment in animal wastewater which have not yet been developed [5, 33]. In order to address these issues, numerous methods and strategies have been developed and adopted. More details about algal cultivation systems can be found in [5, 11, 10, 28].

Harvesting of microalgal culture has been considered a major bottleneck to the industrial-scale processing of microalgae for biofuel production. The cost of biomass recovery from the culture medium can make up to 20–30% of the total cost of producing the biomass [49]. Various methods are currently used for harvesting algae, which include physical-based, chemical-based, mechanical-based, biological-based or, to a lesser extent, electrical-based operations, i.e. flocculation, centrifugation, filtration, ultrafiltrations, air flotation, autoflotation, etc. [10]. However, various combinations or sequences of these methods are also commonly in use. The cell size of algae is very small. Therefore, chemical flocculation is often performed as a pretreatment to increase the particle size of algae before using another method such as flotation to harvest the algae. Multivalent metal salts like

(a) Coil/helical bioreactor (b) Tubular bioreactor

(c) Flat-panel bioreactor (d) Floating photo bioreactor

(e) Column photobioreactor (f) Multi-layer bioreactor

Fig. 4 Algae culture in photobioreactor system. Adapted from [5]

ferric chloride, aluminium sulphate and ferric sulphate are commonly used as flocculants in the flocculation process. In mechanical-based processes (the centrifugation process), centrifugal force is used for recovering suspended algae. Once separated, microalgae can be removed from the culture medium by simply draining the excess medium. This method is recognised as the most reliable and rapid method [10, 50]. The filtration method of harvesting has proved to be the most competitive compared to other harvesting options. There are many different forms of filtration available, such as dead-end filtration, microfiltration, ultrafiltration, pressure filtration, vacuum filtration. In this method, suspended culture is passed through filters on which the algae accumulate and allow the culture medium to pass through the filter. The culture medium continually runs through the microfilters until the filter contains a thick algae paste [10]. In the electrical-based method, negative charge properties of algal cells are used for separating the cells [51]. These cells can be concentrated by the movement in an electric field. However, low-cost algal harvesting options for biofuels applications do not currently exist.

4 Gross Composition of Microalgae and Biofuel Option

Macroalgae generally contains only 10–15% dry matter [52]. The main constituents of microalgae are lipids (2–23%), carbohydrates (5–23%), proteins (6–52%) and some fat [53]. Table 5 shows gross composition of several freshwater and marine microalgal species. However, the chemical composition of algal biomass can greatly be modified by environmental factors, such as temperature, salinity, illumination, pH value, mineral content, CO_2 supply, population density, growth phase and physiological status [54]. Microalgae can produce biodiesel, bioethanol, biogas, biohydrogen and many other fuel types [27] via thermochemical and biochemical methods. Table 6 shows different types of biofuel production from different microalgae.

Table 5 Gross composition of several freshwater and marine microalgal species (% of dry matter). Adapted from [33]

S.N.	Microalgae	Protein	Carbohydrates	Lipids
	Freshwater microalgae species			
1	*Anabaena cylindrica*	43–56	25–30	4–7
2	*Aphanizomenon flosaquae*	62	23	3
3	*Arthrospira maxima*	60–71	13–16	6–7
4	*Chlamydomonas rheinhardii*	48	17	21
5	*Chlorella pyrenoidosa*	57	26	2
6	*Chlorella vulgaris*	51–58	12–17	14–22
7	*Chlorella zofingiensis*	11–20	25–28	26–46
8	*Euglena gracilis*	39–61	14–18	14–20
9	*Scenedesmus dimorphus*	8–18	21–52	16–40
10	*Scenedesmus obliquus*	50–56	10–17	12–14
11	*Scenedesmus quadricauda*	47	–	1.9
12	*Spirogyra* sp.	6–20	33–64	11–21
13	*Spirulina maxima*	60–71	13–16	6–7
14	*Spirulina platensis*	46–63	8–14	4–9
	Marine microalgal species			
15	*Dunaliella bioculata*	49	4	8
16	*Dunaliella salina*	57	32	6
17	*Porphyridium cruentum*	28–39	40–57	9–14
18	*Prymnesium parvum*	28–45	25–33	22–38
19	*Synechococcus* sp.	63	15	11
20	*Tetraselmis maculate*	52	15	3

Table 6 Different types of biofuel production from different microalgae. Adapted from [35]

S.N.	Microalgae	Algae type	Biofuel	Productivity of biofuel
1	*Arthrospira maxima*	Green	Hydrogen,	40–69%
			Biodiesel	
2	*Chlamydomonas reinhardtii*	Green	Hydrogen	2.5 mL h^{-1}/11.73 g/L
3	*Chlorella* sp.	Green	Biodiesel	
4	*Chlorella biomass*	Green	Ethanol	22.6 g/L
5	*Chlorella minutissima*	Green	Methanol	
6	*Chlorella protothecoides*	Green	Biodiesel	15.5 g/L
7	*Chlorella regularis*	Green	Ethanol	
8	*Chlorella vulgaris*	Green	Ethanol	
9	*Chloroococcum humicola*	Green	Ethanol	7.2 g/L or 10 g/L
10	*Chloroococcum infusionum*	Green	Ethanol	0.26 g ethanol/g algae
11	*Chloroococcum* Sp.	Blue green	Biodiesel	10.0 g/L
12	*Chloroococcum* Sp.	Blue green	Ethanol	3.83 g/L
13	*Dunaliella* sp.	Green	Ethanol	11.0 mg/g
14	*Haematococcus pluvialis*	Red	Biodiesel	420 GJ/ha/yr
15	*Neochloris oleoabundans*	Green	Biodiesel	56.0 g/g
16	*Platymonas subcordiformis*	Green	Hydrogen	
17	*Scenedesmus obliquus*	Green	Methanol,	
			Hydrogen	
18	*Spirogyra*	Green	Ethanol	
19	*Spirulina platensis*	Green	Hydrogen	1.8 μmol/mg
20	*Spirulina platensis* UTEX 1927	Blue green	Methane	0.40 m^3/kg
21	*Spirulina* Leb 18	Blue green	Methane	0.79 g/L

Microalgal biodiesel contains no sulphur and can replace diesel in most vehicles with little or no modification of vehicle engines while, on the other hand, the use of it can decrease the emissions of particulate matters, CO, hydrocarbons and SOx [40, 55].

Bioethanol can be produced from microalgae through fermentation of microalgal biomass. Carbohydrates in microalgae can be hydrolysed to sugars [56] and then fermented to bioethanol by yeast; the final product 'ethanol' can be obtained after a distillation process. Microalgae have some unique properties that play positive roles for bioethanol production. The properties are as follows: (i) cell walls of microalgae are largely made up of polysaccharides with low/no percentage of lignin and hemicelluloses; (ii) the composition of microalgae is generally nearly identical and consistent; and (iii) some microalgae contain more than 50% (dry weight) of starch and glycogen, which are useful ingredients for bioethanol production [57]. Bioethanol production from microalgae is still in the preliminary research phase, and further research about its advantages and disadvantages needs to be undertaken.

The microalgae or their residues after lipid extraction have a huge potential for biogas production, which can be achieved by anaerobic digestion, and mainly contain methane and carbon dioxide. Microalgae can directly use sunlight and water to generate biohydrogen in the absence of oxygen in a closed culture system [58]. In addition to biodiesel, bioethanol, biogas and biohydrogen, microalgae can also be converted into biobutanol, bio-oil, syngas, jetfuel, etc., via appropriate conversion methods, including thermochemical, chemical and biochemical conversion processes.

5 Biofuel Production Processes from Microalgae

A number of ways the microalgae biomass can be converted into energy sources. These include the following: (a) biochemical conversion, (b) chemical reaction, (c) direct combustion and (d) thermochemical conversion. Figure 5 illustrates a schematic of biodiesel and bioethanol production processes using microalgae feedstock [4, 6, 10, 40, 59]. Apart from biofuel production, other compounds can also be extracted based on microalgae species. A more detailed information can be found in [10, 37, 40, 43, 49, 60, 61, 62, 63, 64, 65].

The production of biofuel is a complex process. A schematic of biofuel production processes from microalgae is shown in Fig. 6. The process consists of the following stages: (a) stage 1—microalgae cultivation, (b) stage 2—harvesting (cell separation from the growth medium), drying and cell disruption, (c) stage 3—lipid extraction for biodiesel production through transesterification and (d) stage 4—starch hydrolysis, fermentation and distillation for bioethanol production.

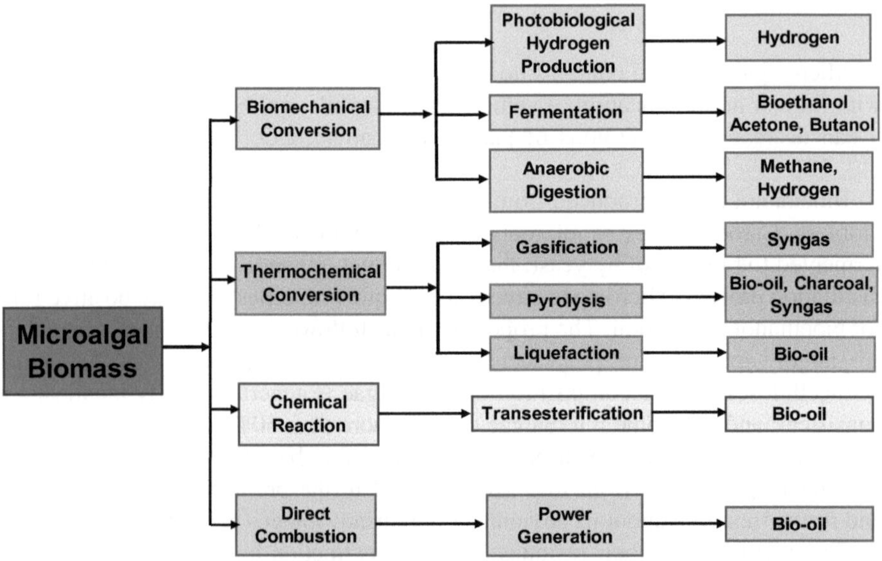

Fig. 5 Biofuel production processes from microalgae biomass. Adapted from [10, 59]

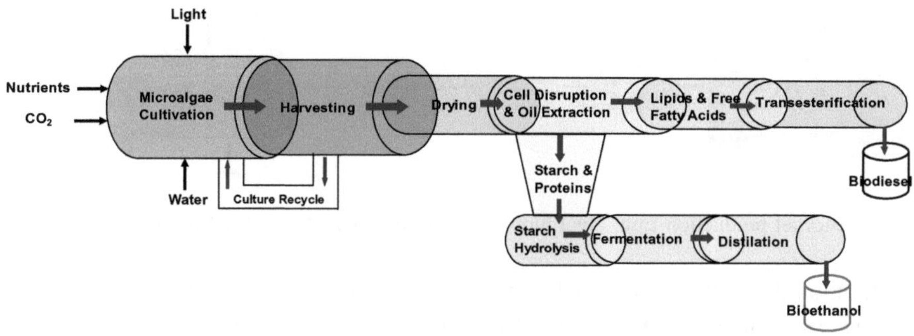

Fig. 6 Biodiesel and bioethanol production processes from microalgae. Adapted from [10]

The sun-drying method is probably the cheapest drying method that has been employed for the processing of microalgal biomass. However, this method takes a long drying time and requires a large drying surface. More efficient but more costly drying methods are drum drying, spray drying, fluidised bed drying, freeze drying and refractance window drying. Cell disruption methods that have been used successfully include high-pressure homogenisers, autoclaving, and the addition of hydrochloric acid, sodium hydroxide or alkaline lysis [10]. Numerous methods for extraction of lipids from microalgae have been applied; but most common methods are expeller/oil press, liquid–liquid extraction (solvent extraction), supercritical fluid extraction (SFE) and ultrasound techniques. Solvent extraction has proved to be successful in extracting lipids from microalgae. In this approach, organic solvents are added to algae paste. Solvents destroy algal cell walls and extract oil from the aqueous medium. Solvent extract can then be subjected to distillation to separate oil from solvent. After the extraction processes, the resulting microalgal oil can be converted into biodiesel through a process called transesterification. The transesterification reaction consists of transforming triglycerides into fatty acid alkyl esters, in the presence of an alcohol, such as methanol or ethanol, and a catalyst, such as an alkali or acid, with glycerol as a by-product [10, 66]. However, these processes are complex, technologically challenging and economically expensive. A significant challenge lies ahead for devising a viable biofuel production process [10, 64, 65, 66].

6 Potential of Algal Biofuel Production Using CO_2 from Thermal Power Plant

CO_2 emission is rapidly increasing with the rising demand for power (electricity). Rapid economic growth in developing nations leads the increasing demand for power generation. For example, power consumption grew from 375.39 TWh in 2000 to 600.65 TWh in 2008, at an annual growth rate of 6.67%, while electric power generation grew from 529.12 TWh in 2000 to 835.27 TWh at an annual growth rate of 5.78% [67]. At present, most power generation in the world is

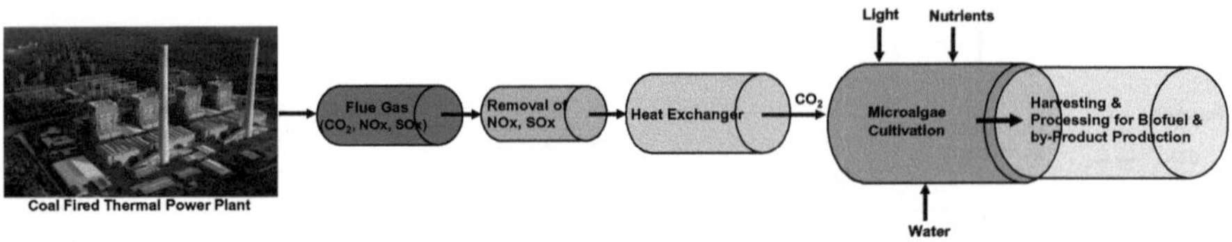

Fig. 7 Biofuel production from algal biomass using flue gas of coal-fired power plant

predominantly coal-based, which generates large amounts of CO_2. Microalgae are capable of utilising CO_2 emitted by thermal power plants with photosynthetic efficiency of greater than 10%, compared to 0.5% of terrestrial plants [68]. One kilogram of microalgal biomass has the capacity to capture 4.8 kg of CO_2 [69].

Coupling microalgal cultivation with a thermal power plant can not only reduce carbon dioxide emissions but also produce value-added products including biodiesel. Baral et al. [69] reported that a 146 MW coal-fired thermal power plant produces 1.1 mega tonnes (Mts) of CO_2 per year. They also reported that out of 3,000 tonnes CO_2 released by a coal-fired thermal power plant, 5.4 tonnes can be sequestered by an algal plant producing around 3 tonnes (dry weight) algal biomass. This algal biomass can produce approximately 1.19 tonnes of biofuel, which is a cleaner fuel than diesel [69]. After lipid extraction, the remaining dry biomass/cake of microalgae can be used as livestock feed, organic fertiliser due to its high nitrogen phosphorous ratio (N:P), or simply burned for energy cogeneration [70].

In addition, the algal biomass produced can also be used to decrease the consumption of coal in power plants. Algal biomass has a higher calorific value than coal. The calorific value of algal biomass is approximately 28,000 kJ/kg, whereas typical high thermal quality coal has calorific value of 27,000 kJ/kg [67, 71]. An average coal-fired power plant of 126 MW capacity uses 430 tonnes of coal per day; co-firing with 3 tonnes of algae can reduce coal usage by 3.11 tonnes per day or 1,026 tonnes per year. Furthermore, the process will reduce SOx content in the flue gas, as biomass burns without releasing any SOx in the atmosphere [69]. A schematic of algal biomass and biofuel production processes using flue gas generated by the coal-fired thermal power plant is shown in Fig. 7.

Several pilot studies on CO_2 utilisation from coal-fired thermal power plants for microalgal biomass production are currently underway around the world including Australia (e.g. the Loy Yang Power Plant in Victoria, Stanwell Power Plant in Queensland and Eraring Power Plant in New South Wales).

7 Microalgal Biofuel Research in Australia

Microalgae research and commercial production in Australia is limited despite having one of the largest microalgal companies for bioactive products in Western Australia [72]. The global challenges of health and nutrition issues, food productivity and security and sustainable energy production have increased the interest in algal research in Australia recently. Several Australian universities (including RMIT University, Adelaide University, University of Queensland, Murdoch University, Flinders University, University of Tasmania, University of Sydney, James Cook University, Wollongong University and Monash University) and the Commonwealth Scientific and Industrial Research Organisation (CSIRO) have been conducting research on microalgae to produce biofuel, fish feed, pharmaceutical and cosmetic products. The CSIRO has established the Australian National Algae Culture Collection Centre. The Centre holds more than 1000 microalgae strains of marine (saltwater) and freshwater species from tropical Australia to Antarctica.

Microalgal biofuel research at RMIT University is cross-disciplinary, involving researchers from the School of Applied Sciences, School of Aerospace, Mechanical and Manufacturing Engineering and other external organisations. The research group is working on developing cost-effective algal cultural, harvesting and extraction techniques. Presently, the Group focusses on typical microalgae such as *Nannochloris* sp. *Isochrysis* sp. *Chlorella* sp., *Pavlova salina* with the collaboration of other algal research groups (Fig. 3a and c). Figure 8 illustrates some representative microalgal collections [73]. These algal species have high lipid contents as shown in Table 3. The biofuel research group at RMIT University utilises expertise and knowledge from aquaculture, biotechnology, engineering and chemical

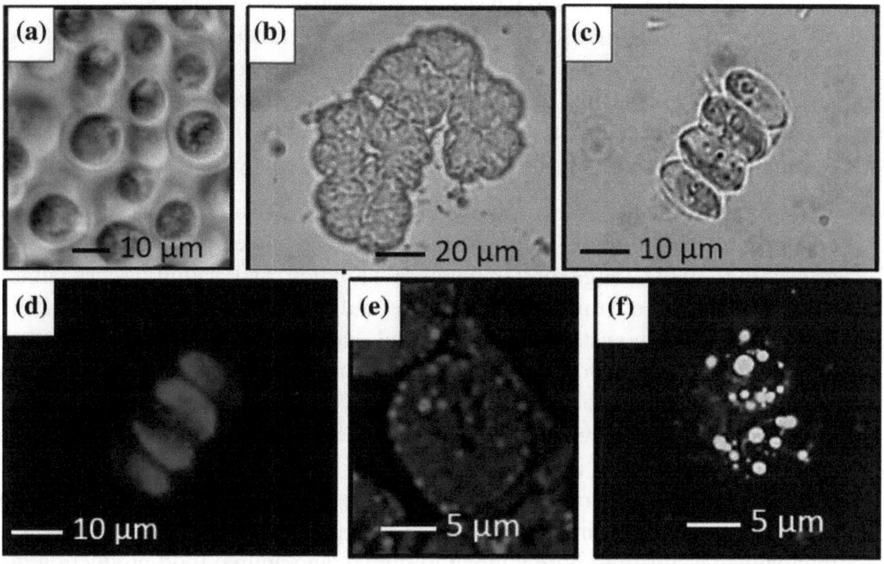

Fig. 8 Microalgal collection in the School of Applied Sciences, RMIT University: **a** *Chlamydomonas reinhardtii*, CC-125; **b** *Botryococcus braunii*, race L; **c** *Scenedesmussp*, bright field and **d** UV filter. *Red* colour represents chloroplasts; **e** Lipid bodies stained with Nile *Red* in *C. reinhardtii* wild type and **f** starch mutant, CC-4334 under nitrogen-limited condition. Adapted from [73]

sciences. The multidisciplinary approach creates a unique environment to investigate sustainable biofuel production for automotive, aviation, food and pharmaceuticals applications. In order to enhance algal biofuel production, it is important to augment algae cellular lipid yield. Hence, the redirection of energy storage compounds from starch to lipid metabolic pathways in starch mutants and/or reduction of lipid catabolism such as β-oxidation are required. The RMIT Research Group is actively working on recombinant proteins in microalgae in enhancing metabolic reprogramming of fatty acid biosynthesis pathways through recombinant enzymes to tailor oil composition. The Group is also researching to develop therapeutic proteins and antigens. Additionally, the Group is working on microalgal culture and harvesting infrastructure and extraction techniques' improvement.

8 Major Challenges

Current global energy demand has been met by mainly fossil fuels (80%) and to a lesser extent by biofuels (11.3%), followed by nuclear fuel (5.5%), and other renewable energy sources, including PV cells and wind (2.2%) [74]. Worldwide initiatives to reduce greenhouse gas emission and increase renewable energy utilisation compel the reduction of our fossil fuel dependency. Hence, we need to increase renewable energy utilisation for growing demand for power and energy. Solar and wind energy utilisation has limitations and significant costs, and technological challenges are involved in rapid increases in its share of the global energy mix. Therefore, third-generation algal biofuel can play a role in filling the partial gap left by reduced fossil fuel utilisation. However, there are significant challenges to be addressed in order to utilise the full potential of third-generation biofuel derived mainly from microalgal masses. Some major challenges are as follows:

(a) Technological challenges in microalgal selection, cultivation, harvesting, processing and distribution. There is a challenge to develop high-yielding (hybridisation), wide range of temperature variation and salinity (euryhaline)-tolerant microalgae strains. Cost-effective and technologically reliable cultivation and harvesting systems are considered one of the significant constraints in commercial-scale microalgal production. Current mechanical and chemical processes used for biofuel extraction are economically costly. The cost of associated infrastructure for the distribution and retailing of biofuels is also a challenge compared to conventional systems.
(b) Land use and agricultural input challenges for microalgae cultivation. Farming land and freshwater cannot be fully utilised in commercial-scale production of microalgae, as this will pose a threat to human and animal food production due to increased farming land reduction as a result of housing, industrialisation and infrastructure expansion, and global sea-level rise. Excessive use of inorganic and organic fertiliser in algal mass production can affect the ecological balance (eutrophication) due to the excessive use of phosphate and nitrate.

(c) A comprehensive understanding of microalgal biology is still a challenge, as in-depth knowledge can facilitate gene cloning and manipulation. Genetic modification is also a challenge by which we can achieve higher photosynthetic efficiency, temperature toleration ability, improved metabolic activity, and higher productivity of algae biomass and its oil content.
(d) Mechanisms and processes of wastewater nutrient utilisation for microalgal cultivation remain a challenge to date. There is no well-developed and cost-effective filtration process to remove the harmful elements that can affect algal growth and its lipid content.
(e) There is a significant challenge to the development of the full value chain of microalgae products and by-products. Production of biofuel alone, without considering co-utilisation of by-products such as animal feed, biofertiliser, cosmetics, chemical and pharmaceutical products, colouring agents, biogas masses, will not be commercially attractive.
(f) At current fossil fuel prices, it is extremely challenging for biofuel to be competitive in the market because the production cost of biofuel remains significantly higher.

9 Concluding Remarks and Future Prospects

In the light of the need for sustainable energy sources, biofuels from microalgae have gained worldwide public and political support, as biofuels have significant potential to play a notable role in combating energy shortages, complementing the shortfall of fossil fuel, reducing global warming and the effects of climate change. The biofuel produced from microalgae is generally carbon neutral, as the CO_2 emitted from burning biofuel is utilised back from the atmosphere by microalgal growth. Hence, the utilisation of biofuels does not add new carbon to the atmosphere.

Microalgae have broad bioenergy potential, since they can flexibly produce multiple biofuels, such as biodiesel, bioethanol, biogas and biohydrogen. However, there is a number of questions that need to be addressed with respect to commercially viable and sustainable microalgal biofuel production at a large scale, such as (i) what leading technology(s) for microalgal production, harvesting, and processing for biofuel production and other valuable by-product development will be, (ii) whether the process of biofuel production will be based on photosynthetic or heterotropic microorganism, (iii) if heterotropic microorganism is used how the CO_2 produced by heterotropic microorganism will be sequestrated and (iv) whether recycling and reuse of CO_2 and various types of nutrients are possible in a cost-effective way.

The main hurdle for economically viable biofuel production is the cost gap due to high cost of cultivation, harvesting and processing technologies. The current cost of infrastructure facilities and the energy required for microalgal cultivation and harvesting are high. Hence, comprehensive research is required to make biofuel industry sustainable and cost-effective. At present microalgal Research and Development (R&D) remains in its infancy and a major R&D initiative needs to be undertaken to:

a. Enhance the yield of microalgal production using genetic engineering technology. The basic understanding of microalgal biology can facilitate gene cloning and manipulation. Genetic modification can be used to increase photosynthetic efficiency, temperature toleration ability and productivity of biomass and oil content of microalgae.
b. Improve metabolic activity of microalgae to obtain efficient strains.
c. Understand required physicochemical parameters of cultured water for mono- and/or multiculture of microalgal species.
d. Improve the environment and reduction of production cost of biofuel by using wastewater from various sources, seawater, fermented liquids and agricultural wastes as a nutrient source.
e. Innovate cost-effective harvesting and processing technologies.
f. Enhance yield of biofuel products and explore the feasibility of an algal high-value co-product strategy and at the same time reduce the overall operating cost.

Microalgae have immense potential for biofuels production although there are some obstacles to the large-scale commercial production of biofuels using microalgal feedstock at present. However, this potential largely depends on the utilisation of technology and inputs such as CO_2, wastewater, salt water, natural light, barren land and aquatic environment. As algae possess significant potential for producing biofuel and valuable by-products, for successful operation of the industry both these options need to be explored efficiently and sustainably. Microalgae fully address the concerns created by first- and second-generation biofuel utilisation.

Biofuel production from microalgae utilising the CO_2 emitted by thermal power plants (coal, gas and liquid fuels) can usher in a new vista for mitigation of greenhouse gas emission and effects of climate change. This will also allow low cost and renewable carbon capture instead of currently used high-cost technology. Carbon capture by microalgae and co-burning of microalgae by-products with coal will assist both power generators and consumers to have affordable power.

References

1. Scaife, M. A., Merkx-Jacques, A., Woodhall, D. L., & Armenta, R. E. (2015). Algal biofuels in Canada: Status and potential. *Renewable and Sustainable Energy Reviews, 44,* 620–642.
2. Quéré, C. L., Raupach, M. R., Canadell, J. G., Marland, G., Bopp, L., Ciais, P., et al. (2009). Trends in the sources and sinks of carbon dioxide. *Nature Geoscience, 2,* 831–836.
3. Morello, L. (2012). Fuel burning rise into high-risk zone, climatewire, 3 December, Retrieved on April 24,2016 from http://www.scientificamerican.com/article/global-co2-emissions-from/.

4. Alam, F., Date, A., Rasjidin, R., Mobin, S., Moria, H., & Baqui, A. (2012). Biofuel from algae- is it a viable alternative? *Procedia Engineering, 49,* 221–222.
5. Mobin, S., & Alam, F. (2014). Biofuel production from algae utilizing wastewater. In *Proceedings of 19th Australasian Fluid Mechanics Conference,* Article No 27.
6. Alam, F., Mobin, S., & Chowdhury, H. (2015). Third generation biofuel from algae. *Procedia Engineering, 105,* 763–768.
7. Li, W. W., & Yu, H. Q. (2011). From waste water to bioenergy and biochemicals via two-stage bioconversion processes: A future paradigm. *Biotechnology Advances, 29,* 972–982.
8. Demirbas, A. (2009). Political, economic and environmental impacts of biofuels: A review. *Applied Energy, 86,* S108–S117.
9. Nigam, P. S., & Singh, A. (2011). Production of liquid biofuels from renewable resources. *Progress in Energy and Combustion Science, 37*(1), 52–68.
10. Dragone, G., Fernandes, B., Vicente, A. A. & Teixeira, J. A. (2010). Third generation biofuels from microalgae in current research. In Mendez-Vilas, A. (Ed.) *Technology and Education Topics in Applied Microbiology and Microbial Biotechnology* (pp. 1355–1366.) Formatex.
11. Chisti, Y. (2007). Biodiesel from microalgae. *Biotechnology Advances, 25,* 294–306.
12. Singh, A., Nigam, P. S., & Murphy, J. D. (2011). Renewable fuels from algae: An answer to debatable land based fuels. *Bioresource Technology, 102,* 10–16.
13. Graham-Rowe, D. (2011). Agriculture: Beyond food versus fuel. *Nature, 474,* S6–S8.
14. Dyer, J. A., Vergé, X. P. C., Desjardins, R. L., Worth, D. E., & McConkey, B. G. (2010). The impact of increased biodiesel production on the greenhouse gas emissions from field crops in Canada. *Energy for Sustainable Development, 14,* 73–82.
15. Groom, M. J., Gray, E. M., & Townsend, P. A. (2008). Biofuels and biodiversity: Principles for creating better policies for biofuel production. *Conservation Biology, 22,* 602–609.
16. Schenk, P., Thomas-Hall, S., Stephens, E., Marx, U., Mussgnug, J., Posten, C., et al. (2008). Second generation biofuels: High efficiency microalgae for biodiesel production. *BioEnergy Research, 1,* 20–43.
17. Sims, R. E. H., Mabee, W., Saddler, J. N., & Taylor, M. (2010). An over view of second generation biofuel technologies. *Bioresources Technology, 101,* 1570–1580.
18. Singh, B., Guldhe, A., Rawat, I., & Bux, F. (2014). Towards a sustainable approach for development of biodiesel from plant and microalgae. *Renewable and Sustainable Energy Reviews, 9,* 216–245.
19. Anex, R. P., Aden, A., Kazi, F. K., Fortman, J., Swanson, R. M., Wright, M. M., et al. (2010). Techno-economic comparison of biomass-to-transportation fuels via pyrolysis, gasification, and biochemical pathways. *Fuel, 89,* S29–S35.
20. Wyman, C. E. (2003). Potential synergies and challenges in refining cellulosic biomass to fuels, hemicals, and power. *Biotechnology Progress, 19,* 254–262.
21. Brennan, L., & Owende, P. (2010). Biofuels from microalgae–A review of technologies for production, processing, and extractions of biofuels and co-products. *Renewable and Sustainable Energy Reviews, 14,* 557–577.
22. Mabee, W. E., McFarlane, P. N., & Saddler, J. N. (2011). Biomass availability for lignocellulosic ethanol production. *Biomass Bioenergy, 35,* 4519–4529.
23. Melillo, J. M., Reilly, J. M., Kicklighter, D. W., Gurgel, A. C., Cronin, T. W., Paltsev, S., et al. (2009). Indirect emissions from biofuels: How important? *Science, 326,* 1397–1399.
24. Fargione, J., Hill, J., Tilman, D., Polasky, S., & Hawthorne, P. (2008). Land clearing and the biofuel carbon debt. *Science, 319,* 1235–1238.
25. Searchinger, T., Heimlich, R., Houghton, R. A., Dong, F., Elobeid, A., Fabiosa, J., et al. (2008). Use of U.S. Croplands for biofuels increases greenhouse gases through emissions from land-use change. *Science, 319,* 1238–1240.
26. Tilman, D., Socolow, R., Foley, J. A., Hill, J., Larson, E., Lynd, L., et al. (2009). Beneficial biofuels: The food, energy, and environment trilemma. *Science, 325,* 270–271.
27. Li, Y., Horsman, M., Wu, N., Lan, C. Q., & Dubois-Calero, N. (2008). Biofuels from microalgae. *Biotechnology Progress, 24,* 815–820.

28. Scott, S. A., Davey, M. P., Dennis, J. S., Horst, I., Howe, C. J., Lea-Smith, D. J., et al. (2010). Biodiesel from algae: Challenges and prospects. *Current Opinion in Biotechnology, 21,* 277–286.
29. Richmond, A. (2004). *Handbook of microalgal culture: Biotechnology and applied phycology*. Oxford, UK: Blackwell Science Ltd.
30. Graham, L. E., Graham, J. M., & Wilcox, L. W. (2009). *Algae*. California, USA: Benjamin-Cummings publishing Company. ISBN-13: 978-0321559654.
31. Ozkurt, I. (2009). Qualifying of safflower and algae for energy. *Energy Education Science and Technology Part A, 23,* 145–151.
32. Aresta, M., Dibenedetto, A., & Barberio, G. (2005). Utilization of macro-algae for enhanced CO_2 fixation and biofuels production: Development of a computing software for an LCA study. *Fuel ProcessingTechnology, 86,* 1679–1693.
33. Zhu, L. D., Hiltunen, E., Antila, E., Zhong, J. J., Yuan, Z. H., & Wang, Z. M. (2014). Microalgal biofuels: Flexible bioenergies for sustainable development. *Renewable and Sustainable Energy Reviews, 30,* 1035–1046.
34. Najafi, G., Ghobadiana, B., & Yusaf, T. F. (2011). Algae as a sustainable energy source for biofuel production in Iran: A case study. *Renewable and Sustainable Energy Reviews, 2011* (15), 3870–3876.
35. Maity, J. P., Bundschuh, J., Chen, C.-Y., & Bhattachary, A. P. (2014). Microalgae for third generation biofuel production, mitigation of greenhouse gas emissions and wastewater treatment: Present and future perspectives—a mini review. *Energy, 78,* 104–113.
36. Ramaraj, R., & Dussadee, N. (2014). Biological purification process for biogas using algae culture: A review. *International Journal of Sustainable and Green Energy, 4,* 20–32.
37. Um, B.-H., & Kim, Y.-S. (2008). Review: A chance for Korea to advance algal-biodiesel technology. *Journal of Industrial and Engineering Chemistry, 15,* 1–7.
38. Sydney, E. B., Sturm, W., de Carvalho, J. C., Thomaz-Soccol, V., Larroche, C., Pandey, A., et al. (2010). Potential carbon dioxide fixation by industrially important microalgae. *Bioresource Technology, 101,* 5892–5896.
39. Koh, L. P., & Ghazoul, J. (2008). Biofuels, biodiversity, and people: Understanding the conflicts and finding opportunities. *Biological Conservation, 141,* 2450–2460.
40. Mata, T. M., Martins, A. A., & Caetano, N. S. (2010). Microalgae for biodiesel production and other applications: A review. *Renewable and Sustainable Energy Reviews, 14,* 217–232.
41. Tredici, M. R., & Materassi, R. (1992). From open ponds to vertical alveolar panels: The Italian experience in the development of reactors for the mass cultivation of phototrophic mircroorganisms. *Journal of Applied Phycology, 4,* 221–231.
42. Lundquist, T. J., Woertz, I. C., Quinn, N. W. T., & Benemann, J. R. (2010). *A realistic technology and engineering assessment of algae biofuel production*. Berkeley, CA: Energy Biosciences Institute.
43. Carvalho, A. P., Meireles, L. A., & Malcata, F. X. (2006). Microalgal reactors: A review of enclosed system designs and performances. *Biotechnology Progress, 22,* 1490–1506.
44. Kunjapur, A. M., & Eldridge, R. B. (2010). Photobioreactor design for commercial biofuelproduction from microalgae. *Industrial and Engineering Chemmistry Reserch, 49,* 3516–3526.
45. Darzins, A., Pienkos, P. T., & Edye, L. (2010). Current status and potential for algal biofuels production. *IEA Bioenergy Task, 39,* 1–131.
46. Lam, M. K. & Lee, K. T. (2014). Scale-up and commercialisation of algal cultivation and biofuel production. In: Pandey, A., Lee, D-J., Chisti, Y. & Soccol, C. R. (Eds.), *Biofuel from algae*. Elsevier, (pp 261–285), ISBN: 978-0-444-59558-4.
47. Zemke, P. E. (2016). Mass cultivation of phototrophic microalgae in recent Advances. In J. Liu, Z. Sun, & H. Gerken (Eds.), *Microalgal biotechnology* (pp. 1–16). Foster City, USA: OMICS eBooks Group.
48. Hopman, J. P. (1998). Wastewater treatment with suspended and nonsuspended algae. *Journal of Phycology, 34,* 757–763.

49. Molina, G. E., Belarbi, E. H., Acién, F. G., Robles, M. A., & Chisti, Y. (2003). *Biotechnology Advances, 20,* 491–515.
50. Christenson, L., & Sims, R. (2011). Production and harvesting of microalgae for wastewater treatment, biofuels, and bioproducts. *Biotechnology Advances, 29,* 686–702.
51. Kumar, H., Yadava, P., & Gaur, J. (1981). Electrical flocculation of the unicellular green alga *Chlorella vulgaris* Beijerinck. *Aquatic Botany, 11,* 187–195.
52. Roesijadi, G. (2010). Macroalgae as a biomass feedstock: A preliminary analysis. U.S. Department of Energy under contract, Pacific Northwest National Laboratory.
53. Brown, M. R., Jeffrey, S. W., Volkman, J. K., & Dunstan, G. A. (1997). Nutritional properties of microalgae for mariculture. *Aquaculture, 151,* 315–331.
54. Trivedi, J., Aila, M., Bangwal, D. P., Kaul, S., & Garg, M. O. (2015). Algae basedbiorefinery —howtomakesense? *Renewable and Sustainable Energy Reviews, 47,* 295–307.
55. Delucchi, M. A. A. (2003). Lifecycle Emissions Model (LEM): Lifecycle emissions from transportation fuels; motor vehicles, transportation modes, electricity use, heating and cooking fuels. Main report UCD-ITS-RR-03-17. University of California, Davis, CA.
56. Liu, C. H., Chang, C. Y., Cheng, C. L., Lee, D. J., & Chang, J. S. (2012). Fermentative hydrogen production by *Clostridium butyricum* CGS5 using carbohydrate-rich microalgal biomassas feedstock. *International Journal of Hydrogen Energy, 37,* 15458–15464.
57. John, R. P., Anisha, G. S., Nampoothiri, K. M., & Pandey, A. (2011). Micro and macroalgal biomass: A renewable source for bioethanol. *Bioresource Technology, 102,* 186–193.
58. Melis, A., & Happe, T. (2001). Hydrogen production. Green algae as a source of energy. *Plant Physiology, 127,* 740–748.
59. Wang, B., Li, Y., Wu, N., & Lan, C. (2008). CO_2 bio-mitigation using microalgae. *Applied Microbiology and Biotechnology, 79,* 707–718.
60. Spolaore, P., Joannis-Cassan, C., Duran, E., & Isambert, A. (2006). Commercial applications of microalgae. *Journal of Bioscience and Bioengineering, 101,* 87–96.
61. Benemann, J. R., Tillett, D. M., & Weissman, J. C. (1987). Microalgae biotechnology. *Trends in Biotechnology, 5,* 47–53.
62. Eriksen, N., Poulsen, B., & Lønsmann, I. J. (1998). Dual sparging laboratory-scale photobioreactor for continuous production of microalgae. *Journal of Applied Phycology, 10,* 377–382.
63. Tredici, M. R. (1999). Bioreactors, photo. In M. C. Flickinger & S. W. Drew (Eds.), *Encyclopedia of Bioprocess Technology: Fermentation, Biocatalysis, and Bioseparation* (pp. 395–419). New York, NY: Wiley.
64. Harun, R., Singh, M., Forde, G. M., & Danquah, M. K. (2010). Bioprocess engineering of microalgae to produce a variety of consumer products. *Renewable and Sustainable Energy Reviews, 14,* 1037–1047.
65. Mendes-Pinto, M. M., Raposo, M. F. J., Bowen, J., Young, A. J., & Morais, R. (2001). Evaluation of different cell disruption processes on encysted cells of Haematococcus pluvialis: Effects on astaxanthin recovery and implications for bio-availability. *Journal of Applied Phycology, 13,* 19–24.
66. Vasudevan, P., & Briggs, M. (2008). Biodiesel production—current state of the art and challenges. *Journal of Industrial Microbiology and Biotechnology, 35,* 421–430.
67. Annual Energy Outlook 2010 with Projections to 2035. The Energy Information Administration, National Energy Information Centre, Forrestal Building, Washington, DC (https://www.eia.doe.gov, 2010).
68. Khan, S. A. & Rashmi. (2008). Algae: An ovelsource of renewable energy and carbon sequestration. *Renewable Energy (Akshayurja), 2,* 14–8 September–October.
69. Baral, S. S., Singh, K., & Sharma, P. (2015). The potential of sustainable algal biofuel production using CO_2 from thermal power plant in India. *Renewable and Sustainable Energy Reviews, 49,* 1061–1074.
70. Jain, S. K., Prakash, V., & Kapoor, S. (2004). Flue gas treatment alternatives or enhancing ESP collection efficiency: NTPC experience, workshop on ESP performance: Role of fly ash resistivity. *IIT-DandNTPC, 23–24,* 1–5.

71. Sudhakar, K., Premlata, M., & Sudarshan, K. (2012). Energy balance and exergy analysis of large scale algal biomass production. In: Indonesia workshop & international symposium on bioenergy from biomass.
72. Winberg, P. (2011). Scaling up for new opportunities in the practical use of algae (Applied Phycology), An Australian Government Report, RIRDC Publication No. 11/174, ISBN 978-1-74254-350-5, (pp. 1-23).
73. Mouradov, A. & Stevenson, T. (2012 September). Algae: An essential link between our past and future. *Microbiology Australia*, pp. 125–127.
74. Koppelaar, R. (2012). Word energy consumption beyond 500 exajoules. Santa Rosa, CA: Post Carbon Institute. http://www.resilience.org/stories/ 2012-02-16/world-energy-consumption-beyond-500-exajoules.

Performance Assessment of an Electrostatic Precipitator of a Coal-Fired Power Plant—A Case Study for Collecting Smaller Particles

A.S.M. Sayem, M.M.K. Khan, M.G. Rasul and N.M.S. Hassan

Abstract Producing sustainable clean energy is one of the key challenges in modern power generation systems. Coal-fired power plants are one of the main sources of electrical energy due to the low cost of coal compared to other fossil fuels. However, one of the major problems of the coal-fired power plant is the exhaust emission of fine particulate matter. Most of the coal power plants and other process industries generally use electrostatic precipitators (ESPs) because of their effectiveness and reliability in controlling particulate matter. The dust particles from the flue gas are separated using flow dynamics and an electrical force induced by the ESP. Baffles and plates are used to obstruct the flue gas flow and to increase residence time to force particle deposition. ESPs are the most reliable control devices to capture the fine particles and their efficiency is also high. However, the precipitator has some serious limitations when capturing smaller size dust particles, especially those less than 2.5 micron. Another drawback is the collection of dust from low-temperature flue gas. In this chapter, a computational fluid dynamics (CFD) model of flow distribution inside the ESP has been discussed which can be useful for collecting smaller particles regardless of operating temperature. A case study is presented showing a wide variety of flow simulation by inserting different shapes of baffles inside the ESP and their effect on particle collection. The collection efficiency of the particles affected by different flow distributions and the possible modifications in the existing ESPs used in the power plants are also discussed.

Keywords Computational fluid dynamics (CFD) · Electrostatic precipitator (ESP) · Collection efficiency · Flow distribution

A.S.M. Sayem (✉) · M.M.K. Khan · M.G. Rasul
School of Engineering and Technology, Higher Education Division,
Central Queensland University, Rockhampton, QLD 4702, Australia
e-mail: a.sayem@cqu.edu.au

N.M.S. Hassan
School of Engineering and Technology, Higher Education Division,
Central Queensland University, Cairns, QLD 4870, Australia

M.M.K. Khan et al. (eds.), *Application of Thermo-fluid Processes in Energy Systems*,
Green Energy and Technology, https://doi.org/10.1007/978-981-10-0697-5_5

1 Introduction

Coal has been a major source of electricity generation for Queensland's electricity supply for decades because it is abundant and cheap. In 2013–14, coal accounted for 61% of total power generation in Australia [1] and 41% of the global electricity generation [2]. This huge amount of electricity generation from coal indicates the future coal consumption scenario for power generation industry. But one of the biggest challenges associated with the coal-based power plant is to control the emission of particulate matter as well as carbon emissions. Because of strict rules and regulations of the Environmental Protection Agency (EPA), particulate emissions have remained one of the major concerns to power industry. Particulates contain materials that can affect our health severely because they may go into the deeper parts of the respiratory tract [3]. Local power stations always desire to control the particulate emissions at a minimum cost by performance optimization of the emission control devices [4] in spite of having high capture capability by the electrostatic precipitators. Currently, the particles of size PM (Particulate Matter) 2.5 micron or more may escape the ESP; however, it is anticipated that new EPA regulations will soon be imposed for mandatory capture of these particles. Use of ESPs has become popular due to many industrial advantages. For example, ESPs can handle large volumes of flue gases and can operate at high temperature with small pressure drop and can remove micronized particles. The main advantage of this device is that it can collect a wide range of particulate matter in both dry and wet states. Table 1 shows the particulate capture range for various available capture devices. Comparing an ESP with all other devices, it is seen that the ESP is capable of collecting a wide range of particulate matter, starting from 0 micron up to 1000 micron. In an ESP, there is no significant amount of pressure drop and collection efficiency is also high compared to other methods. Therefore, the ESP is viewed as a convenient choice as a particulate emission control device.

2 An Overview of Particle-Capturing Technique and Associated Problems

For separation of dust particles from the exhaust gas in the coal-fired plants, several technologies are available commercially. The particles can be separated by applying different working principles. For example, inertial separation is used in a cyclone separator, diffusion is applied in fabric filtration, an electrical field is used in an electrostatic precipitator and contacting, impingent and impaction are applied with wet scrubbers. Each type of separator exhibits different characteristics on collection efficiency, process and application suitability. A short description of these processes is provided below:

1 Dust spectrum captured by various control devices (Parker, 1997a)

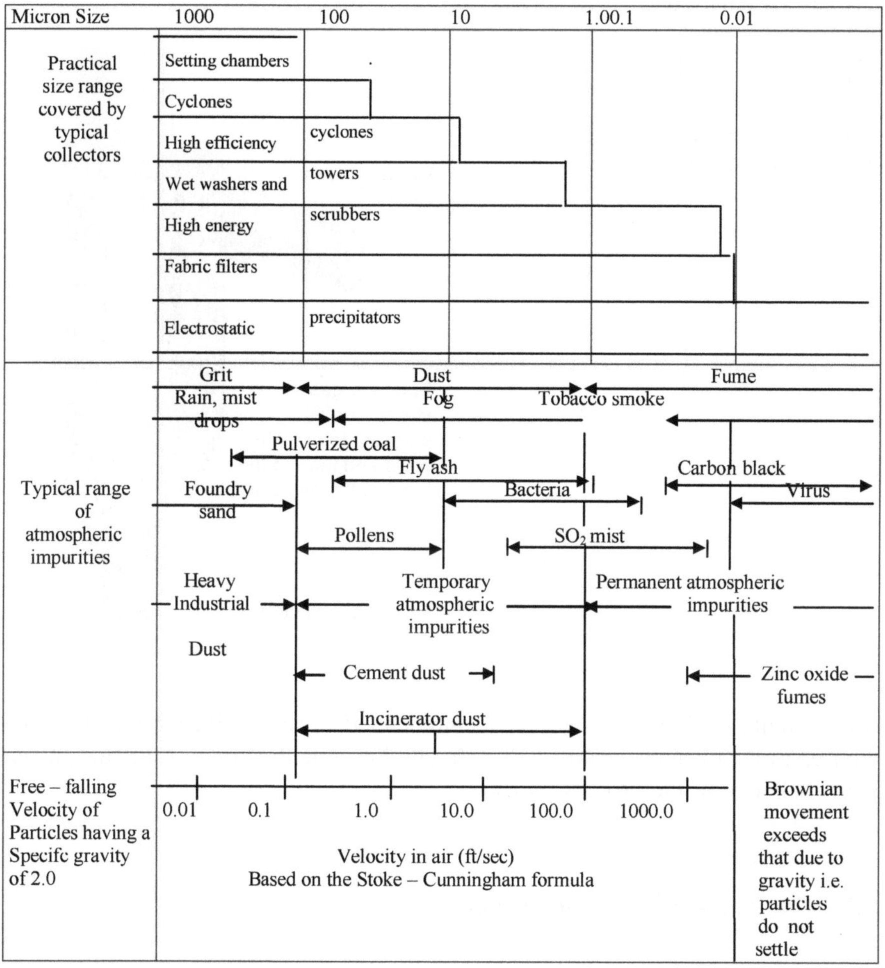

2.1 Cyclone Separator

A cyclone separator as shown below in Fig. 1 works on the principle of centrifugal force which creates a high-speed spiral flow of air. The spiral flow exerts a centrifugal force on the particles and, due to the inertia force, the particles move towards the wall of the cyclone separator and then to the base of the cyclone.

The efficiency of a cyclone separator depends on the velocity of air flow and cyclone diameter. The smaller the diameter, the more efficiently the particles are removed. However, if the diameter is smaller, the pressure drop is higher. In that

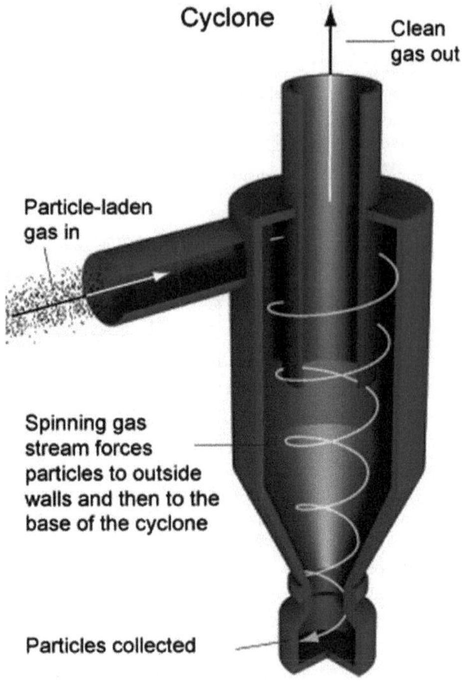

Fig. 1 Cyclone separator. Adopted with permission from © The State of Queensland 2017 Controlling Air Pollution. Available at: https://www.qld.gov.au/environment/pollution/monitoring/air-pollution/controlling

case, a multi-cyclone chamber is recommended. The significant pressure drop required in the cyclone separator to maintain the same flow rate of the flue gas is a great challenge in the process industry.

2.2 *Packed-Bed Wet Scrubber*

A wet scrubber is designed to remove gaseous or contaminated particles from a gas flow. Figure 2 shows the diagram of a packed-bed wet scrubber.

Here, the contaminated gas flows through the inlet duct situated in the lower portion and water is sprayed from the upper right portion. The gaseous pollutants are absorbed by the absorber liquid and the particles are removed in the form of droplets. Gaseous pollutants can be removed by using specific scrubbing liquid. For example, to remove acidic pollutants an alkaline liquid is used. So, for the packed-bed wet scrubber, a specific absorber fluid based on the type of gaseous pollutants is used. This is also problematic for continuous operation since further effluent treatment is necessary for handling the dust pollutants.

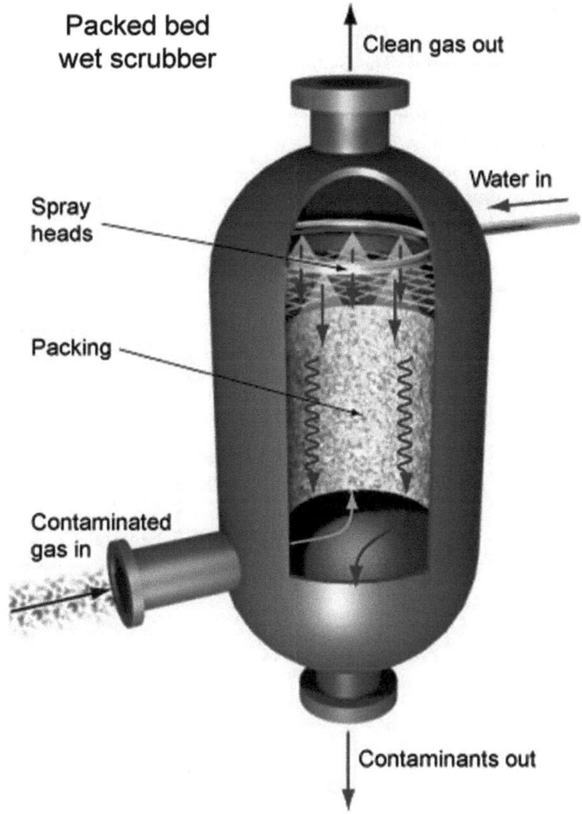

Fig. 2 Packed-bed wet scrubber. Adopted with permission from © The State of Queensland 2017 Controlling Air Pollution. Available at: https://www.qld.gov.au/environment/pollution/monitoring/air-pollution/controlling

2.3 Fabric Filters or Bag Filters

A fabric filter as shown in Fig. 3 is a widely accepted method for particulate emission control. In fabric filtration, particles go through a number of filter bags placed in parallel leaving the dust retained by the fabric. The type of fabric cloth used in the operation determines its operating temperature. For cotton fabric, the temperature limit is 355 °K while that for fibre glass is 530 °K.

The difficulty with bag filters is that most of the energy is needed to overcome the gas pressure drop across the bag, dust cakes and associated duct work. Besides, this device will require a change of the fabric after a designated amount of working hours which are associated with an increased operational cost.

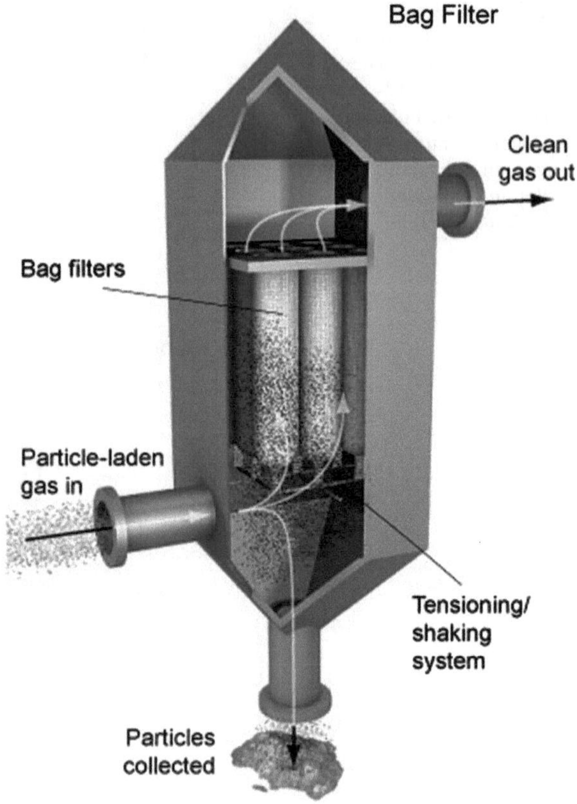

Fig. 3 Diagram of typical bag filter. Adopted with permission from © The State of Queensland 2017 Controlling Air Pollution. Available at: https://www.qld.gov.au/environment/pollution/monitoring/air-pollution/controlling

2.4 *Electrostatic Precipitator*

An electrostatic precipitator (ESP) is one of the most widely used devices for particle removal from the gas stream and uses an induced electrostatic charge. It is a high-precision particle-capturing device. An ESP is generally placed on the flow path of flue gas after the furnace and before the chimney. As there are no significant obstacles on the flow path, an ESP is a good option for maintaining continuous flow. Besides, an ESP offers a wide range of particulate capture ability as well as maintaining a smooth run of exhaust gas flow, especially in industrial applications. However, it is difficult to capture smaller size particulate matter, especially less than 2.5 micron, and enhancing the performance of the ESP for long-term use still remains a challenge.

3 Working Principle of the ESP

A simple ESP consists of a discharge wire and two collector plates. A high voltage is applied to the discharge wire to form an electric field to ionize the flue gas surrounding the plates.

Dust particles in the flue gas are charged either positively or negatively. Then, these charged particles move to oppositely charged plates, commonly referred to as collector plates. By this procedure, a dust layer is created on the plate. The layer of dust particles can be removed either by wrapping or spraying a liquid on the plate. Basically, three steps are followed for ESP. These are: (a) charging, (b) collecting and (c) removing. A short overview of those steps is given below:

(a) Charging: High voltage at an electrode creates a corona effect which will ionize the atmosphere surrounded by the electrode. The particles in the flue gas, having been charged, are then subjected to transverse electrostatic forces that pull them towards the oppositely charged collector plates.
(b) Collecting: Charged particles then move to the collector plates and accumulate on the plate surfaces.
(c) Removing: After formation of a dust cake in the collector plates, plates are periodically rapped for dust collection. Dust layer thickness is determined by the dust resistivity, strength of charges and other factors such as dielectric constants, conductivities of dust and air associated with the process.

After removing dust particles from the collector plates, they enter into the hopper and are removed from the hopper as quickly as possible; otherwise, particles will combine with each other, making them difficult to remove from the hopper.

4 ESP Performance and Improvement Options

The performance of an ESP depends on a number of factors. Some of them are related to the nature of the ash particles and some of them depend on the design and operation of the ESP [5]. Factors that are related to particles basically consist of electrical characteristics, size distribution, shape, particle cohesivity, operating temperature, flue gas conditioning, bipolar agglomeration, pre-charger, pulsed energization, gas cooling and humidification, hybrid particulate collector, electro-core separator, flow optimization, sulphur content etc. Precipitation-related problems include mechanical problems, re-entrainment, non-uniform gas distribution, fouling of the plate (Figs. 4 and 5).

From a literature survey, it is seen that various technologies exist that can be applied to enhance the performance of an ESP. Some of these technologies are stated below and are briefly discussed for relevance.

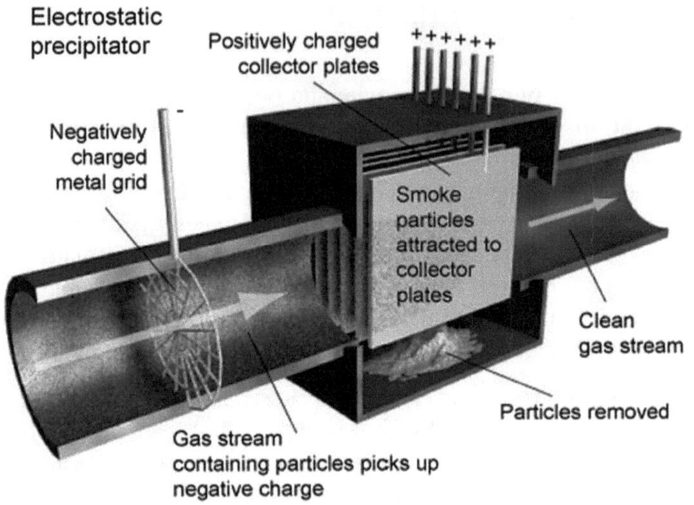

Fig. 4 Typical ESP. Adopted with permission from © The State of Queensland 2017 Controlling Air Pollution. Available at: https://www.qld.gov.au/environment/pollution/monitoring/air-pollution/controlling

Fig. 5 Schematic view of fluid flow inside an ESP [44]

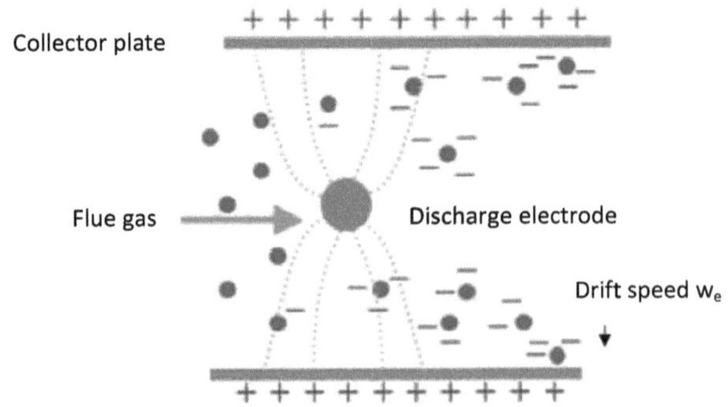

(a) Flue gas conditioning,
(b) Gas cooling and humidification,
(c) Bipolar agglomeration,
(d) Pre-charger,
(e) Pulsed energization,
(f) Electrocore separator,
(g) Hybrid particulate collections systems,
(h) Membrane-based collection surface,
(i) Wet ESP and
(j) Skewed gas flow technology and flow optimization.

Application of a specific technology depends on the material and operating conditions of the ESP. For example, to collect high resistive fly ash, the effective technology is flue gas conditioning [6, 7]. For solving resistivity-related problems, gas cooling and humidification can also be used [8]. Bipolar agglomeration is a technique where the charged particles are attracted by oppositely charged particles [9, 10]. Pre-charger technology [8, 11], pulsed energization [8, 12], electrocore separator [6, 11, 13], hybrid particulate collection systems [6] and membrane-based collection surfaces [6] are also efficient for collecting fine particles. For controlling particle emissions, wet ESP can replace dry ESP [6, 14]. For improving the ESP performance, skewed gas flow technology can be considered as it reduces particle re-entrainment losses [15–17]. For further improvement, the performance of an ESP can be increased by optimizing flow distribution [18–23].

(a) **Flue gas conditioning (FGC)**

The performance of an ESP can be increased by a technology called flue gas conditioning (FGC) for collecting fly ashes whose resistivity is very high. This process also involves conditioning agents which are injected into the flue gas. With the advancement of technologies, many new agents such as urea and elemental sulphur are applied instead of old ones in spite of producing SO_2 and aqueous ammonia which help to regenerate SO_3 and NH_3 [6]. The presence of SO_3 in the flue gas can reduce resistivity in the electrostatic precipitation process. In combustion of coal, most of the sulphur is converted into SO_2. However, approximately 1% of the sulphur is converted into SO_3. The resistivity of the particles decreases as the sulphur content of the coal increases as shown in Fig. 6.

Depending on combustion conditions, 0.5–3% of sulphur dioxide is further oxidized to SO_3. When the temperature rises to the sulphuric acid dew point, the

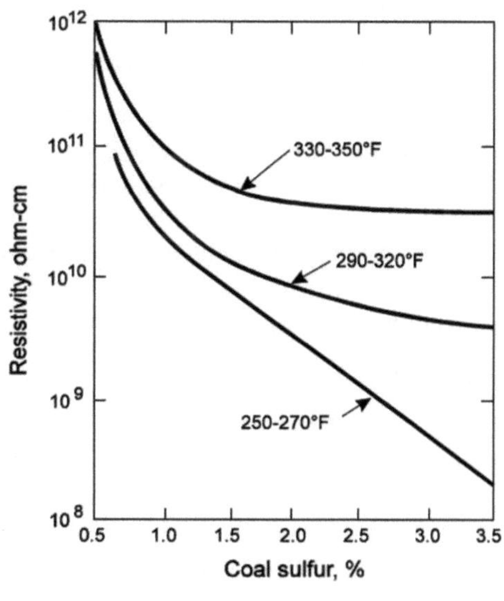

Fig. 6 Fly ash resistivity versus coal sulphur content for several flu gas temperature bands (*Source* [45])

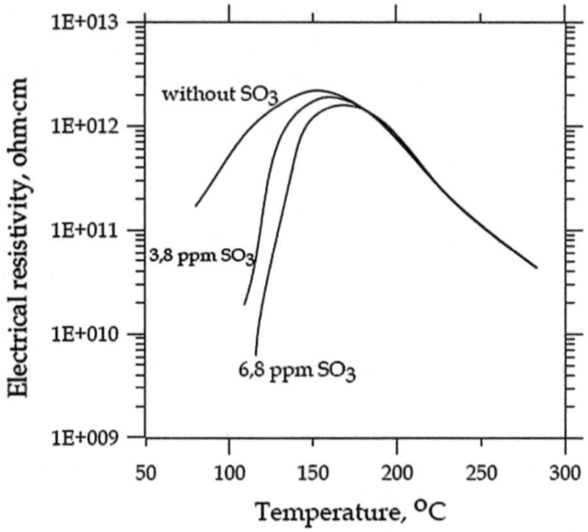

Fig. 7 Electrical resistivity of a fly ash as a function of SO_3 injection [24]

condensation of SO_3 on particle surfaces takes place which helps to reduce electrical resistivity of the fly ash. Figure 7 shows the electrical resistivity as a function of SO_3 injection.

To overcome the low sulphur content coal resistivity problems, SO_3 of 10 to 30 ppm can be injected into the gas stream. The SO_3 is injected into the duct before the precipitator.

Other research showed that a high-alumina content in coal reduces the efficiency of ESP [25]. To overcome this problem, SO_3—flue gas conditioning (FGC) was recommended [26]. An experimental study was done by [26] explaining that injecting SO_3 of 34.3 mg/m^3 increased ESP efficiency of a typical power plant in China from 96.9 to 99.5%.

(b) **Gas cooling and humidification**

When ESPs are operating at temperatures around 175 °C, resistivity-related problems may arise and gas cooling and humidification can overcome these problems. By increasing absorption of SO_3 and moisture, the resistivity of the dust layer can be increased. This is one of the basic advantages of the gas cooling and humidification process. Another advantage is that it reduces treated flue gas volume and velocity which will in turn reduce re-entrainment. Gas cooling can also foster condensation and collection of the vapour phase of mercury, arsenic and selenium species [8].

(c) **Bipolar agglomeration by acoustic or electric waves**

In an ESP, fine particles should be collected in certain amount. To improve the efficiency, a method is used which is known as bipolar agglomeration. The flue gas is first divided into two streams. Bipolar chargers are used for oppositely charging the particles in each stream. In a mixing section, these particles are brought together to form larger particle clusters by attaching the fine particles to the coarse particles.

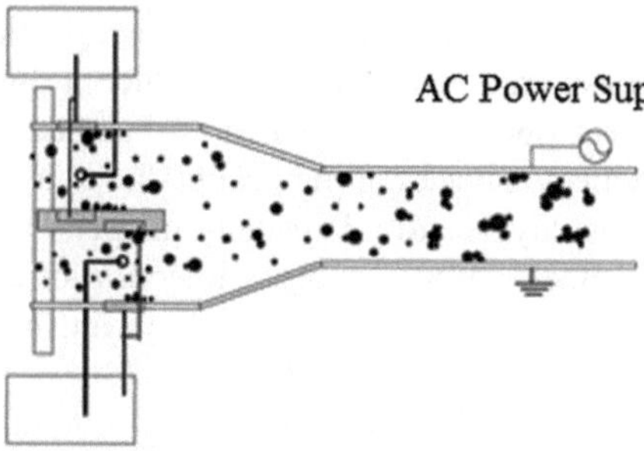

Fig. 8 Schematic of electrical agglomeration [9]

A bipolar agglomerator enhances fly ash particle agglomeration by combining electrostatic and turbulent mixing processes before entering into the ESP. By applying AC fields of 60–500 Hz, submicron particles are reduced by 25–29% [9]. The average number of elementary charges per particle was measured when the polarities of two corona chargers were either positive or negative. For the high-mass loading condition, the corona chargers' current/voltage characteristics were obtained. A comparison was made between the high-mass loading condition and no-mass loading condition. In the negative corona charger, dioctyl sebacate particles intercepted corona-induced ions of about 70–90% [9]. With the increase of the mass loading, there is a decrease in particle charges.

For promoting the chances of collision, intense acoustic waves are applied. These waves have tuned frequency and cause relative motion between the coarse particles and the submicron particles. Acoustic streaming which helps to entrain the submicron particles towards the charged coarse particles is also created by these waves. This effect is also created by applying an AC field by vibrating charged coarse particles [9, 10]. A schematic of the electrical agglomeration principle is shown in Fig. 8.

(d) **Pre-charger**

Fine particle collection efficiency can be improved by one of the ESP elements known as a pre-charger. For efficient charging of the dust upstream of the main ESP, a section is specially designed. Pre-chargers are normally used to agglomerate fine particulate matter into larger sizes for easy collection [27]. Long treatment times and efficient gas mixing resulting from travelling an intense corona towards the gas flow are suitable for ensuring fine particle charging. A pre-charger is mainly used for reducing corona current and operating the main ESP with higher electrical fields to attain efficient dust precipitation [8]. In general, the dust layer has high resistivity which can be reduced by a cooling system. It is easier to charge dust particles by use of a small pre-charger. A relatively small pre-charger is used in temperature-controlled electrodes [11].

(e) **Pulsed energization**

There is another technique available for improving collection efficiency of an ESP which is known as pulsed energization [12]. Normally, ions are generated at the discharge electrode and produce a dust layer on the collection electrodes. For efficient operation of an ESP, a balance should be maintained between these two activities. A back corona occurs due to an excessive build-up of negative charges in the dust layer, and this usually happens when the resistivity of the dust layer is too high. Back corona involves injection of collected ash into the flue gas stream and generation of positive ions which are responsible for discharging the negatively charged particles in the flue gas.

Therefore, back corona can be minimized by pulsed energization by regulating the generation of ions at the discharge electrode. The supply voltage is first rectified with a switched integrated rectifier, and then, it is chopped at line frequency instead of transforming and rectifying at normal frequency (50/60 Hz). By applying varying voltage correctly to high resistivity fly ashes, power consumption and emissions can be reduced.

Evaluation of the various modes of electrical operation was done in the pilot-scale unit, which includes varying current settings, different types of charging and the pulse characteristics.

While performing all these tests, it is found that emissions tend to decrease from 35 mg/Nm^3 in the continuous charge mode to 15 mg/Nm^3 in the pulsed mode [8].

(f) **Electrocore separator**

For separating fly ashes from any type of coal material effectively, there is another particle control technology called an electrocore separator. For achieving higher removal efficiency at lower capital and operating costs, an upgrading of an electrocore was developed to replace a conventional ESP [6]. An array of cylindrical separators is used in the electrocore which consists of a tangential inlet, a tangential outlet and two axial outlets.

A temperature-controlled pre-charger is used for charging incoming particles. These pre-charged particles then enter into the separators along their outer wall. For creating an electric field to impart an electric charge of the same polarity as the particles, a large-diameter electrode is used. This electrode is placed along the separator's central axis. To keep the charged particles in the near-wall region, electrical and centrifugal forces act simultaneously. In the near-wall region, particles are collected with the bleed flow. From the end of the separators, the clean flow is extracted axially. The bleed flow, which is about 10% of the clean flow, is directed either back to the inlet of the upstream ESP or to a separate collector. These arrangements are shown schematically in Figs. 9 and 10.

There is no need for rapping because separators operate in a continuous mode without collecting particles. As this process does not use hoppers, re-entrainment is eliminated. Back corona is eliminated in the separators as the central electrode operates without corona. Back corona in the pre-charger is eliminated by cooling

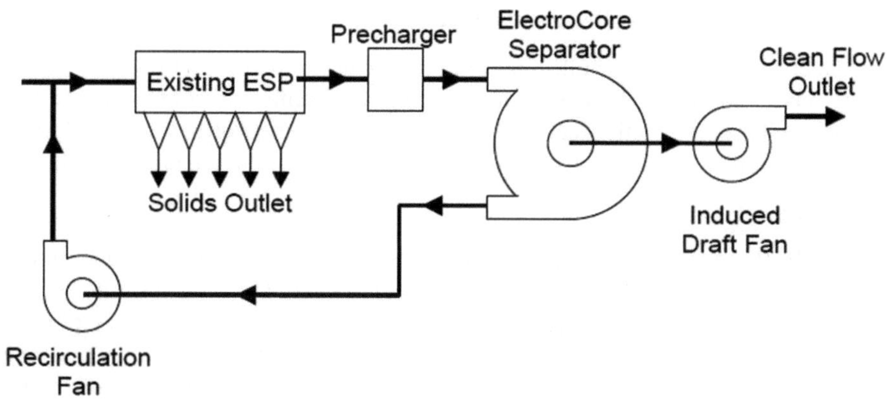

Fig. 9 Electrocore configuration with bleed flow directed back to ESP inlet [11]

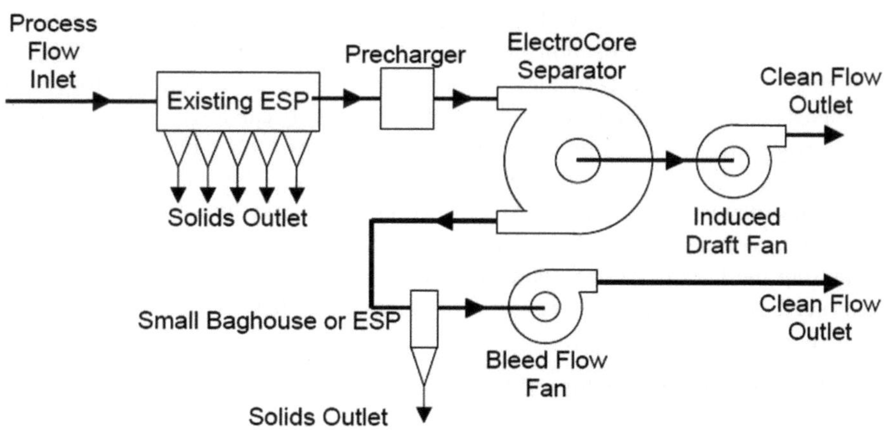

Fig. 10 Electrocore configuration with bleed flow directed back to separate collector [11]

the ash layer on the grounded electrode, thereby reducing the ash temperature and electrical resistivity [11, 28].

(g) **Hybrid particulate collection systems**

For attaining higher removal efficiency of particulate matter at lower cost, hybrid particulate collection systems can be used which combine electrostatic precipitation with fabric filtration. This system can upgrade the existing conventional ESPs by combining with activated carbon powder injection. It is also a way for controlling mercury. Advanced technology can have a collection efficiency of 99.99% for all particles from 0.01 to 50 μm. Advantages of an advanced hybrid particulate collector (AHPC) compared to conventional ESPs are that it requires less space and lower capital cost. AHPCs can be designed for retrofits for traditional ESPs [6].

(h) **Membrane-based collection surface**

Under wet conditions, collection efficiency can be increased by use of woven membranes. But under dry conditions, rapping re-entrainment can be lowered by a membrane-based collection surface which might increase collection efficiency. If successful, this work may lead to low-cost membrane-based wet and/or dry ESPs [6].

(i) **Wet ESP**

Wet ESPs are used for removing fine particles. These particles include various types of air toxins such as mercury particles of 0.01 μm, but wet ESPs have gained much popularity in coal-fired power plants due to their removal efficiency of about 99.9%. It can be used either in the downstream of a wet scrubber as a final polishing system or for replacing a traditional dry ESP at the very last function of electric field. It is used for attaining higher efficiency for removing fine particles. For full-scale demonstration, wet ESPs of different configurations and water delivery systems have been installed on coal-fired utility boilers [6]. A wet ESP was described by Peukert and Wadenpohl (2001) which used a liquid film with the continuously irrigated collection electrodes. The basic advantage of this technology was that it avoided the formation of dust cake on the collection electrode, thus problems of re-entrainment or arcing did not exist. Breakdown occurs only for spray-irrigated ESPs. As this was continuously irrigated, no breakdown occurred. Liquid film was also used for separating particles and the collection electrode; hence, the danger of clogging was low. Grounded liquid film could be used for conducting corona current. For cost-effective design, this technology allows manufacturing the ESP from non-conductive materials such as polypropylene, PVC or reinforced plastics. This also prevents corrosion even when the gas streams contain corrosive components.

(j) **Skew gas flow technology (SGFT)**

Skew gas flow technology (SGFT) involves altered flow distribution throughout the collection chamber which is different from that of uniform flow to an inlet and outlet [29]. Gas velocities in some regions become lower than their average velocities due to this distribution. This happens where the dust distribution is the highest. This technology reduces re-entrainment losses, especially during the rapping cycle. This also increases overall ESP performance [15–17]. Figure 11 shows the dust mass flow distribution with skewed flow which differs from the assumed dust concentration (Fig. 12) and actual dust concentration with uniform flow standards (Fig. 13).

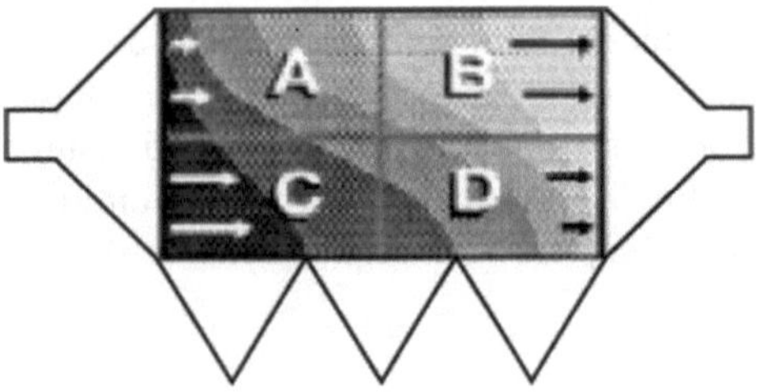

Fig. 11 Actual dust mass flow distribution with skewed gas flow [17]

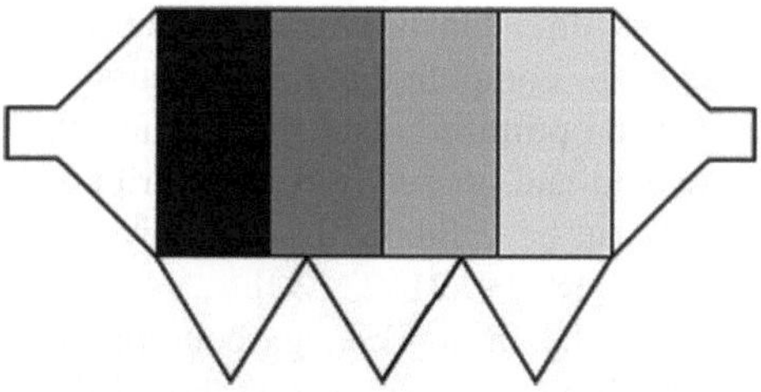

Fig. 12 Assumed dust concentration on the basis for uniform gas flow standards [17]

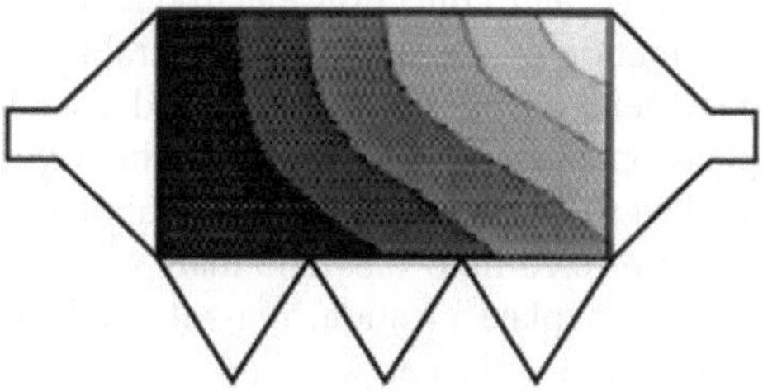

Fig. 13 Actual dust concentration with uniform gas flow [17]

For determining the optimal skew configurations, Lockhart and Weiss [15] used a CFD model for the precipitators of the Israel Electric Corporation. They took dust re-entrainment and non-uniform vertical dust profiles into consideration for their model. Basically, modification was made at both the inlet and outlet faces of the precipitators. They considered negligible re-entrainment with the existing hopper baffles. So, according to their concept, no modification was needed at the hopper baffles. But Boyd [16] introduced additional hopper baffles for reducing re-entrainment by performing tests on ATCO Power's Battle River Generating Station's flue gas flow.

Again, inlet skews seemed to be less pronounced to Ojanpera and Hein [17] because, when the precipitator length in direction of flow increases, the average gas velocity decreases. They preferred outlet skews for their higher velocity.

Considering all of the technologies discussed above, skew gas flow in an ESP appears to be an attractive option for a further study to enhance ESP efficiency. Therefore, a numerical investigation of skew flow and its effect on particle capture has been undertaken and discussed in the following section.

5 Flow Optimization

Computational fluid dynamics is used to predict the flow distribution inside the ESP which helps to show the flow behaviour of flue gas, particulate distribution behaviour and also helps to calculate the required time to settle the particles on the collector plates. Flow field characteristics and particle trajectories can be optimized for the maximum capture of the finest particles since optimization is a much cheaper and less time-consuming technique than other improvement techniques. As it is based on computational application, repeated simulations can be done until expected optimized results are achieved. To improve the ESP performance, suitable flow field and proper velocity distribution are selected [18–20]. The computational model can be validated by the flow field data obtained from the different research projects [18, 19, 21–23, 30].

Typically in an ESP, gaseous particles flow slowly through the ESP chamber which should be evenly distributed throughout the whole unit. By expanding or diverging, the velocity of the gas stream is reduced. To ensure usage of all the discharge and collection electrodes across the entire width of the ESP, an even distribution of the flue gas is desirable. To evenly distribute the gas flow into the chambers, an inlet plenum is used containing perforated openings called diffuser plate openings. For controlling the velocity distribution emerging from a wide angle, tests were performed by Sahin and Ward-Smith [19] on the use of perforated plates of area ratio 6.8. The main effects on the flow in wide angle diffusers were caused by plate location, porosity and the number of plates. High-flow velocities were found in the central region of the diffuser exit plane for plates of high porosity of 58%. The best results were obtained using 50% porous plates with one of the plates positioned a short distance downstream of the diffuser entry plane with the second plate just upstream of the exit plane. The 2D simple model developed by Skodras [20] with discharge electrode and collection electrode describes ESP operation. The model explains that increasing the residence time by lowering inlet velocity could increase the collecting performance, but there is no validation for this result with experimental or on site measured data.

A numerical flow model of an ESP was developed by Schwab and Johnson [31], but they did not design any physical geometry for the collection plates. For representing geometry, they accounted for flow resistance. Instead of creating any physical collecting plates, a porous region was introduced by Varonos [32] in their numerical model. By inserting smoothing grids and flaps, they also optimized the fluid flow. A comparative study was done by Dumont and Mudry [21] on flow simulation results from different precipitator CFD models with actual field measurements of velocity patterns. From the literature review, it is seen that it is necessary to learn and know the flow distribution inside the ESPs to seek any improvement in ESP efficiency and capability. This research introduces a new approach to optimize the velocity distribution inside the ESP.

A case study of the CFD flow optimization which has been conducted for a laboratory-scale ESP model is presented in the following section.

5.1 Description of Experimental Set-up

A laboratory-scale ESP model, geometrically similar to an industrial ESP, was designed and fabricated by Shah et al. [33–38] at CQ University to examine the flow behaviour inside the ESP. This laboratory-scale ESP consisted of a rectangular

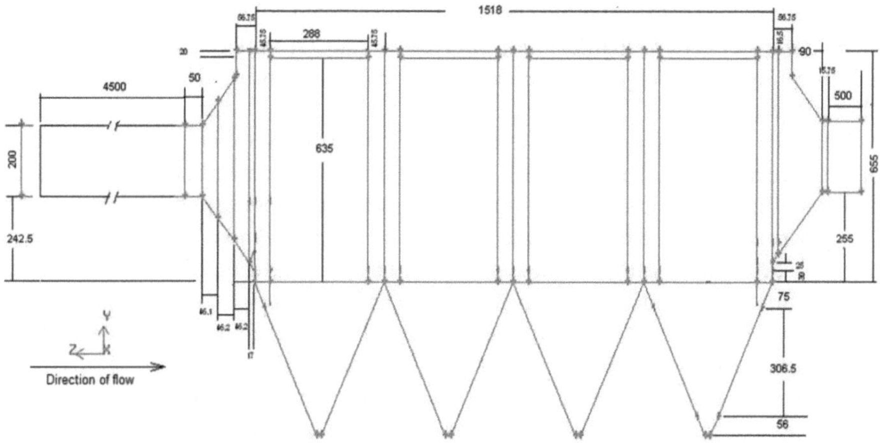

Fig. 14 Side elevation of laboratory-scale ESP (dimensions are in mm) [38]

collection chamber and an inlet evase and an outlet evase. A side view of the laboratory-scale ESP is presented in Fig. 14 which is geometrically similar to the ESP of a local power station. For dimensional analysis, the scale ratio between a real ESP and the laboratory-scale ESP was considered as 20 meaning the laboratory

Fig. 15 Inlet evase (dimension are in mm) [38]

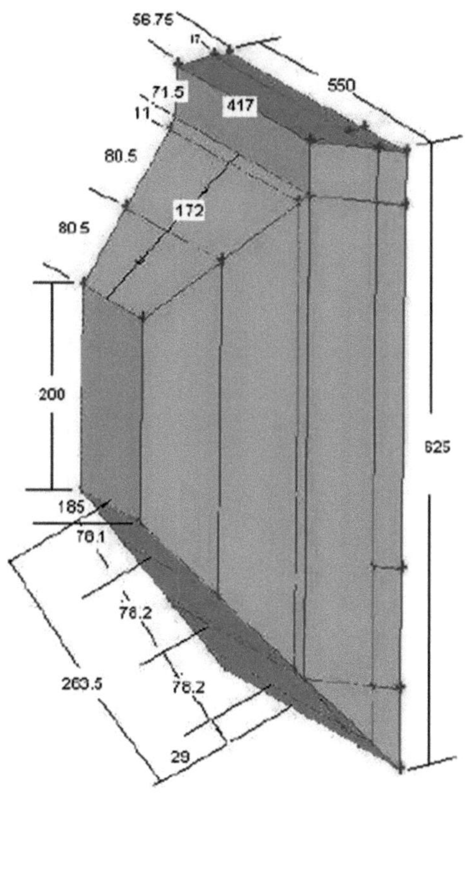

scale was reduced by 20 times from the full-scale ESP. The model includes a rectangular collection chamber, an inlet evase and an outlet evase. The effective length, width and height of the collection chamber are 1.518 m, 0.655 m and 0.55 m, respectively.

The inlet evase shown in Fig. 15 has a pyramidal diffuser with approximately 50° divergence angle which is located in front of the collection chamber. The outlet evase shown in Fig. 16, which is a convergent duct, is located after the collection chamber. One perforated plate with 62% opening and 1.5 mm thickness is placed inside the inlet evase and two more perforated plates with 69% opening and 1 mm thickness are placed after the first one inside the inlet evase satisfying the geometrical similarity of the full-scale ESP. Another perforated plate with 62% opening is placed inside the outlet evase, which is mainly used for flow variation inside the ESP.

Figures 17 and 18 show the hopper for the first and the fourth zones and second and third zones, respectively. Four sets of collection plates are placed inside the collection chamber. Each set consists of 24 vertical collection plates which are placed 2 cm apart.

Fig. 16 Outlet evase (dimension are in mm) [38]

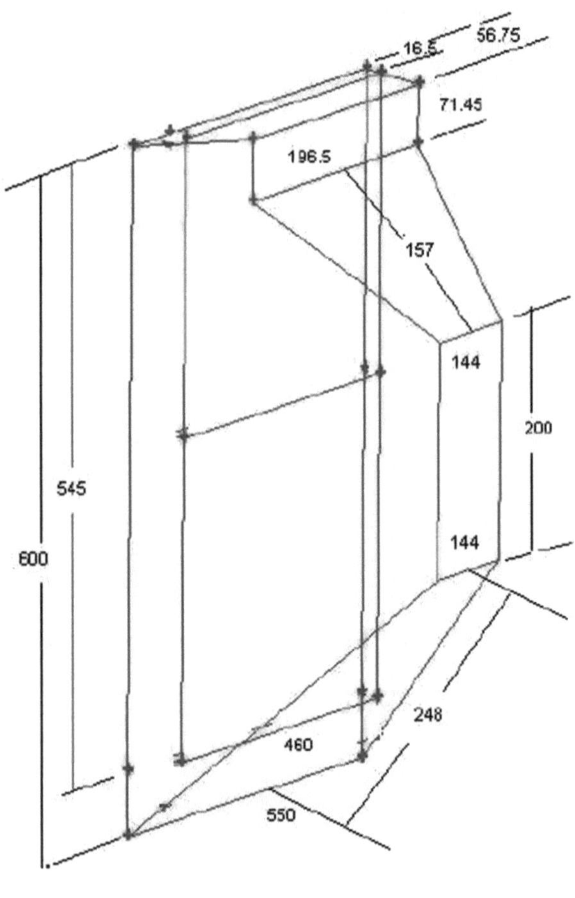

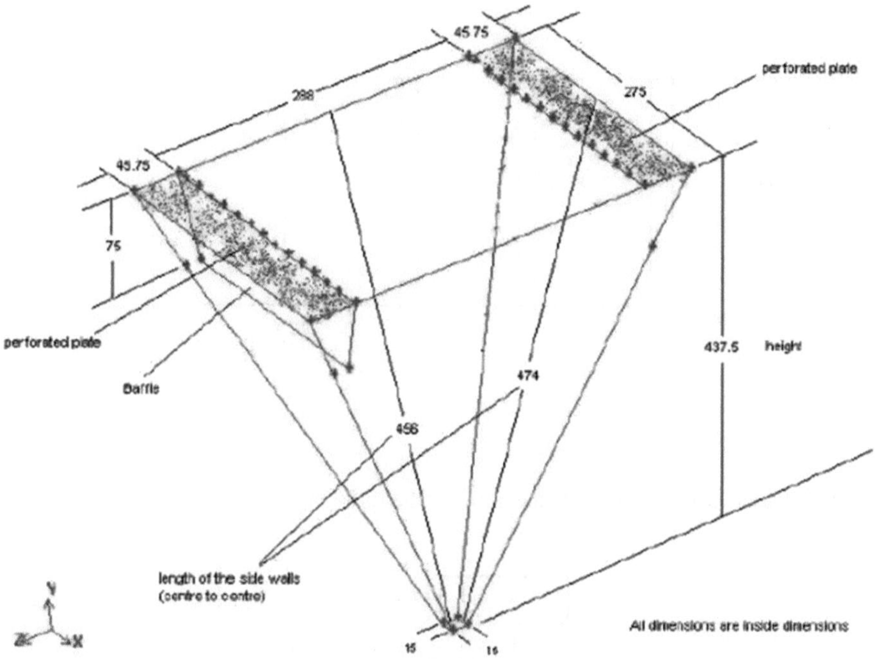

Fig. 17 Hopper 1 and 4 (dimension are in mm) [38]

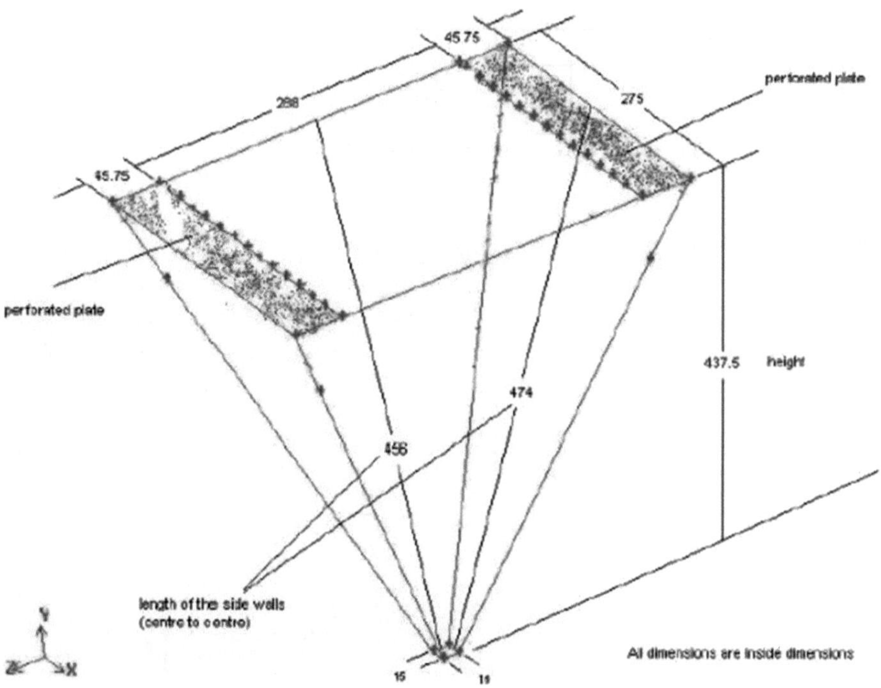

Fig. 18 Hopper 2 and 3 (dimension are in mm) [38]

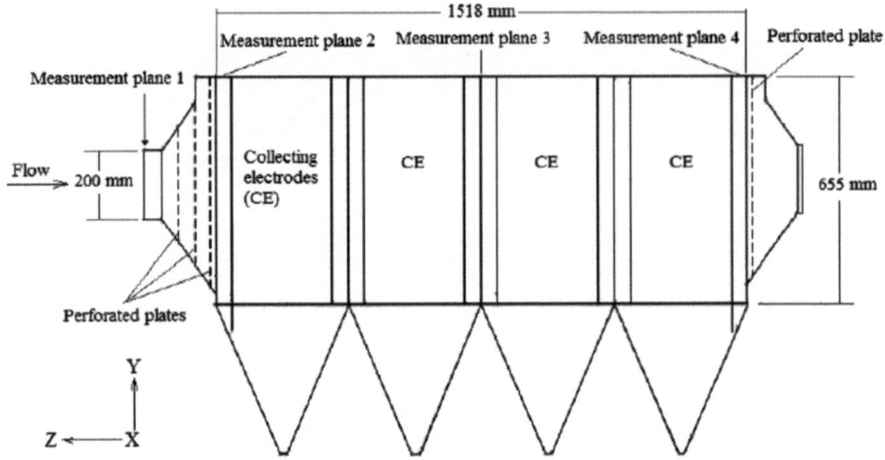

Fig. 19 Measurement planes inside the duct and the ESP [38]

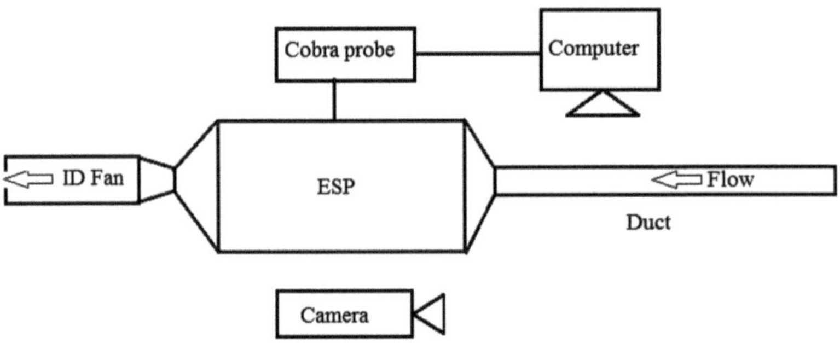

Fig. 20 Schematic diagram of the experimental set-up [38]

Fig. 21 ESP in running condition

Fig. 22 *Top view* shows the baffles used to redirect the flow

The collection chamber is made of transparent acrylic material to visualize the flow phenomena. A long duct was placed in front of the ESP and one induced draft fan was used downstream to suck the air through the ESP. One measurement plane inside the duct upstream of the ESP and three measurement planes inside the ESP are presented in Fig. 19. The schematic diagram of the experimental set-up is presented in Fig. 20 and photographs of the experimental set-up are presented in Figs. 21 and 22.

6 Simulation and Analysis

This study examines the flow behaviour inside the ESP with a view to optimizing the flow for improved performance. In particular, the flow behaviour resulting from changing the baffle shape has been investigated [39]. For simplicity of the modelling and simulation, the rectangular collection chamber has been considered as a rectangular duct including the inlet evase, the duct section and the outlet evase. This is because the dust particle separation from the flue gas and the collection of dust particles usually occur inside the rectangular duct [40]. Two different types of baffles in ESP duct have been considered. The first sets of baffles are arrow-shaped and the second sets are circular-shaped. The Geometry is drawn in Design Modeller of ANSYS Fluent, and further processing is done for meshing and refinement. ANSYS code 'FLUENT' was used for numerical simulation of fluid flow behaviour.

The ESP considered in this chapter has a single chamber. Six circular- and arrow-shaped baffles are inserted in two opposite sidewalls of two separate ducts. Figure 23 shows the ESP containing arrow-shaped baffles and Fig. 24 shows the ESP containing circular-shaped baffles. The length and width of the ESP are 151.8 cm and 55 cm, respectively. The baffles are equally spaced and the distance between the two baffles is 17.4 cm. The width of each baffle is 5 cm and length is 7.5 cm from wall to pick point, respectively [41].

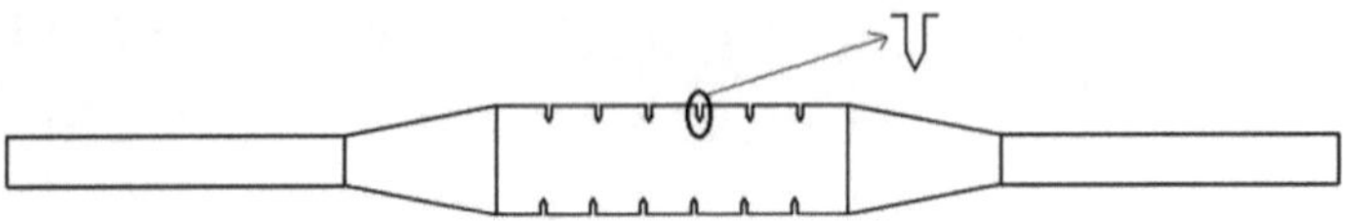

Fig. 23 ESP containing *arrow-shaped baffles*

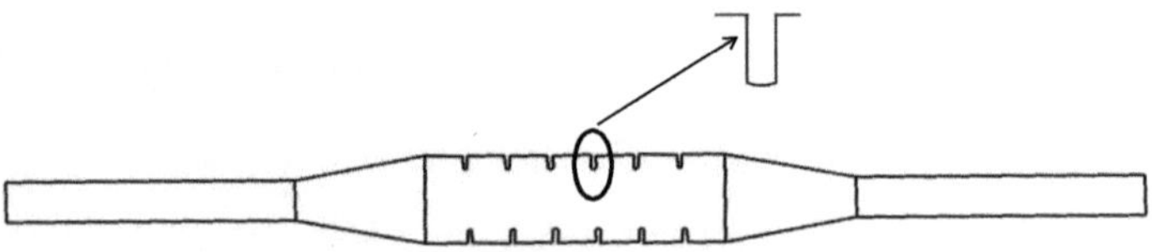

Fig. 24 ESP containing *circular-shaped baffles*

6.1 Numerical Approach and Simulation Procedure

The mesh created by Design Modeller is exported to ANSYS to discretize the fluid domain into small cells to form a volume mesh or grid and set-up appropriate boundary conditions. Numerical computation of fluid transport includes continuity, momentum and turbulence model equations. The flow properties and equations are solved and analysed by the CFD code FLUENT [42].

Continuity equation:

$$\frac{\partial \bar{u}}{\partial x} + \frac{\partial \bar{v}}{\partial y} + \frac{\partial \bar{w}}{\partial z} = \frac{\partial (u_i)}{\partial x_i} = 0. \tag{1}$$

Momentum Equation:

$$\frac{\partial}{\partial t}(\rho u_i) + \frac{\partial}{\partial x_j}(\rho u_i u_j) = -\frac{\partial p}{\partial x_j} + \frac{\partial \tau_{ij}}{\partial x_j} + \rho g_i + F_i. \tag{2}$$

In this equation, p is static pressure and self-defined source terms are contained in Fi. The stress tensor is determined by the following equations:

$$\tau_{ij} = \left[\mu\left(\frac{\partial u_i}{\partial x_j} + \frac{\partial v_j}{\partial x_i}\right)\right] - \frac{2}{3}\mu\frac{\partial u_i}{\partial x_i}\delta_{ij}. \tag{3}$$

$$\frac{\partial}{\partial t}(\rho k) + \frac{\partial}{\partial x_j}(\rho k u_j) = \frac{\partial}{\partial x_j}\left[\left(\mu + \frac{\mu t}{\sigma k}\right)\frac{\partial k}{\partial x_j}\right] + (G_k + G_B - Y_M) - \rho \in + S_k. \tag{4}$$

$$\frac{\partial}{\partial t}(\rho \in) + \frac{\partial}{\partial x_j}(\rho \in u_i) = \frac{\partial}{\partial x_j}\left[\left(\mu + \frac{u_t}{\sigma \in}\right)\frac{\partial \in}{\partial x_j}\right] + C_{1\in}\frac{\in}{k}(G_k + C_{3\in}G_b) - C_{2\in}\rho\frac{\in^2}{k} + S_{\in}, \tag{5}$$

where u, v, w is each component of gas velocity (m/s); and μ, μ_t are molecular viscosity and kinetic viscosity (Pa.s), respectively; ρ (kg/m^3) is fluid density; G_k (J/Kg) represents turbulent energy generated by mean velocity gradient; G_B is turbulent energy generated by buoyancy; Y_M represents pulsation expansion in the turbulent model of compressible flow; $C_{1\in}$, $C_{2\in}$ and $C_{3\in}$ are empirical constants; σk and $\sigma \in$ are corresponding turbulent Prandtl numbers in k-equation and ε-equation; S_k and S_ε are self-defining source terms; $C_{1\in} = 1.44$, $C_{2\in} = 1.92$, $C_{3\in} = 0.09$, $\sigma k = 1.0$ and $\sigma \in = 1.3$ [43].

The realizable k-∈ model was selected in this study as it differs from other k-∈ models in two important ways. Firstly, the realizable k-∈ model contains a new formulation for the turbulent viscosity. Secondly, a new transport equation for the dissipation rate, ε (dissipation rate of turbulent kinetic energy, m^2/s^3), has been derived from an exact equation for the transport of the mean-square velocity fluctuation [42]. The turbulent kinetic energy k (turbulent kinetic energy, m^2/s^2) and its dissipation rate ∈ for realizable k-∈ model are obtained from the transport Eqs. (4) and (5), respectively [42, 43].

6.2 Mesh Generation and Boundary Conditions

Two-dimensional simplified models of the ESP were employed in this study to investigate the flow properties. ANSYS 15 was used to establish models for calculating regions where unstructured grids were generated, and finally, after resizing and refinement, a structured grid was obtained. A mesh independency test was conducted by selecting and plotting 3 sets of data and, as shown in Fig. 25, sample 3 mesh was chosen because it is very close to sample 2 and allows better mesh quality.

The number of nodes and cells obtained for sample 3 were 3,5897 and 6,9604, respectively, for arrow-shaped baffles and 3,6289 and 7,0364, respectively, for the circular-shaped baffles. Air was used as the fluid and its acquiescent properties were maintained with a constant velocity. The boundary conditions were applied as follows: inlet velocity was considered as constant at 7 m/s, the outlet boundary condition was pressure outlet and 'no slip' (velocity at the wall to be zero) boundary condition was imposed on the sidewalls including the baffle's side face and front face.

The finite volume methods were used to discretize the partial differential equations of the model. The semi-implicit method for pressure–velocity coupling equation (COUPLED) scheme was used for pressure–velocity coupling, and the first-order upwind scheme was used because of its accuracy and stability and because this scheme also interpolates the variables on the surface of the control volume. Turbulent kinetic energy k and turbulent dissipation rate ε were considered as a first-order upwind for better simulation accuracy.

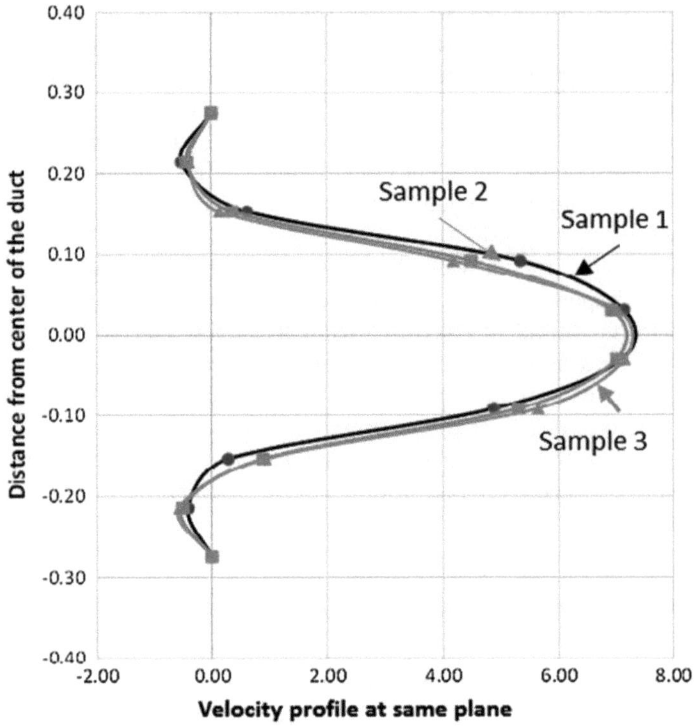

Fig. 25 Mesh independency test

In order to compute the results, all simulations were carried out on an Intel Core i5 processor computer that has a 2.80 GHz processor and 8.00 GB of RAM, 64-bit operating system.

7 Results and Analysis

The results of the impact of the baffles on the flow are discussed on a qualitative basis since more results and their experimental validation are required for a quantitative analysis.

The assembled graphs below are the simulation results of velocity distribution for circular-shaped and arrow-shaped baffles, respectively. Six baffles were inserted into each wall in air flow distribution plates. In the following figures, the influence of baffle on forming a skewed air flow pattern is considered only (Figs. 26 and 27).

Figures 28 and 29 show the contour of the velocity distribution for arrow-shaped baffles and circular-shaped baffles, respectively, and Figs. 30 and 31 are the corresponding velocity vectors for those baffles. It is seen from these figures

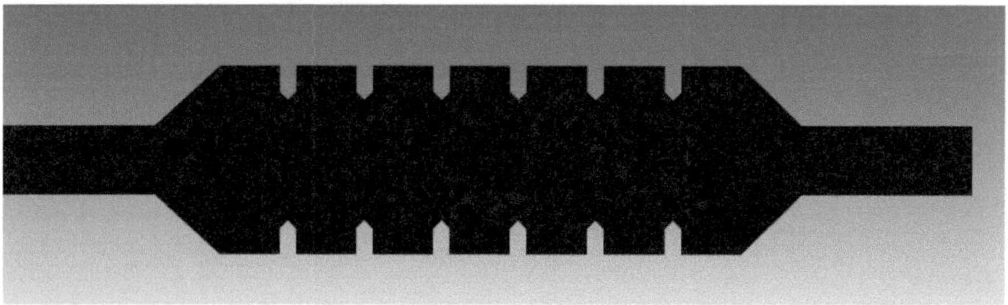

Fig. 26 Structured mesh for *arrow-shaped baffles* (2D)

Fig. 27 Structured mesh for *circular-shaped baffles* (2D)

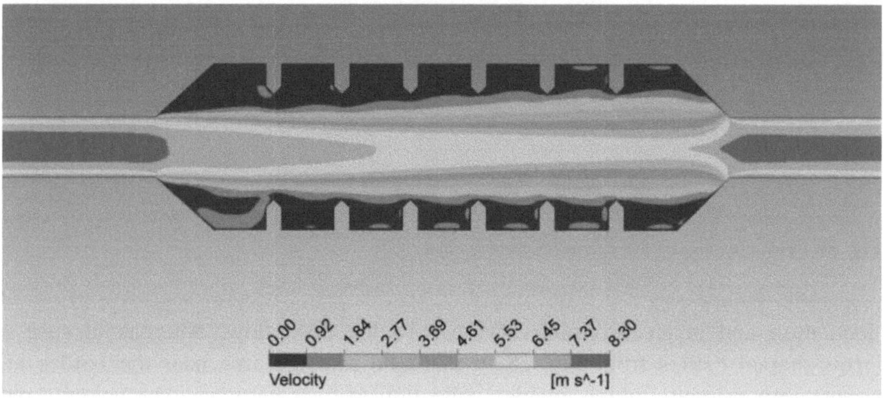

Fig. 28 Velocity distributions for *arrow-shaped baffles*

that the velocity distribution pattern is different because of the changing shapes of the baffles. For the circular-shaped baffles, the velocity streamline changed more frequently compared to the arrow-shaped baffles. In the case of arrow-shaped baffles, the velocity distribution maintains a smooth streamline. From the view point of velocity pattern, circular-shaped baffles have greater potential to capture particulate matter since the velocity change is more frequent causing pressure

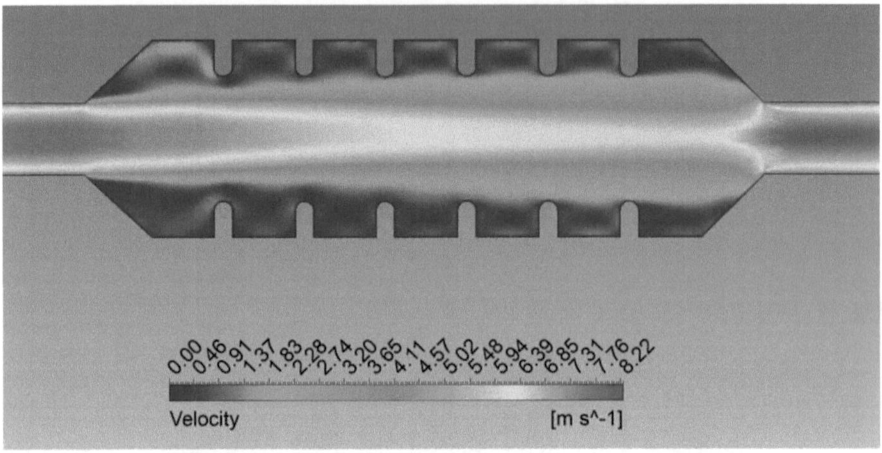

Fig. 29 Velocity distributions for *circular-shaped baffles*

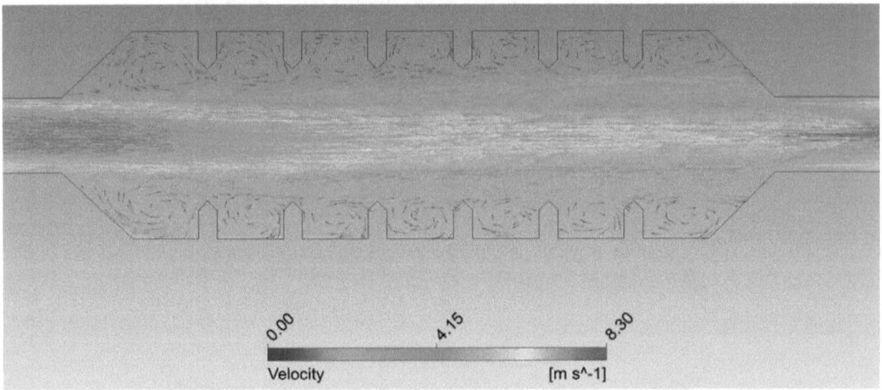

Fig. 30 Velocity vector for *arrow-shaped baffles*

fluctuation and hence dropping of the particles from flow, whereas in case of arrow-shaped baffles from Fig. 28, it is found that the area near the baffles has almost zero velocity, which enables collection of finer particles. The velocity pattern of arrow-shaped baffles keeps a smooth centreline velocity stream which reduces drag force for upcoming flue gas and also decreases residence time due to smooth velocity pattern shown in Fig. 28.

The flow pattern and the CFD analysis of the two different shapes of baffles in the ESP show some encouraging and favourable results. The summary that can be drawn is as below:

- Qualitative analysis of the two different flow patterns shows that insertion of baffles in the ESP forms skewed gas flow whose formation is influenced by the

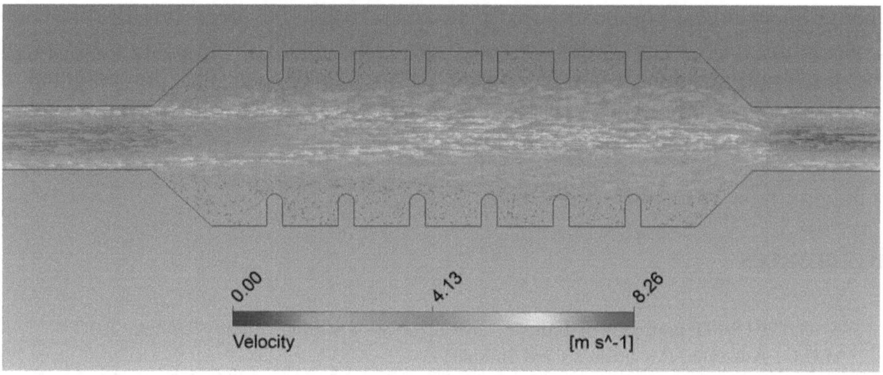

Fig. 31 Velocity vector for *circular-shaped baffles*

baffles' shape and the interval between two adjacent baffles. Skewed gas flow also increases the residence time of the flue gas inside the duct, leading to more dust collection. Further study will be conducted on the gap between two opposite baffles or placing the baffles on an incline inside the duct.

- Velocity distribution indicates that a smooth velocity pattern is obtained in the case of arrow-shaped baffles, whereas circular-shaped baffles have greater potential to capture particulate matter because of frequent velocity changes causing pressure fluctuation and hence more particle dropping from the flow. Further study is needed to examine the residence time and drag force on these baffles.
- Arrow-shaped baffles offer lower drag force whereas circular-shaped baffles offer more residence time. Therefore, it is necessary to optimize both the models to get the desired flow and residence time to enable dust collection.
- The dynamics of formation of the skewed flow is complicated and is not quantitatively analysed. However, qualitatively, it shows the creation of vortex flow near the baffles and the existence of a near-zero velocity close to the wall of the baffles, which suggests an improvement of the dust collection efficiency.
- Further investigation is required to determine the strength of vortex flow and the time for dust collection from the wall. This is because, as the thickness of dust collection is increased, further interaction with the spinning motion of flow may wipe out the accumulated dust in the wall.

8 Conclusion

In this chapter, different technologies for capturing smaller size particles have been discussed and a flow optimization technique has been examined as it is the most cost-effective and less time-consuming option. The simulation results showed a significant impact on improvement of the capture of particles. The model of flow optimization by inserting baffles which will not change the flow rate or cause any change in chemical composition of particulate matter has been briefly discussed. Considering all the circumstances and from a sustainable clean energy viewpoint, the flow optimization technique discussed in this chapter has the potential to increase the capability and performance of the existing ESPs which would result in a huge reduction in investment in power plants.

References

1. A. G. (2015). Department of Industry and Science Australian Energy Update 2015.
2. M.C.O. Australia, Australia's Coal Industry.
3. Morawska, L., Agranovski, V., Ristovski, Z., & Jamriska, M. (2002). Effect of face velocity and the nature of aerosol on the collection of submicrometer particles by electrostatic precipitator. *Indoor Air, 12,* 129–137.
4. Jaworek, A., Krupa, A., & Czech, T. (2007). Modern electrostatic devices and methods for exhaust gas cleaning: A brief review. *Journal of Electrostatics, 65,* 133–155.
5. K. M. S. C A J Paulson, Electrostatic precipitation of fly ash from australian bituminous coals in January 1998.
6. Zhu, Q. (2003). *Developments in particulate control.* IEA Coal Research.
7. Chemithon. (2011). Chemithon Enterprise Inc. Retrieved from http://www.chemithon.com/Enviro_fluegas.html.
8. Srinivasachar, S., Pease, B. R., Porle, K., Mauritzson, C., & Haythornthwaite, S. (1997). Ultra high efficiency ESP for fine particulate and air toxics control. In *National Energy Technology Lab, Pittsburgh, PA, and Morgantown, WV (US).*
9. Ji, J.-H., Hwang, J., Bae, G.-N., & Kim, Y.-G. (2004). Particle charging and agglomeration in DC and AC electric fields. *Journal of Electrostatics, 61,* 57–68.
10. Nakajima, Y., & Sato, T. (2003). Electrostatic collection of submicron particles with the aid of electrostatic agglomeration promoted by particle vibration. *Powder Technology, 135–136,* 266–284.
11. Altman, R. F., Easom, B. H., Box, C., & Harrison, W. A. (2002). Results of electrocore™ pilot testing at EC gaston steam plant. In *Proceedings of Air Quality III: Mercury, Trace Elements, and Particulate Matter Conference, Arlington, VA.*
12. Elayyan, H. S. B., Bouziane, A., & Waters, R. T. (2002). Theoretical and experimental investigation of a pulsed ESP. *Journal of Electrostatics, 56,* 219–234.
13. LSR Technologies Report, Integrated system to control primary pm 2.5 from electric power plants. In Other Information: PBD: 30 June 2000, Medium: ED; Size: 9 pp.
14. Peukert, W., & Wadenpohl, C. (2001). Industrial separation of fine particles with difficult dust properties. *Powder Technology, 118,* 136–148.
15. Lockhart, J., Weiss, O., & Hadera, I. (2001). The application of skewed gas flow technology at the Israel electric corp. MD-A Station. In *8th International conference on electrostatic precipitation.*
16. Boyd, M. (2001). Skewed gas flow technology offers antidote to opacity derates. *Power Engineering, 105.*
17. Ojanpera, R., & Hein, A. G. (1999). Skewed gas flow technology-a method to improve precipitator performance. In *Annual Meeting-Technical Section Canadian Pulp And Paper Association, Canadian Pulp & Paper Assn-Technical Section* (pp. A261–A264).
18. Grainger, C. (2001). Ductwork changes improve ESP performance. *Clean Air and Environmental Quality, 35,* 43–44.
19. Sahin, B., & Ward-Smith, A. J. (1987). The use of perforated plates to control the flow emerging from a wide-angle diffuser, with application to electrostatic precipitator design. *International Journal of Heat and Fluid Flow, 8,* 124–131.

20. Skodras, G., Kaldis, S. P., Sofialidis, D., Faltsi, O., Grammelis, P., & Sakellaropoulos, G. P. (2006). Particulate removal via electrostatic precipitators—CFD simulation. *Fuel Processing Technology, 87,* 623–631.
21. Dumont, B. J., & Mudry, R. G. (2003). Computational fluid dynamic modeling of electrostatic precipitators. In *Proceedings of Electric Power Conference.*
22. Gan, G., & Riffat, S. B. (1997). Pressure loss characteristics of orifice and perforated plates. *Experimental Thermal and Fluid Science, 14,* 160–165.
23. Kim, S. H., & Lee, K. W. (1999). Experimental study of electrostatic precipitator performance and comparison with existing theoretical prediction models. *Journal of Electrostatics, 48,* 3–25.
24. Parker, K. R. (1997). *Applied Electrostatic Precipitation.* Chapman & Hall.
25. Qi, L., & Yuan, Y. (2013). Mechanism of the effect of alkali metal on the electrostatic precipitability of fly ash. *Fuel, 107,* 848–851.
26. Qi, L., & Yuan, Y. (2013). Influence of SO3 in flue gas on electrostatic precipitability of high-alumina coal fly ash from a power plant in China. *Powder Technology, 245,* 163–167.
27. Thonglek, V., & TanongkiatKiatsiriroat. (2013). Improvement of electrostatic precipitator for submicron particle collection by non-thermal plasma pre-charger. *International Journal of Emerging Technology and Advanced Engineering, 3*(10).
28. Botros, M. P., & Altman, R. Integrated System to Control Primary PM 2.5 from Electric Power Plants.
29. Sayem, A. S. M., Khan, M. M. K., Rasul, M. G., Hassan, N. M. S., & Amanullah, M.T.O. (2015). Modelling of baffles in electrostatic precipitator (ESP) to achieve optimum flow distribution. In *The 11th International Conference on Heat Transfer, Fluid Mechanics and Thermodynamics (HEFAT 2015), Kruger National Park, South Africa* (p. Paper ID 1570073667).
30. Haque, S. M. E., Rasul, M. G., Deev, A. V., Khan, M. M. K., & Subaschandar, N. (2009). Flow simulation in an electrostatic precipitator of a thermal power plant. *Applied Thermal Engineering, 29,* 2037–2042.
31. Schwab, M., & Johnson, R. (1994). Numerical design method for improving gas distribution within electrostatic precipitators. In *Proceedings of the American Power Conference* (pp. 882–882). Illinois Institute of Technology.
32. Varonos, A. A., Anagnostopoulos, J. S., & Bergeles, G. C. (2002). Prediction of the cleaning efficiency of an electrostatic precipitator. *Journal of Electrostatics, 55,* 111–133.
33. Haque, S. M. E., Rasul, M. G., Khan, M. M. K., Deev, A. V., & Subaschandar, N. (2009). Influence of the inlet velocity profiles on the prediction of velocity distribution inside an electrostatic precipitator. *Experimental Thermal and Fluid Science, 33,* 322–328.
34. Haque, S. M. E., Rasul, M. G., Deev, A. V., Khan, M. M. K., & Zhou, J. (2006). Numerical simulation of turbulent flow inside the electrostatic precipitator of a power plant. In *International Conference on Fluid Mechanics, Miami, Florida, USA* (pp. 25–30).
35. Haque, S. M. E., Rasul, M., Khan, M. M. K., Deev, A., & Rao, S. (2007). Numerical modelling for optimizing flow distribution inside an electrostatic precipitator. *International Journal of Mathematics and Computers in Simulation, 1.*
36. Haque, S. M. E., Rasul, M. G., & Khan, M. M. K. (2008). Modelling and simulation of particle trajectory inside an electrostatic precipitator. In *4th BSME-ASME International Conference on Thermal Engineering, Dhaka, Bangladesh.*
37. Haque, S. M. E., Rasul, M., & Khan, M. M. K. (2010). Fine participate emission control by optimizing process parameters of an electrostatic precipitator. In *Proceedings of the WSEAS International Conference, Mechanical Engineering Series, World Scientific and Engineering Academy and Society.*
38. Haque, S. M. E. (2009). *Performance study of the electrostatic precipitator of a coal fired power plant: Aspects of fine particulate emission control.* Ph.D. Thesis, Faculty of Sciences, Engineering and Health, CQ University.
39. Sayem, A. S. M., Khan, M. M. K., Rasul, M. G., & Hasan, N. M. S. (2015). Fluid flow analysis in electrostatic precipitator of a coal fired power plant considering electrode with two different

shape of baffles. In *International Conference on Mechanical Engineering and Renewable Energy 2015 (ICMERE2015), Chittagong, Bangladesh* (pp. ICMERE2015-PI-2285).
40. Deev, A. V., Rasheed, T., Welsh, M. C., Khan, M. M. K., & Rasul, M. G. (2009). Measurement of instantaneous flow velocities in a concentric reducer using particle image velocimetry: Study of scale deposition. *Experimental Thermal and Fluid Science, 33,* 1003–1011.
41. Sayem, A. S. M., Khan, M. M. K., Rasul, M. G., Amanullah, M. T. O., & Hassan, N. M. S. (2015). Effects of baffles on flow distribution in an electrostatic precipitator (ESP) of a coal based power plant. In: *6th BSME International Conference on Thermal Engineering (ICTE 2014), Procedia Engineering, Dhaka, Bangladesh* (p. 8).
42. ANSYS FLUENT 12.0 Theory Guide.
43. Dubois, F., & Huamo, W. (2001). *New advances in computational fluid dynamics—theory, methods and applications [M], in.* Beijing: Higher Education Press.
44. William W. Nazaroff & Lisa Alvarez-Cohen. (2017). "Electrostatic Precipitators." Dartmouth. http://engineering.dartmouth.edu/~d30345d/courses/engs37/esps.pdf. Accessed 1.2.2017.
45. White, H. J. (1977). Fly Ash and Furnace Gas Characteristics. *Journal of the Air Pollution Control Association*, *27*, 114–120. http://www.tandfonline.com/doi/pdf/10.1080/00022470.1977.10470386

Part II
Thermofluid Process Applications

Experimental Investigation and Molecular-Based Modeling of Crude Oil Density at Pressures to 270 MPa and Temperatures to 524 K

Isaac K. Gamwo, Babatunde A. Bamgbade and Ward A. Burgess

Abstract A predictive crude oil density model reliable over a wide range of temperature and pressure conditions is increasingly important for the safe production of oil and accurate estimation of oil reserves. While hydrocarbon density data at low-to-moderate temperatures and pressures are plentiful, data and validated models that have reasonable predictive capability for crude oil at extreme temperatures and pressures are limited. In this investigation, we present new experimental density data for crude oil sample obtained from the Gulf of Mexico region. Density data are measured at pressures to 270 MPa and temperatures to 524 K. These conditions simulate those encountered from ultra-deep formations to platforms. These density data points are then used to validate both empirical-based and molecular-based equations of state models. Results show that the molecular-based perturbed-chain statistical associating fluid theory (PC-SAFT) models, without the use of any fitting parameters, predict the crude oil density within 1% of the experimental data. These results are superior to the density predictions obtained with the high-temperature, high-pressure, volume-translated cubic equations of state.

Keywords Crude oil · Density · PC-SAFT · High-temperature · High-pressure · Equation of state

Nomenclature

Latin symbols

a_i, b_i	constants in Tait equation
B, P_0	pressure terms in Tait equation, MPa
CN	carbon number calculated using expression of Huang and Radosz

I.K. Gamwo (✉) · B.A. Bamgbade · W.A. Burgess
United States Department of Energy, National Energy Technology Laboratory, Research & Innovation Center, Pittsburgh, PA 15236-0940, USA
e-mail: gamwo@netl.doe.gov

M.M.K. Khan et al. (eds.), *Application of Thermo-fluid Processes in Energy Systems*, Green Energy and Technology, https://doi.org/10.1007/978-981-10-0697-5_6

CN_p	carbon number calculated assuming species of given M_w has 100% paraffin character
CN_{pna}	carbon number calculated assuming species of given M_w has 100% polynuclear aromatic character
H/C	molecular hydrogen-to-carbon ratio
m	segment number parameter for PC-SAFT equation
M_w	molecular weight, g/mol
P	pressure, MPa
V	molar volume, cm^3/mol
T	temperature, K
asph	subscript to signify the asphaltene phase
A + R	subscript to signify the aromatics and resins
w	weight fraction

Greek symbols

β	isothermal compressibility, MPa^{-1}
δ	mean absolute percent deviation for a set of numbers, %
ε/k	energy parameter for PC-SAFT equation, K
λ	standard deviation associated with a given value of δ
ρ	density, g/cm^3
σ	size parameter for PC-SAFT equation, Å
γ	aromaticity parameter

1 Introduction

Rapid worldwide economic growth over the past century has led the petroleum industry to search for oil and gas in harsh conditions, including in ultra-deep reservoirs several miles beneath the deep seafloor where high temperature and pressure conditions prevail. While the first oil well drilled in the USA had a depth of approximately 100 feet, prospectors today must often drill several miles deep to tap oil reserves. The hydrocarbons within these wells are at increasingly under extreme conditions [1] of temperature and/or pressure (such as the oil located in deep wells in the Gulf of Mexico). These high-temperature, high-pressure (HTHP) wells can exhibit reservoir temperatures in excess of 423 K and/or reservoir pressures in excess of 10,000 psi (69 MPa). Between 1982 and 2012, there were 415 such wells established around the world [2]. This trend is expected to accelerate in the future as existing oil reserves are depleted [3]. From a thermodynamic and thermophysical property modeling perspective, the increased tendency for new reservoirs to exist at HTHP conditions is problematic as the predictive capability of many models at such conditions is unproven. Development of HTHP reservoirs is capital-intensive and therefore represents a significant financial risk, particularly when they are offshore. A proper evaluation of such reservoirs typically requires accurate pressure–volume–temperature (*PVT*) data. In addition, the safe transport of hydrocarbon liquids in pipelines from reservoir to processing facility requires knowledge of second derivative properties [4, 5], such as the isothermal compressibility β. Accurate prediction of β is also necessary to estimate the rate of production decline during the primary recovery phase. Accurate knowledge of bubble point and asphaltene onset pressures in oil wells and dew point pressures in gas wells is critical to the design of processes that maintain a steady flow of hydrocarbons from the well. Ideally, these properties are measured experimentally. However, often such experimentation is not feasible for technical and/or economic reasons. Therefore, the United States Department of Energy is interested in the identification and development of thermodynamic and thermophysical property models that give accurate predictions even at extreme conditions (defined in this work as temperatures to 533 K and pressures to 240 MPa).

Experimental data for crude oil systems at extreme HTHP conditions are necessary in order to properly validate model performance at such conditions. Limited high-pressure data are available for the density and isothermal compressibility of some crude oil systems. However, to our knowledge, no published data are available for crude oil systems at temperatures higher than $\sim$400 K and 155 MPa [6]. Therefore, we present new experimental density data for a crude oil sample measured from ambient conditions to HTHP conditions at temperatures to 523 K and pressures to 270 MPa.

To date, several types of methods have been employed to describe hydrocarbon systems in petrochemical applications. A popular method involves a number of empirical correlations [4, 7–16] that are used in the estimation of crude oil properties, including *PVT* data, gas/oil ratio, and β. Such "black oil" correlations are typically only valid over a given range of temperature and pressure values; however, more HTHP wells are starting to exceed these ranges. Oftentimes, correlations developed for crude oil in one particular region of the world do not provide reliable predictions when applied to oil from another part of the world. The reason is mainly attributed to the significant compositional differences encountered between crude oils produced in various regions. For these reasons, investigators are increasingly turning to equation of states (EoSs) to obtain such HTHP crude oil properties.

While empirical "black oil" correlations are typically created for whole crude oil systems, EoSs rely on the use of suitable characterization procedures. A typical crude oil is comprised of numerous components, and it would not be computationally efficient to consider each component individually. However, characterization of the oil as a mixture of well-defined fractions greatly simplifies the computation. Several popular characterization procedures, based on different properties of the oil, have been reported in the literature [17–19] since the earliest studies reported by Katz and Firoozabadi [20], who used *n*-paraffin boiling point temperatures to represent the crude oil composition as separate carbon number fractions. The characterization procedure proposed by Whitson [19], widely used in upstream applications, also involves the subdivision of crude oil into single carbon number fractions. These methods are continually being applied and improved [21, 22].

Equation of state (EoS) methods, while more complex, are more versatile and can be divided into two general classes. Firstly, there are semiempirical methods based on cubic EoSs originating from the van der Waals EoS, such as the Peng–Robinson [23] (PR) and Soave–Redlich–Kwong (SRK) [24] equations. Cubic EoSs are often used in petrochemical applications and in the oil industry because of their relative simplicity and reliability at predicting the phase equilibrium of nonpolar hydrocarbon systems. However, their performance has been demonstrated to be poor for modeling thermophysical HTHP properties of asymmetric hydrocarbons [25] and their mixtures [26, 27].

Therefore, there is interest in a second, newer type of EoS method—the molecular-based statistical associating fluid theory (SAFT). These SAFT-based EoSs are typically constituted as a sum of residual Helmholtz energy terms, which account for hard-sphere repulsion between molecules, dispersion interactions between molecules, the acentric nature of a molecule as conceptualized by a chain comprising m number of hard-sphere segments, and associations between molecules (such as hydrogen bonding). For each component, three pure-component parameters must be defined: the segment number m, the segment size parameter σ, and the segment–segment energy parameter ε/k. Of the SAFT-based EoSs, the perturbed-chain SAFT (PC-SAFT) EoS has been widely evaluated for its ability to model hydrocarbon mixtures. The PC-SAFT EoS [28], in conjunction with an appropriate characterization method, is a promising modeling prospect for predicting crude oil properties.

Two general types of methods of crude oil characterization using PC-SAFT exist. The first such method of crude oil characterization is the paraffins–naphthenes–aromatics (PNA) method [29], which is based on refractive index data. The method used in this work is the second type of characterization method, which was used by Punnapala and Vargas [30]. This characterization method for crude oil is based on oil composition in terms of the saturate, aromatic, resin, and asphaltene (SARA) components. This method originates from the work proposed by Ting et al. [31], used to model crude oil bubble points, as well as asphaltene onset precipitation. In order to obtain PC-SAFT parameters for the SARA components, correlations have been derived using the parameter sets published by Gross and Sadowski [26].

2 Materials and Methods

2.1 Materials

A dead crude oil sample from the Gulf of Mexico was supplied by a company that has operations in the region. Carbon disulfide (CS_2, HPLC grade, CAS number 75-15-0) was obtained from Aldrich and used in the preparation of the dead crude oil samples for GC analysis. Asphaltene precipitation measurements were performed by dissolving the dead crude oil sample in *n*-pentane (>99%, CAS 109-66-0), obtained from Sigma-Aldrich. For the solid-phase extraction (SPE) analysis of the crude oil sample, *n*-pentane, toluene (HPLC grade, CAS number 108-88-3), and ethyl acetate (LC-MS grade, CAS number 141-78-6) were used. Mallinckrodt Co. and Sigma-Aldrich supplied the toluene and ethyl acetate, respectively.

2.2 Experimental Method

The high-pressure cell used in this study is based on the design of McHugh and colleagues [32], but with some alterations. Figure 1 shows a schematic of the high-pressure density cell used in this study. The body of the cell shown in Fig. 1 is constructed of a high-nickel content alloy, Inconel 625. The alteration involves the replacement of the elastomeric o-rings previously used for sealing the sapphire window to the front end of the view cell with a two-piece window holder made out of Inconel 625. As the name implies, the window holder encases the window and then mates with the front end of the view cell to provide a metal-to-metal seal with the view cell. In addition, the floating piston that also required the use of an elastomeric o-ring for sealing has been replaced with a bellows, which is also constructed from Inconel 625. The o-rings were replaced due to their potential degradation by any of the numerous components of crude oil.

The cell contents are compressed to the desired pressures by inflating the bellows (from BellowsTech) with water using a high-pressure syringe pump (Vinci Corp.) capable of generating pressures up to 40,000 psi (276 MPa). The cell pressure is measured using a Viatran pressure transducer (Viatran Corporation, model no. 245, pressure range 0–345 MPa and accurate to ±0.35 MPa) situated on the water side of the apparatus. The temperature of the cell contents is measured using a type-K thermocouple (Omega Corporation, calibrated at four temperatures from ambient to 528 K to within 0.2 K using methods traceable to NIST standards). Temperature variation for each experimental isotherm is within 0.2 K. The internal volume of the density cell is determined from positional information provided by a

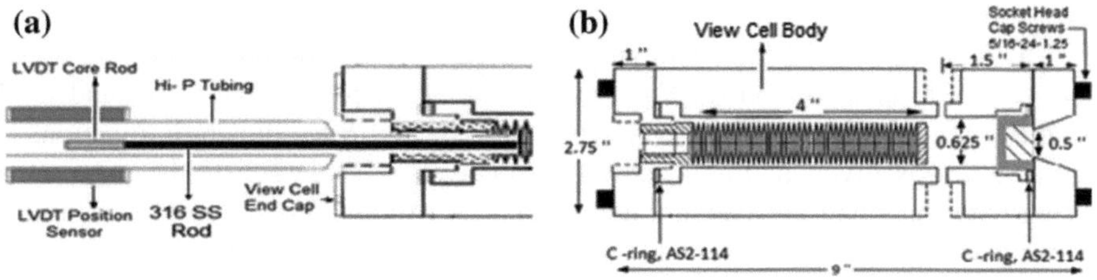

Fig. 1 Schematic diagram indicating **a** linear variable differential transformer and **b** the high-pressure view cell used in this work

linear variable differential transformer (LVDT, Schaevitz Corp, Model 2000 HR). The LVDT positional information is correlated to the internal volume of the cell by calibrations against known *n*-decane density values obtained at 323, 423, and 523 K. The reference density values were obtained from NIST WebBook, which are available at pressures to 800 MPa and temperatures to 673 K for maximum density values to 770 $kg{\cdot}m^{-3}$ [33]. The estimated uncertainty of this volume calibration is within 0.7% of the calculated volume, while the accumulated uncertainty in the reported density data is estimated to be less than 1%.

Before loading the cell with the crude oil sample, air must be purged from the cell as it does not dissolve readily in hydrocarbons. Therefore, the cell is flushed three times with propane at pressures to about 0.5 MPa and then vacuumed. Following this, approximately 16 g of the crude oil sample is charged into the cell, and the cell is heated and allowed to stabilize at a desired temperature. For a particular isotherm, the cell contents are compressed to pressure points that are selected in a nonlinear fashion in order to minimize any experimental peculiarities.

3 Experimental Results and Correlations

For the dead crude oil studied, experimental data points are obtained for three isotherms and approximately two dozen pressures. If a suitable correlation is used to represent the experimental data points, then reliable reference density values can be interpolated for temperature and pressure conditions at intermediate conditions (e.g., the 373 K isotherm) for which experimental values were not recorded. Thus, the modified Tait equation [34] is fitted to experimental data measured from ambient conditions to 525 K and 270 MPa for the dead crude oil sample. Equation 1 gives the Tait equation where ρ_0 is the density at the reference pressure P_0, which is chosen as 0.1 MPa, while B and C are adjustable parameters.

$$\frac{\rho - \rho_0}{\rho} = C \log_{10} \frac{P + B}{P_0 + B} \tag{1}$$

$$\delta = \frac{1}{n} \sum_{1}^{n} \left| \frac{\rho_{i,Tait} - \rho_{i,experimental}}{\rho_{i,experimental}} \right| \bullet 100 \tag{2}$$

The Tait equation parameters are optimized by the minimization of the mean absolute percent deviation (δ, Eq. 2) while adjusting parameters ρ_0, B, and C. A single value of C is used for the system, while ρ_0 and B are quadratic functions of temperature as given in Eqs. 3 and 4, respectively. Table 1 lists the parameters along with the δ and the standard deviation, λ, for the fit. The value of C is similar to what was obtained for the well-characterized hydrocarbon pure compounds and their binary mixtures previously investigated [23–25].

Table 1 Tait equation parameter fits, mean absolute percent deviations, δ, and standard deviation values, λ, for δ, obtained for the crude oil sample

Parameters	Optimized values
C	0.2275
a_0	1123.0
a_1	−0.8258
a_2	2.195×10^{-4}
b_0	370.00
b_1	-1.091×10^{-4}
b_2	8.526×10^{-4}
δ	0.14
λ	0.16

$$\rho_0/\left(kg \cdot m^{-3}\right) = \sum_{i=0}^{2} a_i (T/K)^i \quad (3)$$

$$B/MPa = \sum_{i-0}^{2} b_i (T/K)^i \quad (4)$$

4 Modeling of Experimental Data

4.1 Volume-Translated Cubic EoS Modeling Methods for Crude Oil

Although cubic EoSs, such as the Peng–Robinson (PR) and Soave–Redlich–Kwong (SRK) equations, give accurate phase equilibrium predictions for hydrocarbon systems, they fail to give reliable density predictions. In an earlier study, promising results have been obtained with the model of Pedersen et al. [35] for the prediction of crude oil density at pressures to ~80 MPa. Pedersen et al. adapted the volume-translated SRK method of Peneloux and Rauzy [36] in order to model crude oil systems. Larsen et al. [6] used the model of Pedersen et al. [37] for predicting bottom-hole reservoir fluid density at temperatures to 400 K and pressures to 155 MPa that was within 3% of the experimental values. Therefore, in this work, we extended the capability of the model to reliably predict HTHP density of crude oils at pressures to 270 MPa and temperatures to 523 K.

The molecular weight distribution of the crude oils is obtained from the results of gas chromatography (GC) analyses. Specific single carbon number (SCN) fraction molecular weights are defined from either the GC chromatogram or in the manner put forth by Katz and Firoozabadi [20]. Predictions obtained with the model are within 3% of the experimental density data. However, as shown in Fig. 2a, a noticeable drop-off in performance is seen at pressures above ~70 MPa. Therefore, we investigated the performance of the volume-translated method of Baled et al. [38],

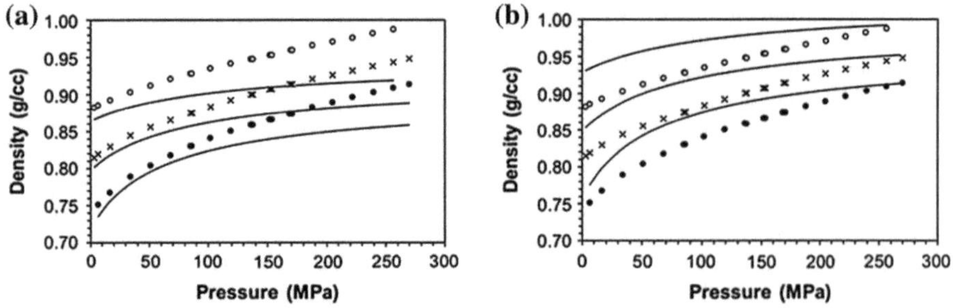

Fig. 2 Density predictions (*solid lines*) for the dead crude oil are made using the SRK EoS applying volume translation methods of **a** Pedersen et al. [33] and **b** Baled et al. [36]. Experimental data points were obtained at temperatures of 324.5 K (o), 424.0 K (x), and 523.6 K (●)

which was specifically developed to be applicable within the full temperature and pressure range of interest—that is, to 533 K and 276 MPa. While density predictions in Fig. 2b indicate that this model yields improved density predictions at higher pressures, poorer predictions are obtained at pressures below 150 MPa. Therefore, on average, there is no significant difference between the volume translation methods of both Baled et al. [36] and Pedersen et al. [33].

4.2 PC-SAFT EoS Modeling Methods for Crude Oil

The perturbed-chain statistical associating fluid theory (PC-SAFT) EoS has been widely evaluated for its ability to describe hydrocarbon mixtures. Most previous PC-SAFT crude oil modeling studies have focused on modeling of asphaltene precipitation, but PC-SAFT has also been used to model reservoir hydrocarbon density [6]. In recent PC-SAFT modeling studies of crude oils, it has been established that dispersion force and not polarity is the major intermolecular force that dominates the phase behavior calculations [28, 39, 40]. Thus, for nonpolar molecules, only three characteristic parameters are required when using PC-SAFT: the number of segments, m; the temperature-independent segment diameter, σ; and the segment–segment dispersion interaction energy, ε/k. The parameters are obtained for pure compounds by fitting the PC-SAFT EoS to experimental vapor pressure and density data, and these parameters are reported for n-alkanes by Gross and Sadowski [26]. However, crude oils contain numerous compounds, and it is not feasible to define pure-component parameters for all the compounds. Therefore, the crude oil is divided into saturate, aromatic, and resin (SARA) pseudocomponents. The PC-SAFT EoS parameters have been demonstrated to be a smooth function of molecular weight for several chemical families [26]. Researchers have obtained the PC-SAFT parameters m, σ, and ε/k for SARA components from such correlations [28, 29, 41–46]. Several sets of correlations are available in the literature,

Table 2 PC-SAFT pure-component parameter correlations for each dead oil fraction [28]

Correlations for saturates	Correlations for aromatics + resins and asphaltenes
$m = 0.0257 * M_w + 0.8444$	$m = (1-\gamma)(0.0257 * M_w + 0.8444) + \gamma(0.0101 * M_w + 1.7296)$
$\sigma = 4.047 - 4.8013\frac{\ln(M_w)}{M_w}$	$\sigma = (1-\gamma)\left(4.047 - 4.8013\frac{\ln(M_w)}{M_w}\right) + \gamma\left(4.6169 - \frac{93.98}{M_w}\right)$
$\varepsilon/k = \exp\left(5.5769 - \frac{9.523}{M_w}\right)$	$\varepsilon/k = (1-\gamma)\exp\left(5.5769 - \frac{9.523}{M_w}\right) + \gamma\left(508 - \frac{234100}{M_w^{1.5}}\right)$

from which we chose the correlations recently proposed by Punnapala and Vargas [28] that are listed in Table 2. Note that the γ parameter listed in this table represents the aromaticity of a SARA pseudocomponent.

This parameter, first proposed by Ting et al. [29], has been employed in a number of PC-SAFT modeling studies concerning asphaltene precipitation [29, 40, 41, 43, 44]. In this work, we define γ in the manner of Punnapala and Vargas [28], who assumed that a value of $\gamma = 1$ represents components having 100% character of a polynuclear aromatic (PNA) such as naphthalene or anthracene. Conversely, a value of $\gamma = 0$ means that the component has 100% saturate character (100% alkane behavior). We chose to use the definition of Punnapala and Vargas because it greatly facilitates the development of a predictive method for the γ values of the aromatics + resins and asphaltenes pseudocomponents.

The organic contents of crude oils are predominantly present as saturated and unsaturated hydrocarbons. Therefore, the dead oils used in this study can be characterized based on the SARA composition of the dead oil. The SARA method has been applied on several occasions to facilitate PC-SAFT modeling of asphaltene precipitation in crude oil systems [28, 39, 41–44]. The saturate fraction contains linear, branched, and cyclic hydrocarbons. Aromatics comprise any molecule that contains one or more aromatic rings. Resins, or polar aromatics that contain at least one polar heteroatom such as oxygen, comprise the fraction of the stock tank oil that is insoluble in propane yet soluble in *n*-pentane, *n*-heptane, and toluene. Asphaltenes are defined as the portion of the oil that is soluble in aromatic solvents such as toluene but insoluble in an excess of light alkane solvents such as *n*-pentane [28].

Asphaltenes are precipitated from the dead crude oil by mixing the oil with n-pentane on a 1 g oil to 40 mL n-pentane basis. Fractional SARA composition results for the crude oil are obtained via solid-phase extraction technique once the asphaltene components are filtered off. The saturate contents are extracted with pentane, followed by the extraction of the aromatic contents with toluene. Lastly, the resin fractions are obtained by extraction with ethyl acetate. Table 3 shows the SARA analysis results as relative mass distribution of saturates, aromatic and resin, and asphaltene fractions for the dead crude oil sample.

The molecular weight distribution of the crude oils is obtained from the results of gas chromatography (GC) analyses. Specific single carbon number (SCN) fraction molecular weights are defined from either the GC chromatogram (carbon number 6 or lower) or in the manner put forth by Katz and Firoozabadi [20].

As discussed previously, the aromaticity γ is needed for the estimation of PC-SAFT parameters for the SARA constituents of crude oil. In this work, γ is estimated by obtaining molecular hydrogen-to-carbon ratios (H/C). Elemental analyses of the dead crude oil and its associated asphaltene fraction are obtained using a Perkin Elmer Series II CHNS/O Analyzer (model 2400). The molecular H/C ratio results are presented in Table 4 along with the calculated aromaticity values, γ, for the crude oil as well as its asphaltene fraction.

The values for γ are calculated using the correlations proposed by Huang and Radosz [45]. Strictly speaking, the correlations are defined for use with hydrocarbons only containing hydrogen and carbon atoms. However, crude oils do contain a small amount of sulfur, oxygen, nitrogen, and other heteroatoms. These heteroatoms are not deemed to be present in sufficient amount to significantly alter the γ calculations and are, thus, neglected. The γ values are calculated using Eq. 5 where CN, defined in Eq. 6, is the carbon number of the hydrocarbon in question. CN_p is the carbon number that a sample of that molecular weight would possess if it were 100% saturate character, and CN_{PNA} is the carbon number that it would possess if it had 100% PNA character. CN, CN_p, and CN_{PNA} are calculated as indicated by Huang and Radosz [45].

$$\gamma = \frac{CN - CN_p}{CN_{PNA} - CN_p} \tag{5}$$

$$CN = \frac{M_w}{12.011 + 1.0079(H/C)} \tag{6}$$

The aromaticity of the aromatic + resin fraction is calculated by taking a weighted average over the dead oil and using the values obtained for the dead oil and the asphaltene fraction according to Eq. 7. Note that in Eq. 7, the aromaticity

Table 3 Saturates, aromatics + resins, and asphaltenes compositional fraction of dead crude oil

Percent mass	Dead crude oil
Saturates	0.666
Aromatics + resins	0.279
Asphaltenes	0.0551

Table 4 Values of H/C molecular ratio in dead crude oil and associated asphaltene fraction. Aromaticity, γ, is calculated using the method of Huang and Radosz [47]. The molecular weights of the crude oils are obtained from its mol wt distribution while that of the asphaltenes are set at 1700 g/mol as has been done in the literature [29, 39, 41–44]

	Mol wt (g/mol)	H/C	Aromaticity γ
Dead crude oil	390	1.67	0.254
Asphaltenes fraction	1700	1.21	0.508

value for the saturate fraction does not appear as it is defined to be equal to zero as per the definition of Punnapala and Vargas [28].

$$\gamma_{A+R} = \frac{\gamma_{oil} - w_{asph}\gamma_{asph}}{w_{A+R}} \tag{7}$$

4.3 PC-SAFT Density Predictions

Once the characterization results are obtained, the PC-SAFT EoS can now be employed to model the crude oil density reported in this study. The PC-SAFT EoS is used with the correlations for the pure-component PC-SAFT parameters as defined in Sect. 4.2 and Table 2. A computer program, written in Visual BASIC, was used to perform all calculations. Necessary inputs to this model include the molecular weight distribution (molecular weight and mole fraction), which is resolved into a manageable number of pseudocomponents using a technique such as the one proposed by Whitson [19]. Other necessary inputs include the mass fractions of saturates, aromatics + resins, and asphaltenes pseudocomponents along with the aromaticity values for the asphaltene, γ_{asph}, and the aromatics + resins, γ_{A+R}, pseudocomponents. The user then specifies the temperature and pressure conditions at which experimental density predictions are desired. The PC-SAFT density modeling results for the crude oil, characterized by the mean absolute percent deviation, δ, and the standard deviation, λ, are presented in Table 5 and in Fig. 3.

As shown in Fig. 3, the PC-SAFT EoS under-predicts the density at low pressures, although the difference between calculated and experimental data decreases as the pressures increase. In fact, at high pressures, the model starts to over-predict the density data. Interestingly, for well-characterized systems, the PC-SAFT EoS typically over-predicts the density at high pressures by 5% or more [23–25, 30]. The deviation values obtained with PC-SAFT EoS are all within 1% and show an improvement on the performance of the volume-translated cubic EoS.

Table 5 PC-SAFT EoS HTHP density modeling results for dead crude oil B. The aromaticity, γ, is 0.810 for the aromatic + resin fraction and 0.508 for the asphaltene fraction

T/K	δ	λ
324.5	0.91	0.77
424.0	0.91	0.78
523.6	0.74	0.53

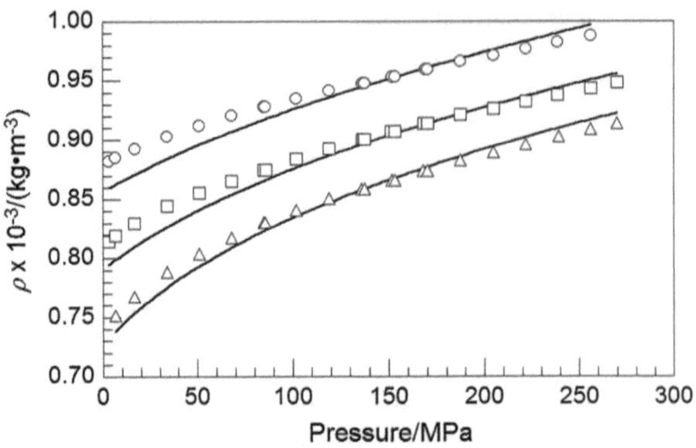

Fig. 3 Experimental (*symbols*) and PC-SAFT EoS modeling results (*solid lines*) of dead crude oil sample. Experimental data points were obtained at temperatures of 324.5 K (o), 424.0 K (□), and 523.6 K (Δ)

5 Conclusion

In order to meet the increasing world energy demand, the search for newer sources for petroleum has led to the necessity of pursuing and recovering petroleum from HTHP ultra-deep reservoirs that are several miles beneath the earth's surface. In light of this fact, HTHP densities are measured at temperatures to 523 K and pressures to 270 MPa for a dead crude oil sample supplied by a company with operations in the Gulf of Mexico. This study covers a significant gap found in the literature by providing crude oil density at temperatures to 523 K and pressures to 275 MPa for the first time. Experimental data are measured with a variable-volume high-pressure view cell that is coupled to a linear variable differential transformer.

PC-SAFT EoS modeling methodology obtained from the literature is applied to model the crude oil density reported here without the use of any fitting parameters, which are typically required when using PC-SAFT to model crude oil properties. Fractional composition information for the saturate, aromatic + resin, and asphaltene fractions is obtained from the SARA analysis of the crude oil. The aromaticity parameters for both the aromatic + resin and the asphaltene fractions, γ_{A+R} and γ_{ASPH}, are estimated from the results of the elemental analysis of the crude oil. Using the results from these analyses, along with information from the molecular weight distribution, the PC-SAFT EoS parameters are calculated from correlations provided in the literature. PC-SAFT predictions obtained from these calculated parameter sets are then compared to the HTHP crude oil density data obtained in this study.

The PC-SAFT EoS provided better HTHP density predictions for the crude oil investigated in this study as compared to the volume-translated cubic EoS predictions. However, PC-SAFT under-predicts the density data at low pressures but over-predicts the data at high pressures, indicating that the slopes of the predicted density versus pressure curves are wrong. Therefore, predictions obtained by the model for the isothermal compressibility and other derivative properties, which depend on the derivative $(dV/dP)_T$, are expected to be erroneous, and a more reliable model should be utilized for their predictions. Further work is required in the development and improvement of EoS models that can accurately capture the thermophysical properties of crude oil and its constituents at various conditions. One possibility is the use of correlations for the pure-component PC-SAFT parameters that are specifically designed for the one-phase, HTHP region.

References

1. Baird, T., Fields, T., Drummond, R., Mathison, D., Langseth, B., Martin, A., & Silipigno, L. (1998). High-pressure, high-temperature well logging, perforating and testing. *Oilfield Review*, *10*(2), 50–67.
2. Avant, C., Daungkaew, S., Befera, B., Danpanich, S., Laprabang, W., De Santo, I., Heath, G., Osman, K., Khan, Z. A., Russell, J., Sims, P., Slapal, M., & Tevis, C. (2012). Testing the limits in extreme well conditions. *Oilfield Review*, *24*(3), 4–19.
3. De Bruijn, G., Skeates, C., Greenaway, R., Harrison, D., Parris, M., James, S., Mueller, F., Ray, S., Riding, M., Temple, L., & Wutherich, K. (2008). High-pressure, high-temperature technologies. *Oilfield Review*, *20*(3), 46–60.
4. Brelvi, S. W. (1998). Generalized correlations for the isothermal compressibility of reservoir fluids and crude oils. *Saudi Aramco Journal of Technology, Spring*, 30–34.
5. Ramage, W. E, Castanier, L. M., & Ramey, H. J. (1987). The comparative economics of thermal recovery projects. DOE/SF/11564-22, SUPRI TR-56, July 1987.
6. Larsen, J., Sorensen, H., Yang, T., & Pedersen, K. S. (2011). EOS and viscosity modeling for highly undersaturated Gulf of Mexico reservoir fluids. In SPE Annual Technical Conference and Exhibition, Society of Petroleum Engineers.
7. Petrosky, G. E., & Farshad, F. F. (1993) Pressure-volume-temperature correlations for Gulf of Mexico crude oils. In SPE 26644, Presented at the 68th Annual Technical Conference and Exhibition of the Society of Petroleum Engineers, Houston, Texas, October 3–6, 1993.
8. Petrosky, G. E., & Farshad, F. F. (1998). Pressure-volume-temperature correlations for Gulf of Mexico crude oils. *SPEREE*, *1*(6), 416–420.
9. Elsharkawy, A. M., & Alikhan, A. A. (1997). Correlations for predicting solution gas/oil ratio, oil formation volume factor, and undersaturated oil compressibility. *Journal of Petroleum Science and Engineering*, *17*, 291–302.
10. Dindoruk, B., & Christman, P. G. (2004). PVT properties and viscosity correlations for Gulf of Mexico oils. *SPEREE*, *7*(6), 427–437.
11. Farshad, F., LeBlanc, J. L., Garber, J. D., & Osorio, J. G. (1996). Empirical PVT correlations for colombian crude oils. In SPE 36105, Presented at the 4th Latin American and Caribbean Petroleum Engineering Conference, Port-of-Spain, Trinidad and Tobago April 23–26, 1996.
12. Hanafy, H. H., Macary, S. M., ElNady, Y. M., Bayomi, A. A., & El Batanony, M. H. (1997). A new approach for predicting the crude oil properties. In SPE 37439, Presented at the 1997 SPE Production Operations Symposium, Oklahoma City, Oklahoma, March 9–11, 1997.
13. Almehaideb, R. A. (1997). Improved PVT correlations for UAE crude oils. In SPE 37691, Presented at the 1997 Middle East Oil Conference and Exhibition held in Manama, Bahrain, March 17–20, 1997.
14. Al-Marhoun, M. A. A New correlation for undersaturated isothermal oil compressibility. SPE 81432-SUM.
15. Vazquez, M., & Beggs, H. D. (1980). Correlations for fluid physical property prediction. *Journal of Petroleum Technology*, *32*(6), 968–970.
16. Brelvi, S. W., & O'Connell, J. P. (1972). Corresponding states correlations for liquid compressibility and partial molal volumes of gases at infinite dilution in liquids. *AIChE Journal*, *18*(6), 1239–1243.

17. Leelavanichkul, P., Deo, M. D., & Hanson, F. V. (2004). Crude oil characterization and regular solution approach to thermodynamic modeling of solid precipitation at low pressure. *Petroleum Science and Technology*, *22*, 973–990.
18. Riazi, M. R. (1997). A continuous model for C7+ fraction characterization of petroleum fluids. *Industrial & Engineering Chemistry Research*, *36*, 4299–4307.
19. Whitson, C. (1983). Characterizing hydrocarbon plus fractions. *SPE J.*, *23*, 683–694.
20. Katz, D. L., & Firoozabadi, A. (1978). Predicting phase behavior of condensate/crude oil systems using methane interaction coefficients. *Journal of Petroleum Technology*, *30*, 1649–1655.
21. AlHammadi, A. A., Vargas, F. M., & Chapman, W. G. (2015). Comparison of cubic-plus-association and perturbed-chain statistical associating fluid theory methods for modeling asphaltene phase behavior and pressure-volume-temperature properties. *Energy & Fuels*, *29*, 2864–2875.
22. He, P., & Ghoniem, A. F. (2015). A group contribution pseudocomponent method for phase equilibrium modeling of mixtures of petroleum fluids and a solvent. *Industrial & Engineering Chemistry Research*, *54*, 8809–8820.
23. Peng, D.-Y., & Robinson, D. B. (1976). A new two-constant equation of state. *Industrial & Engineering Chemistry Fundamentals*, *15*, 59–64.
24. Soave, G. (1972). Equilibrium constants from a modified Redlich-Kwong equation of state. *Chemical Engineering Science*, *27*, 1197–1203.
25. Wu, Y., Bamgbade, B., Liu, K., McHugh, M. A., Baled, H., Enick, R. M., Burgess, W. A., Tapriyal, D., & Morreale, B. D. (2011). Experimental measurements and equation of state modeling of liquid densities for long-chain *n*-Alkanes at pressures to 265 MPa and temperatures to 523 K. *Fluid Phase Equilibria*, *311*, 17–24.
26. Bamgbade, B. A., Wu, Y., Burgess, W. A., Tapriyal, D., Gamwo, I. K., Baled, H. O., Enick, R. M., & McHugh, M. A. (2015). Measurements and modeling of high-temperature, high-pressure density for binary mixtures of propane with *n*-Decane and propane with *n*-Eicosane. *The Journal of Chemical Thermodynamics*, *84*, 108–117.
27. Bamgbade, B. A., Wu, Y., Burgess, W. A., Tapriyal, D., Gamwo, I. K., Baled, H. O., Enick, R. M., & McHugh, M. A. (2015). High-temperature, high-pressure volumetric properties of propane, squalane and their mixtures: Measurement and PC-SAFT modeling. *Industrial & Engineering Chemistry Research*, *54*, 6804–6811.
28. Gross, J., & Sadowski, G. (2001). Perturbed-chain SAFT: An equation of state based on a perturbation theory for chain molecules. *Industrial & Engineering Chemistry Research*, *40*, 1244–1260.
29. Pedersen, K. S., & Sorensen, C. H. (2007). PC-SAFT equation of state applied to petroleum reservoir fluids. SPE 110483.
30. Punnapala, S., & Vargas, F. M. (2013). Revisiting the PC-SAFT characterization procedure for an improved asphaltene precipitation prediction. *Fuel*, *108*, 417–429.
31. Ting, P. D., Hirasaki, G. J., & Chapman, W. G. (2003). Modeling of asphaltene phase behavior with the SAFT equation of state. *Petroleum Science and Technology*, *21*, 647–661.
32. Liu, K., Wu, Y., McHugh, M. A., Baled, H., Enick, R. M., & Morreale, B. D. (2010). Equation of state modeling of high-pressure, high-temperature hydrocarbon density data. *The Journal of Supercritical Fluids*, *55*, 701–711.
33. Linstrom, P. J., & Mallard, W. G. (Eds.). (2005). NIST Chemistry WebBook, NIST Standard Reference Database Number 69, June 2005, National Institute of Standards and Technology, Gaithersburg MD, 20899. http://webbook.nist.gov/chemistry/fluid/.
34. Dymond, J., & Malhotra, R. (1988). *International Journal of Thermophysics, 9*, 941–951.
35. Pedersen, K. S., Milter, J., & Sorensen, H. (2004). Cubic equations of state applied to HT/HP and highly aromatic fluids. *SPE J*, *9*, 186–192.
36. Peneloux, A., & Rauzy, E. (1982). A consistent correction for Redlich-Kwong-Soave volumes. *Fluid phase equilibria, 8*, 7–23.
37. Pedersen, K. S., Milter, J., & Sorensen, H. (2004). Cubic equations of state applied to HT/HP and highly aromatic fluids. *SPE J*, *9*, 186–192.

38. Baled, H., Enick, R. M., Wu, Y., McHugh, M. A., Burgess, W., Tapriyal, D., & Morreale, B. D. (2012). Prediction of hydrocarbon densities using volume-translated SRK and PR equations of state fit to high temperature, high pressure PVT data. *Fluid Phase Equilibria, 317*, 65–76.
39. Buckley, J. S. (1999). Asphaltene precipitation and solvent properties of crude oils. *Energy & Fuels, 13*, 328–332.
40. Buckley, J. S, Hirasaki, G. J., Liu, Y., Von Drasek, S., Wang, J. X., & Gil, B. S. (1998). Asphaltene precipitation and solvent properties of crude oils. *Petroleum Science and Technology, 16*, 251–285.
41. Panuganti, S. R., Vargas, F. M., Gonzalez, D. L., Kurup, A. S., & Chapman, W. G. (2012). PC-SAFT characterization of crude oils and modeling of asphaltene phase behavior. *Fuel, 93*, 658–669.
42. Gonzalez, D. L., Ting, P. D., Hirasaki, G. J., & Chapman, W. G. (2005). Prediction of asphaltene instability under gas injection with PC-SAFT equation of state. *Energy & Fuels, 19*, 1230–1234.
43. Gonzalez, D. L., Hirasaki, G. J., Creek, J., & Chapman, W. G. (2007). Modeling of asphaltene precipitation due to changes in composition using the perturbed chain statistical associating fluid theory equation of state. *Energy & Fuels, 21*, 1231–1242.
44. Vargas, F. M., Gonzalez, D. L., Hirasaki, G. J., & Chapman, W. G. (2009). Modeling asphaltene phase behavior in crude oil systems using the perturbed chain form of the statistical associating fluid theory (PC-SAFT) equation of state. *Energy & Fuels, 23*, 1140–1146.
45. Gonzalez, D. L., Vargas, F. M., Hirasaki, G. J., & Chapman, W. G. (2008). Modeling study of CO_2-induced asphaltene precipitation. *Energy & Fuels, 22*, 757–762.
46. Gonzalez, D. L. (2008). Modeling of asphaltene precipitation and deposition tendency using the PC-SAFT equation of state. Doctoral dissertation. Retrieved from ProQuest Dissertations and Theses. (Accession Order No. 3309879).
47. Huang, S., & Radosz, M. (1991). Phase behavior of reservoir fluids V: SAFT model of CO, and bitumen systems. *Fluid Phase Equilibria, 70*, 33–54.

Heat Transfer Enhancement in a Baffled Attic-Shaped Space

Suvash C. Saha, Y.T. Gu and M.M.K. Khan

Abstract To reduce the natural convection heat loss from attic-shaped spaces, many researchers used convection suppression devices in the past. In this chapter, a single baffle is used under the top tip to investigate numerically the natural convection heat loss in an attic-shaped enclosure, which is a cost-effective approach. The case considered in this chapter is one inclined wall of the enclosure which is uniformly heated while the other inclined wall uniformly cooled with adiabatic bottom wall. The finite volume method has been used to discretize the governing equations, with the QUICK scheme approximating the advection term. The diffusion terms are discretized using central differencing with second-order accuracy. A wide range of governing parameters is studied (Rayleigh number, aspect ratio, baffle length, etc.). It is observed that the heat transfer due to natural convection in the enclosure reduces when the baffle length is increased. Effects of other parameters on heat transfer and flow field are described in this study.

1 Introduction

One of the most important forms of heat transfer and fluid flow in an enclosure is natural convection where the fluid motion is simply induced by density gradients due to temperature differences. Natural convection in enclosures has received extensive attention during past few decades due to its superior importance and applications in nature and industry. Various cavities and spaces such as rectangular shapes, annular spaces, and cylindrical cavities subjected to different thermal and moving boundary conditions have been investigated employing useful working

---

S.C. Saha (✉) · Y.T. Gu
School of Chemistry, Physics and Mechanical Engineering, Queensland University of Technology, GPO Box 2434, Brisbane, QLD 4001, Australia
e-mail: suvash.saha@qut.edu.au; s_c_saha@yahoo.com

M.M.K. Khan
School of Engineering & Built Environment, Central Queensland University, Rockhampton Campus, Rockhampton, QLD 4702, Australia

M.M.K. Khan et al. (eds.), *Application of Thermo-fluid Processes in Energy Systems*, Green Energy and Technology, https://doi.org/10.1007/978-981-10-0697-5_7

fluids such as water and air. The comprehensive review and discussion about the researched topics can be found in [1–4]. However, it is important to note that models attributed to rectangular cavities cannot estimate the complete buoyancy-driven flow within geometries with variable or sloping boundaries. Even though natural convection in a triangular enclosure [5] has many engineering applications in energy transfer in rooms and buildings, convective motion in solar stills, nuclear reactor cooling, and electronic equipment cooling it receives much less attention.

Attic-shaped buildings expose to different thermal boundary conditions during a daytime dependent on the climate situation. Therefore, the natural convection process within the space should be precisely revealed. Since the thermal and fluid flow conditions should provide comfort for the present people according to residential ventilation standards, the space geometry, inside structure, and the insulation are important factors for better design. Moreover, energy consumption for air-conditioning should be considered to achieve higher energy efficient building. According to those literature attended for natural convection within enclosures, fluid flow and heat transfer phenomena are limited for triangular spaces.

One of the earliest studies on natural convection in triangular cross-sectional enclosure was conducted by Probert and Thirst [6, 7]. In their experiment studies, the authors found the optimal pitch angle leading to the minimum rate of heat transfer in the attic of a modeled pitched roof with specified boundary conditions. They showed that the contribution by convection heat transfer is increased rather than conduction with the increase of aspect ratio up to a critical value.

Flack [8, 9] performed several experimental tests using isosceles triangular enclosure filled with air for various aspect ratios and Rayleigh numbers. The enclosure was cooled or heated from base boundary and heated or cooled from inclined walls. Thermal and fluid flow pattern was investigated as well as local and average Nusselt number of heated or cooled walls. It was found that for the case of base heating and inclined cooling, with the increase of Rayleigh number, laminar free convection heat transfer regime became unstable achieving to a critical Rayleigh number, where flow pattern altered to turbulent natural convection regime. Latter, natural convection in a right-angled triangular enclosure with bottom heating and cold side walls for air and distilled water was studied by Poulikakos and Bejan [10, 11] where Rayleigh–Benard type convection occurred. The authors established a new relationship for mean Nusselt number when air filled the enclosure. Considering similar boundary conditions proposed by Flack [5, 6] for their experimental works, Ridouane et al. [12] numerically found a good agreement after comparison of Nusselt number. Latter, Ridouane et al. [13] cut both bottom tips off significantly with exposing thermal insulation boundary condition and reported a much energy efficient space, which consumes less energy to keep the region at desired temperature. There are several studies conducted to deal with the stability of flow cellular structure [14–17]. In their studies, they reported the flow pattern for different boundary conditions with emphasis on the stability of the single cell formed at the core of the region followed by cell division into several smaller pieces and found out some transitional conditions with respect to the aspect ratio and Rayleigh number.

Holtzman et al. [18] performed series of experiments using smoke injected into the triangular enclosure with bottom heated and cold side inclined walls. Symmetry flow assumption was reported for lower Rayleigh number; however, with the increase of Rayleigh number first time they observed asymmetry and multicellular flow progressed inside the enclosure denoting the appearance of pitchfork bifurcation. Predictions of symmetry bifurcation of buoyancy-driven flow within triangular cavity numerically were studied again by Ridouane and Campo [19]. Their numerical reports indicated a critical Rayleigh number of 1.42×10^5 in which for Ra above that critical value, subcritical pitchfork bifurcation is created. The appearance of pitchfork bifurcation is also investigated for diurnal thermal forcing condition [20] where the temperature on inclined surfaces sinusoidally changes with time.

Recently, fluid flow and heat transfer inside an attic space for the boundary conditions of sudden heating/cooling and ramp heating/cooling have been analyzed by Saha et al. [21–24] by employing scaling analysis with numerical verifications. The authors studied the transient flow development inside the enclosure and showed different stages of the flow development.

Since one of the main objectives of designing a residential house is to reduce the heat transfer inside the attic space, several researchers have conducted research by adding adiabatic fin on the walls [25–29]. Varol [25] incorporated single thin adiabatic plate onto a triangular cavity in order to disrupt the heat flow. It was found that for a low Rayleigh number, the main mode of heat transfer is by conduction which is in line with the definition of Rayleigh number. From this research, the maximum and minimum mass flow of the stream function can be decreased by increasing the distance of the plate from the origin. For better understanding on the way of affecting heat transfer in a triangular enclosure, more works have been conducted by adding adiabatic fins on several places of the walls [26]. This is similar to a study conducted by Chamkha [27] with the addition of a single fin to the base of the triangle.

In this chapter, we will focus on a numerical investigation of the steady natural convection in a triangular enclosure with a baffle attached to the top tip. The finite volume method has been used to solve the governing equations. Effects of Rayleigh number, aspect ratio, and baffle length on heat transfer and flow field are described in this study. It is found that the effect of the length of the baffle or interrupter has a great influence on the heat transfer. The results are presented as a form of temperature contours and stream functions. The heat transfer is represented as a form of Nusselt number.

2 Model Theory

Natural convection in an attic space satisfies the conservation equations of mass, momentum, and energy under the assumption of continuous media. The mathematical expressions of these conservation equations may be written together with the equation of state in Cartesian tensor notations as follows.

The equation for conservation of mass, or continuity equation,

$$\frac{\partial \rho}{\partial t} + \frac{\partial (\rho u_j)}{\partial x_j} = 0, \tag{1}$$

where t is the time, ρ is the density of the fluid, u_j are the velocity components, and x_j are the Cartesian coordinates. Conservation of momentum based on Newton's second law,

$$\rho \frac{Du_i}{Dt} = \frac{\partial p}{\partial x_i} + \rho F_i + \frac{\partial}{\partial x_j}\left[\mu\left(\frac{\partial u_i}{\partial x_j} + \frac{\partial u_j}{\partial x_i}\right)\right] - \frac{2}{3}\frac{\partial}{\partial x_i}\left(\mu \frac{\partial u_j}{\partial x_j}\right), \tag{2}$$

where p is the pressure, F_i are body forces, and μ is the viscosity. The conservation of energy equation is:

$$\rho c_p \frac{DT}{Dt} = u_j \frac{\partial p}{\partial x_i} + \frac{\partial}{\partial x_j}\left(k \frac{\partial T}{\partial x_j}\right) + \mu\left[\frac{\partial u_j}{\partial x_j}\left(\frac{\partial u_i}{\partial x_j} + \frac{\partial u_j}{\partial x_i}\right) - \frac{2}{3}\left(\frac{\partial u_j}{\partial x_j}\right)^2\right], \tag{3}$$

where T is the temperature, c_p is specific heat, and κ is the thermal diffusivity. The equation of state is:

$$\rho = \rho(p, T), \tag{4}$$

In Eq. (2), the body force F_i has three components: F_x, F_y, and F_z in x-, y-, and z-directions, respectively. In addition, D/Dt is a derivative operator defined as,

$$\frac{D}{Dt} = \frac{\partial}{\partial t} + u_j \frac{\partial}{\partial x_j}, \tag{5}$$

3 Appropriate Assumptions

The governing Eqs. (1)–(3) are general forms of conservation equations and may be simplified for particular problems such as the one considered in this chapter. Moreover, these equations are highly nonlinear and coupled and are very difficult to solve. Therefore, it is of significance to simplify these equations based on particular features such as follows so that numerical discretization and analyses can be easily performed.

For the steady natural convection through the inclined wall of an attic space, it is understood that, when heat is added to the fluid, the fluid expands and thus changes its density. If the gravity is present, this change in density induces a change in the body force, which may cause the fluid to move by itself without any externally imposed flow velocity. If the temperature difference between the walls and the ambient is not very large, the correlation between the density and the temperature may be considered as a linear relation. As a result, the equation of state (4) may be given by

$$\rho = \rho_0[1 - \beta(T - T_O)], \tag{6}$$

where β is the thermal expansion coefficient. It is generally of an order of magnitude between 10^{-2} and 10^{-4} for different fluids [30].

The natural convection flows often involve a relatively small temperature difference and a low flow velocity, and thus, an incompressible flow assumption is appropriate (see [31]). Accordingly, the continuity equation (1) may be simplified as follows,

$$\frac{\partial u_j}{\partial x_j} = 0. \tag{7}$$

The Boussinesq approximation is assumed for this problem, meaning that all fluid properties such as the viscosity and the thermal conductivity are treated as constants except the density in the buoyancy term. In the present study, a two-dimensional coordinate system has been adopted, where the x-axis and the y-axis are parallel to the horizontal and the vertical, respectively (see Fig. 1). The effect of the gravitational acceleration on the flow field is along the y-axis. Since there is no other volumetric force, the body force in Eq. (2) may be written as

$$\rho F = -g(\rho - \rho_0). \tag{8}$$

Substituting Eq. (6) in Eq. (8) and applying the Boussinesq approximation, we have

$$F = -g\beta(T - T_0) \tag{9}$$

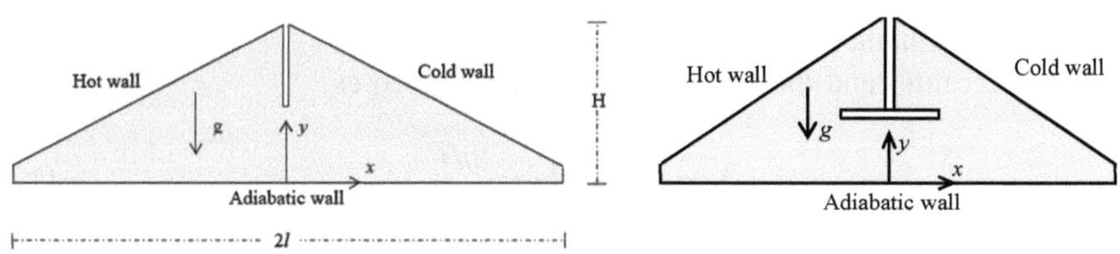

Fig. 1 Schematic of the geometry and the boundary conditions

Now Eq. (2) becomes after incorporating Eqs. (7), (8), and (9)

$$\frac{Du_i}{Dt} = -\frac{1}{\rho}\frac{\partial p}{\partial x_i} + v\frac{\partial^2 u_i}{\partial x_j \partial_j} \pm Fx_i. \tag{10}$$

Similarly, the energy equation can be simplified by assuming that the viscous energy dissipation and the pressure gradient associated with the incompressible assumption may be neglected if the velocity induced by the natural convection flow is lower. Therefore, the energy equation (3) can be simplified as

$$\frac{DT}{Dt} = k\frac{\partial^2 T}{\partial x_j \partial x_j}. \tag{11}$$

The simplified and normalized steady continuity, momentum, and energy equations of a 2*D* model are expressed as follows:

$$\frac{\partial u}{\partial x} + \frac{\partial \mathrm{v}}{\partial y} = 0, \tag{12}$$

$$u\frac{\partial u}{\partial x} + \mathrm{v}\frac{\partial u}{\partial y} = -\frac{1}{\rho}\frac{\partial p}{\partial x} + \mathrm{Pr}\left(\frac{\partial^2 u}{\partial x^2} + \frac{\partial^2 u}{\partial y^2}\right), \tag{13}$$

$$u\frac{\partial \mathrm{v}}{\partial x} + \mathrm{v}\frac{\partial \mathrm{v}}{\partial y} = -\frac{1}{\rho}\frac{\partial p}{\partial y} + \mathrm{Pr}\left(\frac{\partial^2 \mathrm{v}}{\partial x^2} + \frac{\partial^2 \mathrm{v}}{\partial y^2}\right) + \mathrm{Ra}\ \mathrm{Pr}\, T, \tag{14}$$

$$u\frac{\partial T}{\partial x} + \mathrm{v}\frac{\partial T}{\partial y} = \frac{\partial^2 T}{\partial x^2} + \frac{\partial^2 T}{\partial y^2}. \tag{15}$$

The above set of governing equations has been successfully adopted in previous numerical simulations of natural convection in cavity (see, e.g., [32, 33]). It is noted that in the case with a large temperature difference, the fluid does not satisfy the Boussinesq approximation, and thus, cautions must be taken when adopting the above equations.

The Rayleigh number, Ra and the Prandtl number, Pr are defined by

$$\mathrm{Ra} = \frac{g\beta\Delta T H^3}{\kappa \nu} \text{ and } \mathrm{Pr} = \frac{\nu}{\kappa}, \tag{16}$$

where ν is the kinematic viscosity.

The aspect ratio and the Nusselt number are defined as

$$A = \frac{H}{l} \text{ and } \mathrm{Nu} = \frac{qH}{\Delta T k}, \tag{17}$$

where q is the convective heat flux through a boundary and k is the thermal conductivity.

4 Problem Description

Under consideration is the steady flow behavior resulting from heating/cooling a quiescent, isothermal Newtonian fluid of air in a two-dimensional triangular enclosure of height H and horizontal length $2l$ (see Fig. 1). There is a flow interrupter (baffle) attached with the top tip of the enclosure. The baffle with or without a blade attached at the end will affect the flow field. The left inclined surface is heated and the right inclined surface is cooled, whereas the bottom base and the sides of the baffle and the blade are adiabatic as shown in Fig. 1. All boundaries are considered as nonslip. It is also assumed that the flow is laminar. In order to avoid the singularities at the tips in the numerical simulation, the tips are cut off by 5% and at the cutting points (refer to Fig. 1) rigid nonslip and adiabatic vertical walls are assumed. We anticipate that this modification of the geometry will not alter the overall flow development significantly [20, 21].

The initial and boundary conditions are defined as follows:

- Initially, the fluid is quiescent and isothermal.
- On the sloping walls, a rigid nonslip and uniform heating temperature conditions are applied on the left wall and uniform cooling temperature conditions are applied on the right wall (see Fig. 1).
- The bottom horizontal wall, sides of the baffle and blade are maintained as adiabatic and rigid nonslip.
- At the cutting points of the bottom tips, rigid nonslip and adiabatic vertical walls are assumed.

The governing equations (1)–(4) along with the specified initial and boundary conditions are solved using control volume method. The finite volume scheme has been chosen to discretize the governing equations, with the QUICK scheme approximating the advection term. The diffusion terms are discretized using central differencing with second-order accuracy. An extensive grid independence test has been carried out for this study. The results are not shown here for brevity. The suitable numbers of grid nodes adopted for three different aspect ratios of $A = 1.0$, 0.5, and 0.2 are 169×848, 166×744, and 160×532, respectively.

5 Results and Discussions

Results are presented in the form of isotherms and stream functions as well as Nusselt number in this study. Effects of Rayleigh number, aspect ratio, baffle height, and blade length on heat transfer and fluid flow have been presented and discussed in details.

5.1 *Effects of Length of the Interrupter*

Temperature contours and the stream functions are shown in Fig. 2. It is clearly observed that with the addition of an interrupter on the top tip of the enclosure, the heat transfer within the enclosure changes drastically. Due to buoyancy effect, the hot air should travel along the left inclined wall and when it reaches the top tip the thermal stratification starts to happen. As the length of the interrupter increases, the

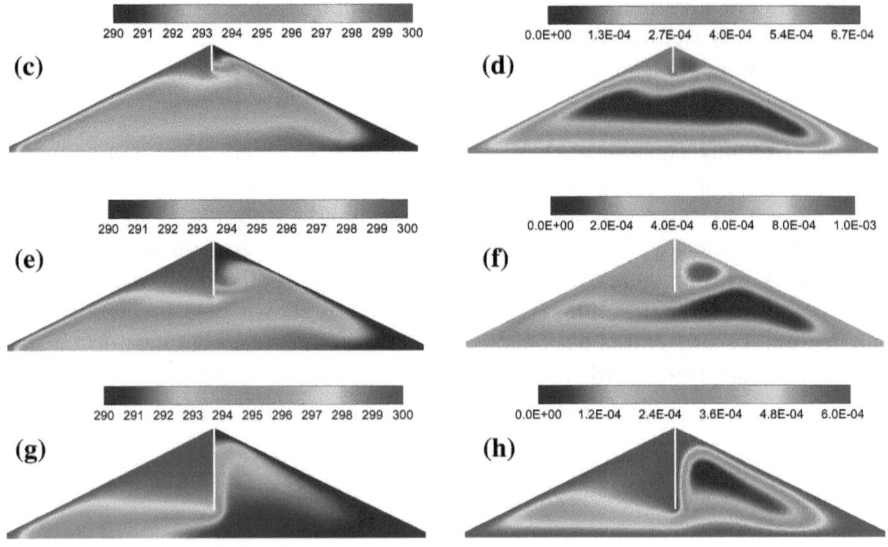

Fig. 2 Temperature contours (*left*) and stream functions (*right*) for different length of the interrupter; **a**, **b** no interrupter, **c**, **d** 25% of *H*, **e**, **f** 50% of *H*, **g**, **h** 75% of *H* when $\mathrm{Ra} = 10^6$ and $A = 0.5$

heat trapped on the hot side of the wall increases. This is due to the nature of heated air which does not flow downward easily due to buoyancy. This is clearly seen in the enclosure for the case when the length of the interrupter is 75% of the cavity height. The buildup of heat layers occurs before the heat begins to flow to the cold wall. This modification of the geometry would be very beneficial during winter conditions as the trapped layer of heated air would play a vital role in heating the house and have a significant saving in heating application.

5.2 *Effect of Rayleigh Numbers*

Figure 3 represents the temperature contours and stream functions for different values of Rayleigh number. It is observed from this figure that for low Rayleigh number single cell is visible; however, as Rayleigh number increases, the solution becomes a two-vortex solution and with further increase in the Rayleigh number a multiple vortex solutions can be observed. It is also observed from the temperature contours that the thermal boundary layer for low Rayleigh number is thicker. However, with increase of Rayleigh number the thermal boundary layer becomes thinner (concentrated near the inclined walls). That means, the flow is dominated by convection.

Effect of Rayleigh number on fluid flow and heat transfer is shown for two Rayleigh numbers in the presence of a blade attached horizontally to the baffle end

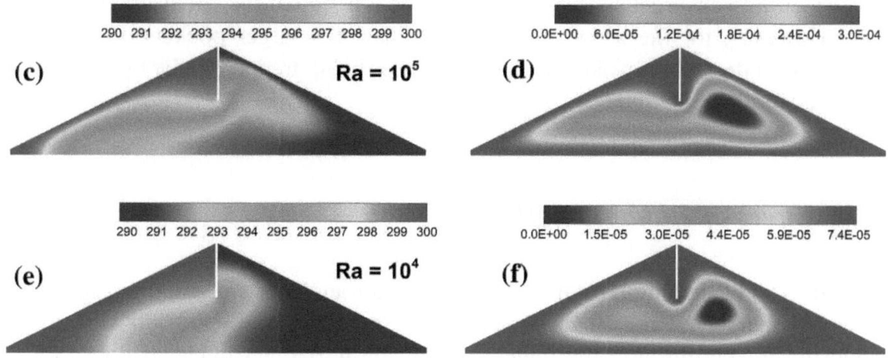

Fig. 3 Temperature contours (*left*) and stream functions (*right*) for different Ra while A = 0.5 and length of the interrupter is 50% of height

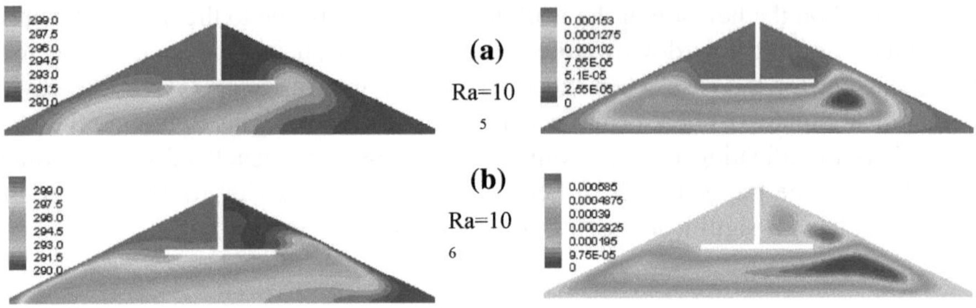

Fig. 4 Temperature contours (*left*) and stream functions (*right*) for different Ra while $A = 0.5$ and length of the interrupter is 50% of height

in Fig. 4. It is observed that the air flow is severely affected by the presence of horizontal blade. As Rayleigh number increases, the flow becomes stronger due to convection dominates the flow. The hot fluid travels underneath the blade and tries to cover the right side of the baffle. We have also observed multicellular flow structure as Ra increases.

5.3 *Effects of Aspect Ratio*

To show the effect of aspect ratio (height to base ratio), temperature contours (a, c, e) and the stream functions (b, d, f) are depicted in Fig. 5. Three aspect ratios have been considered as $A = 1.0$, 0.5, and 0.2.

From the temperature contour in Fig. 5, it is observed that the heat distribution is the quickest when the aspect ratio is the lowest whereas heat distribution is the lowest when the aspect ratio is the highest. These phenomena can be verified by the stream function. For the aspect ratio of 0.2, there are areas of concentrated lines indicating a low pressure, high-velocity zone. With this low pressure, it would act as a vacuum to ensure air distribution from the left side of the enclosure to the right side. However, for the aspect ratio of 1.0, there are larger pockets of cooler air compared to the 0.2 ratio. This is due to the heat being trapped in the upper areas of the enclosure.

5.4 *Effect of Blade Length*

The effect of blade length on fluid flow and heat transfer is presented in Fig. 6. It is observed that when the blade is longer the hot fluid from the left side of the enclosure travels through the underneath of the horizontal blade. It does not get much space to travel to the top right side of the enclosure. This region is controlled

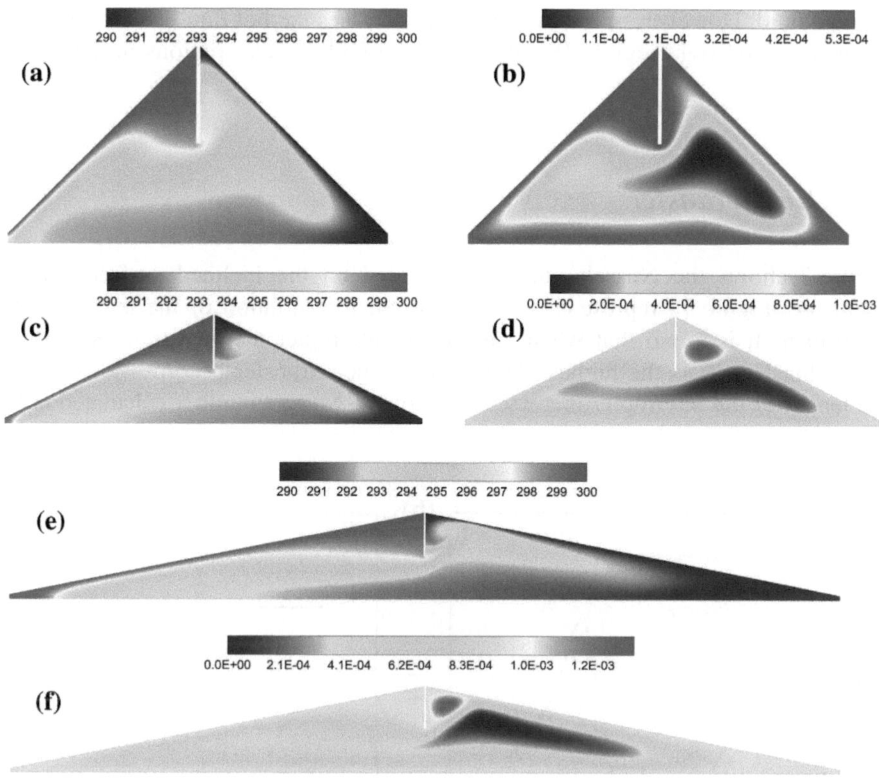

Fig. 5 Temperature contours (**a**, **c**, **e**) and stream functions (**b**, **d**, **f**) for different A while Ra = 10^6 and length of the interrupter is 50% of height

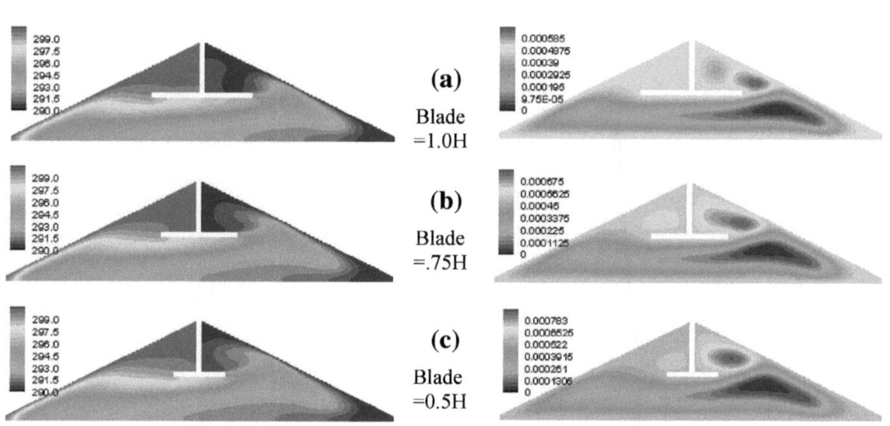

Fig. 6 Temperature contours (*left*) and stream functions (*right*) for different Ra while A = 0.5 and length of the interrupter is 50% of height

by cold fluid. However, as the blade length decreases the hot fluid gets much space to enter the top right region. We also observed multicellular regions in the stream function.

5.5 Heat Transfer

Figure 7 shows the Nusselt number calculated on the inclined surfaces of the enclosure. Figure 7a represents variation of Nu for variation of the length of the interrupter. It is found that when there is no interrupter the heat transfer through the inclined walls is the highest. However, the heat transfer becomes lower when the length of the interrupter increases. The lowest Nusselt number calculated is for

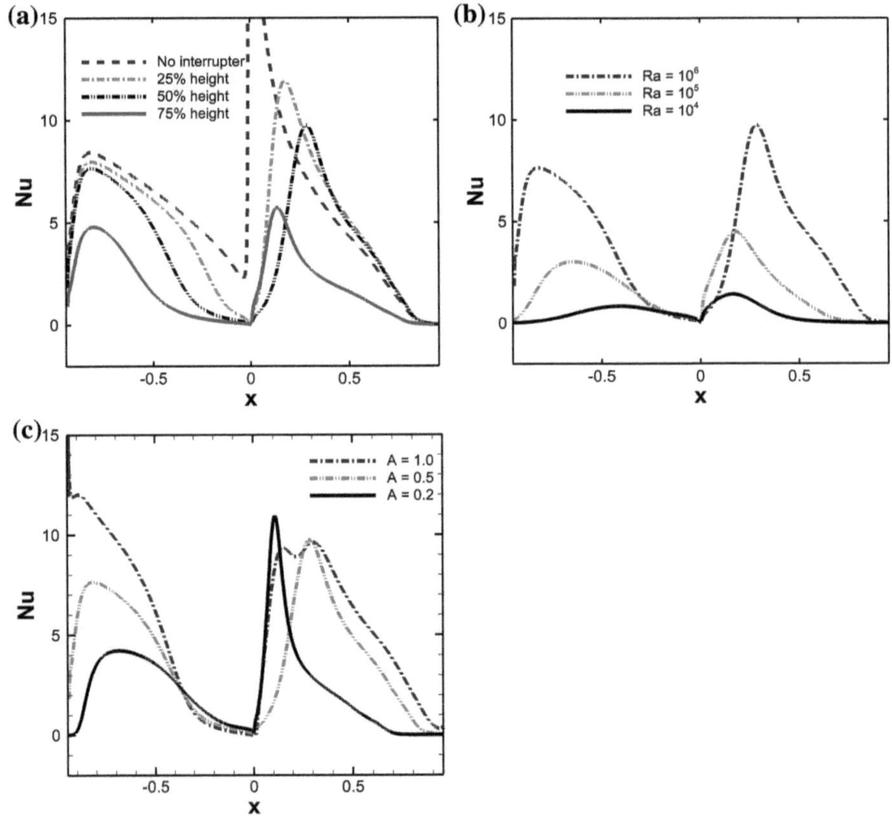

Fig. 7 Nusselt number calculated on the inclined surfaces: **a** effect of interruption for Ra = 10^6 and A = 0.5, **b** effect of Ra when A = 0.5 and length of interrupter is 50% of H, and **c** effect of A while Ra = 10^6 and length of interrupter is 50% of H

75% of the height of the enclosure which is the highest length of the interrupter. In Fig. 5b, the heat transfer is higher for higher Rayleigh number which is expected due to the fact that convection dominates the heat transfer for higher Rayleigh number. Variation of Nu is shown for different aspect ratios and is shown in Fig. 5c. It is found that heat transfer is lower for lower aspect ratio. Since the base of the enclosure is much longer than its height, the heat transfer near the two tips (left and right tips) is dominated by conduction. The mixing of hot and cold air occurs quicker.

Nusselt number is also calculated on the inclined walls for different values of blade length. It is observed that when the blade length is smaller the heat transfer rate is higher on the left inclined wall. As the blade length increases, the heat transfer decreases as the hot fluid from the left side of the enclosure does not get much space to travel to the right top region. The heat transfer through the right inclined wall is bit complex. However, for lower value of blade length, the heat transfer is the highest. When the length of the blade is similar to the interrupter height the heat transfer is the highest near $x = 0.9$. This is because the hot fluid from the left side of the enclosure travels through the underneath of the blade and hits the right inclined wall near $x = 0.9$. Near the top tip, the heat transfer is lower. Therefore, it can be concluded that in the presence of baffle attached to the top tip causes heat transfer reduction through the inclined walls of the enclosure. Heat transfer is also reduced when the baffle height increases or the blade length increases (Fig. 8).

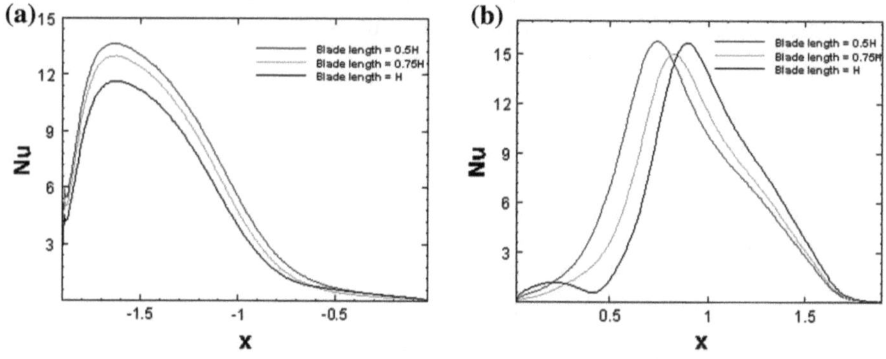

Fig. 8 Nusselt number calculated on the **a** left and the **b** right inclined surfaces for different blade length for baffle height = 0.5, Ra = 10^6, and $A = 0.5$

6 Conclusions

This study shows the enhancement/suppression of natural convection when there is a modification of the geometry, i.e., the aspect ratio, or addition of an interrupter hanging vertically from the apex of the triangular enclosure. From the simulations obtained, it is found that with the addition of an interrupter, the heat transfer is significantly reduced. It is also found that the heat transfer reduces as the height of the interrupter increases. This would be a useful application on houses with winter climates. The trapped heat would be able to heat the interior of the house reducing the power consumption for heating devices. Here, it has also been demonstrated that with a higher aspect ratio, the mass transfer of the air decreases while with a lower aspect ratio the mass transfer increases, thereby increasing the flow and mixing of hot air with the cold air.

For further studies on the effects of an interrupter on the heat flow in an attic space, the geometry of the interrupter has been modified. A horizontal adiabatic blade is attached at the end of the baffle. Inclusion of this blade further reduces the heat transfer. Moreover, the heat transfer depends on the length of the blade. When the blade length is smaller, the heat transfer is higher. However, as the length of the blade increases the heat transfer decreases. To ensure that the interrupter would be viable for both summer and winter conditions, it would be recommended to introduce movable covered slots onto the interrupter itself. The purpose of these slots is to allow heat flow through the interrupter during summer conditions when opened and to act as a solid boundary layer during winter conditions when closed. Even though this is a good design to keep the heat from reaching the house, it is impractical as the cost of materials needed would increase.

References

1. Hyun, J. M. (1994). Unsteady buoyant convection in an enclosure. *Advances in Heat Transfer, 24,* 277–320.
2. Ostrach, S. (1988). Natural convection in enclosures. *Transactions of the ASME: Journal of Heat Transfer, 110,* 1175–1190.
3. Markatos, N. C., & Pericleous, K. A. (1984). Laminar and turbulent natural convection in an enclosed cavity. *International Journal of Heat and Mass Transfer, 27*(5), 755–772.
4. Bejan, A. (2013). *Convection heat transfer*. Hoboken: Wiley.
5. Saha, S. C., & Khan, M. M. K. (2011). A review of natural convection and heat transfer in attic-shaped space. *Energy Building, 43,* 2564–2571.
6. Probert, S. D., & Thirst, T. J. (1977). Thermal insulation provided by triangular sectioned attic spaces. *Applied Energy, 3,* 41–50.
7. Thirst, T. J., & Probert, S. D. (1978). Heat transfer versus pitch angle for nonventilated triangular-sectioned, apex-upward air-filled spaces. *ASTM Special Technical Publication, 660,* 203–210.
8. Flack, R. D. (1980). The experimental measurement of natural convection heat transfer in triangular enclosures heated or cooled from below. *Transactions of the ASME: Journal of Heat Transfer, 102,* 770–772.

9. Flack, R. D. (1979). Velocity measurements in two natural convection air flows using a laser velocimeter. *Transactions of the ASME: Journal of Heat Transfer, 101,* 256–260.
10. Poulikakos, D., & Bejan, A. (1983). Natural convection experiments in a triangular space. *Journal of Heat Transfer, 105,* 652–655.
11. Poulikakos, D., & Bejan, A. (1983). The fluid dynamics of an attic space. *Journal of Fluid Mechanics, 131,* 251–269.
12. Ridouane, E. H., Campo, A., & McGarry, M. (2005). Numerical computation of buoyant airflows confined to attic spaces under opposing hot and cold wall conditions. *International Journal of Thermal Sciences, 44,* 944–952.
13. Ridouane, E. H., Campo, A., & Hasnaoui, M. (2006). Benefits derivable from connecting the bottom and top walls of attic enclosures with insulated vertical side walls. *Numerical Heat Transfer, Part A—Applications, 49,* 175–193.
14. Salmun, H. (1995). The stability of a single-cell steady-state solution in a triangular enclosure. *International Journal of Heat and Mass Transfer, 18,* 363–369.
15. Farrow, D. E., & Patterson, J. C. (1993). On the response of a reservoir sidearm to diurnal heating and cooling. *Journal of Fluid Mechanics, 246,* 143–161.
16. Asan, H., & Namli, L. (2002). Numerical simulation of buoyant flow in a roof of triangular cross-section under winter day boundary conditions. *Energy Buildings, 33,* 753–757.
17. Asan, H., & Namli, L. (2000). Laminar natural convection in a pitched roof of triangular cross-section: Summer day boundary conditions. *Energy Buildings, 33,* 69–73.
18. Holtzman, G. A., Hill, R. W., & Ball, K. S. (2000). Laminar natural convection in isosceles triangular enclosures heated from below and symmetrically cooled from above. *Journal of Heat Transfer, 122,* 485–491.
19. Ridouane, E. H., & Campo, A. (2006). Formation of a pitchfork bifurcation in thermal convection flow inside an isosceles triangular cavity. *Physics of Fluids, 18,* 074102.
20. Saha, S. C., Patterson, J. C., & Lei, C. (2010). Natural convection and heat transfer in attics subject to periodic thermal forcing. *International Journal of Thermal Sciences, 49,* 1899–1910.
21. Saha, S. C. (2011). Unsteady natural convection in a triangular enclosure under isothermal heating. *Energy Buildings, 43,* 701–709.
22. Saha, S. C. (2011). Scaling of free convection heat transfer in a triangular cavity for Pr > 1. *Energy and Buildings, 43*(10), 2908–2917.
23. Saha, S. C., Patterson, J. C., & Lei, C. (2010). Natural convection in attics subject to instantaneous and ramp cooling boundary conditions. *Energy Buildings, 42,* 1192–1204.
24. Saha, S. C., Patterson, J. C., & Lei, C. (2010). Natural convection in attic-shaped spaces subject to sudden and ramp heating boundary conditions. *Heat Mass Transfer, 46,* 621–638.
25. Varol, Y., & OZtop, H. F. (2009). Control of buoyancy-induced temperature and flow fields with an embedded adiabatic thin plate in porous triangular cavities. *Applied Thermal Engineering, 29,* 558–556.
26. Varol, Y., & OZtop, H. F. (2007). Natural convection in porous triangular enclosures with a solid adiabatic fin attached to the horizontal wall. *International Communications in Heat and Mass Transfer, 34,* 19–27.
27. Chamkha, A. J. (2010). Double-diffusive natural convection in inclined finned triangular porous enclosures in the presence of heat generation/absorption effects. *Heat and Mass Transfer, 46*(7), 757–768.
28. Anderson, T. (2009). Convection suppression in a triangular-shaped enclosure. *Computational Thermal Sciences, 1,* 121–309.
29. Ridouane, E. H., & Campo, A. (2007). Effects of attaching baffles onto the inclined walls of attic frames for purposes of energy conservation. *Heat Transfer Engineering, 28,* 1–103.
30. Ostrach, S. (1964). Laminar flows with body forces. In Moore, F. K. (Ed.), *Theory of laminar flows*. Princeton: Princeton University Press.
31. Batchelor, G. K. (1954). Heat transfer by free convection across a closed cavity between vertical boundaries at different temperature. *Quarterly of Applied Mathematics, 12,* 209–233.

32. Saha, S. C., Patterson, J. C., & Lei, C. (2011). Scaling of natural convection of an inclined flat plate: Sudden cooling condition. *Journal of Heat Transfer, 133*(4), 041503.
33. Saha, S. C., Gu, Y. T., & Khan, M. M. K. (2015). A natural convection heat transfer in the partitioned attic space. In M. M. K. Khan, & N. M. S. Hassan (Eds.), *Thermofluids modeling for energy efficiency applications* (1st ed., pp. 59–72). Elsevier (Chapter 3). ISBN-13: 978-0128023976, ISBN-10: 012802397X.

Enhanced Thermo-Fluid Dynamic Modelling Methodologies for Convective Boiling

Tilak T. Chandratilleke and Nima Nadim

Abstract Analytical tools embedded in current thermal design practice for convective boiling systems are traditionally built upon correlated empirical data, which are constrained by the thermo-fluid dynamical complexities associated with stochastic and interactive behaviour of boiling fluid mixtures. These methodologies typically overlook or under-represent key characterising aspects of bubble growth dynamics, vapour/liquid momentum exchange, boiling fluid composition and local phase drag effects in boiling processes, making them inherently an imprecise science. Resulting predictive uncertainties in parametric estimations compromise the optimal design potential for convective boiling systems and contribute to operational instabilities, poor thermal effectiveness and resource wastage in these technologies. This book chapter first discusses the scientific evolution of current boiling analytical practice and predictive methodologies, with an overview of their technical limitations. Forming a foundation for advanced boiling design methodology, it then presents novel thermal and fluid dynamical enhancement strategies that improve modelling precision and realistic processes description. Supported by experimental validations, the applicability of the proposed strategies is ascertained for the entire convective boiling flow regime, which is currently not possible with existing methods. The energy-saving potential and thermal effectiveness underpinned by these modelling enhancements are appraised for their possible contributions towards a sustainable energy future.

Nomenclature

C_d	Drag coefficient
C_p	Specific heat (J/kg-K)
D_w	Bubble departure diameter (m)
E	Energy rate (W/m3)
f	Bubble departure frequency (Hz)

T.T. Chandratilleke (✉) · N. Nadim
Department of Mechanical Engineering, Curtin University, GPO Box U1987, Perth, WA WA6845, Australia
e-mail: t.chandratilleke@curtin.edu.au

M.M.K. Khan et al. (eds.), *Application of Thermo-fluid Processes in Energy Systems*, Green Energy and Technology, https://doi.org/10.1007/978-981-10-0697-5_8

$\vec{F}_{lift}$ Lift force (N)
$\vec{F}^{TD}$ Turbulence drift force (N)
$\vec{F}_{wl}$ Wall lubrication force (N)
g Gravity (m/s^2)
G Mass flow rate (kg/s)
H Enthalpy (kJ/kg)
h_{lg} Latent heat (kJ/kg)
h_{sl} Interfacial heat transfer coefficient (W/m^2-K)
Ja Jacob number
k Turbulent kinetic energy (m^2/s^2)
k_{eff} Effective conductivity (W/m-K)
K_{pq} Interfacial momentum transfer coefficient
L Total length of the channel
L_H Heated length of the channel
$\dot{m}$ Mass flux (kg/m^2-s)
p Pressure (Pa)
$\dot{q}$ Heat flux (W/m^2)
r_c Cavity radius (m)
T Temperature (K)
u^* Frictional velocity on the wall (m/s)
v Velocity (m/s)
$\nabla\vec{V}$ Mean strain rate tensor
$\nabla\vec{V}^T$ Turbulent strain rate tensor
We_s Surface Weber number
y^+ Dimensionless distance from wall
Y^* Dimensionless vertical distance from centre of channel
Z^* Dimensionless axial distance from channel inlet

Subscripts

b Bubble
d Droplet
b,d Bubble or droplet
E Evaporative
L Liquid
m Mixture
p Primary phase
q Secondary phase
Q Quenching
Sat Saturation
Sub Subcooled
Sup Superheated
v Vapour
w Wall

Greek Symbols

α Volume fraction
μ Viscosity (kg/m-s)
ρ Density (kg/m3)
σ Surface tension coefficient (n/m)
$\bar{\tau}$ Stress tensor
τ_D Bubble dwelling time (s)
τ_G Bubble growth time (s)
ω Specific dissipation rate (1/s)
λ Thermal diffusivity $(k/\rho c_p)$

1 Introduction

Owing to high thermal effectiveness associated with fluids undergoing phase change, convective flow boiling mechanisms are widely deployed in thermal energy conversion systems [1] such as steam power plants, industrial boilers, nuclear reactors, refrigerators and electronic cooling heat sinks. Thermal characteristics of flow boiling are crucial design considerations for such technological applications where the overall system performance is fundamentally governed by the dynamics of fluid phase change, flow structures developed and heat transport mechanisms within heated pipes carrying operating fluids. In these, vapour bubble generation and bubble detachment essentially influence the degree of flow turbulence and slip velocity between phases, which in turn determine the thermal and fluid flow characteristics.

Through decades of experimental work and numerical modelling, an extensive knowledge base has been developed in understanding the flow boiling regimes and thermal behaviour in straight pipes. As depicted in Fig. 1, a single phase fluid flow entering a heated pipe undergoes gradual phase change from Bubbly flow to the regimes of Slug, Annular and Mist flow along the pipe with increasing proportions of vapour volume or void fraction. These extreme changes in flow composition

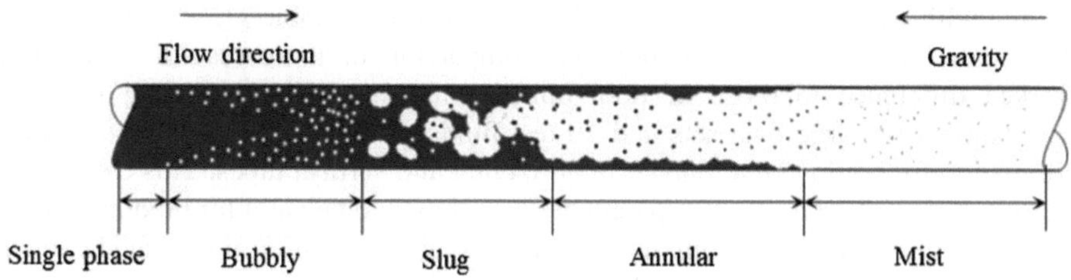

Fig. 1 Variation of flow regimes in upward convection boiling in a straight duct

create widely differing flow mechanisms within boiling regimes and would deter the development of unified thermal and flow simulation methodologies covering all boiling regimes.

The traditional modelling approach for two-phase involves the assumption of independent behaviour in fluid phases and treats the gas-liquid flow domain with slip velocity that contributes to augmented pressure loss. Thus, fluid pressure head losses and average heat transfer rates are computed for the cases of homogenous and non-homogenous mixtures. This approach evidently ignores the interfacial mass and momentum transfer between phases. Thermal and hydrodynamic behaviour in a boiling field has strongly inter-dependent mass and momentum transfers across phase interface due to condensation and/or evaporation processes present. Moreover, these dependencies significantly vary from bubbly flow regime to mist flow regime. Consequently, interfacial hydrodynamic models for gas-liquid systems are inadequate for accurate representation of convective boiling in pipes.

Among the models using gas-liquid interfacial concept, Lockhart and Martinelli [2] method is regarded the pioneering approach, where isothermal two-component flow is analysed for frictional pressure gradient. Another version by Martinelli and Nelson [3] addresses pressure drop during forced circulation boiling and condensation.

In improving accuracy and validity range of analytical solution, empirical and semi-empirical treatment of boiling fields have been suggested and utilised under various operating conditions and materials. Model and correlations suggested by Zuber et al. [4], Bankof [5], Marchaterre and Hoglund [6] and Griffith [7] have made significant contributions to the knowledge of multi-phase hydrodynamics in boiling. For more than five decades, those pioneering models have formed the traditional modelling basis in predicting key parameters of pressure loss, void fraction and slip ratio of phases whilst being the catalyst for further developments. However, heat transfer schemes deployed in traditional boiling analyses generally indicate large uncertainties towards the critical heat flux (CHF). This is due to the assumption of homogenous phase interaction in developing theoretical or empirical models. Moreover, simple mechanisms of fully developed nucleate boiling used would be inadequate for accurate representation of flow boiling phenomena.

For improved understanding of boiling processes, extensive parametric studies have been conducted to examine bubble nucleation, growth and detachment along with pipe wall dry-out, burnout and CHF. Bergles and Rohsenow [8] carried out an experiment observing the characteristics boiling curve for forced convection surface boiling with water at low pressure and investigated the bubble growth process and the requirements for onset of boiling. Comparison of these results indicated a marked difference to those of pool boiling characteristics. Extending his own correlation with large data sets and 10 test fluids, Kandlikar [9] developed a new correlation for boiling heat transfer in horizontal and vertical tubes. This correlation incorporating a fluid dependent parameter was shown to be valid for predicting both nucleate and convective boiling heat transfer. Analysing a uniformly heated coolant channel, Boyd and Meng [10] obtained heat transfer coefficients for both single phase and fully developed boiling (FBD) regimes. These authors suggested an interpolation function to estimate the heat transfer coefficient for nucleate boiling conditions between the single phase and FBD regimes. However as a major drawback, these results were applicable only to the tested fluid in the study.

The next level of enhancement in boiling investigation was to explicitly investigate the dynamics of bubble behaviour at the wall and associated influence on thermal processes. Cole [11] concluded that bubble behaviour and growth at a heated surface are pivotal for accurate determination of heat and mass transfer, and boiling model development. Pioneering such research, Cole [11] conducted a photographic study of boiling phenomenon at heated surfaces up to CHF and identified three stages of dwelling, growth and departure associated with vapour bubble generation. He defined bubble departure frequency based on the fluid conditions and indicated its influence on bubble departure size. Although this study was performed under pool boiling conditions, his bubble departure frequency predictions were shown to provide an acceptable accuracy with flow boiling as well.

Thorncroft et al. [12] visually examined boiling characteristics for convective boiling under subcooled conditions in upward and downward channel flow. Considering heat and mass fluxes in the range of 13–14.6 kW/m^2 and 190–666 kg/m^{2}s, respectively, these authors reported significantly different bubble dynamics, where bubble sliding along nucleation sites was specifically observed. Bubbles were noted to grow partially attached to the nucleation site while most of the growth occurred during the sliding process. Their results were validated for different heat flux conditions by Situ et al. [13] in vertical annular flow using high-speed digital photography. This forced convection boiling analysis indicated bubble departure frequency was proportional to the wall heat flux and reported bubble lift-off diameter, bubble growth rate and bubble velocity after lift-off. In subsequent work, Situ et al. [14] formulated dimensionless form of bubble lift-off diameter as a function of Jacob and Prandtl numbers with acceptable consistency against the experimental data.

Basu et al. [15] have developed a subcooled flow boiling model by separating the wall heat flow into components at the pipe wall and formulated closures for bubble departure diameter, departure frequency, nucleation site density and surface properties. Model predictions were compared with the experimental tests performed for mass flow rates of 124–926 kg/m^2.s, heat fluxes of 25–900 kW/m^2 and contact angles of 30°–90°. These authors concluded that, when boiling approached fully developed conditions, the wall heat transfer was dominated by the transient heat conduction in the superheated liquid film at the wall and consequently, the flow velocity had less influence on the overall heat transfer rate. Compared to other analyses, this model accounted for bubble sliding effect in determining bubble dynamics although the requirement of contact angle value was a major drawback.

Describing bubble ebullition cycle in subcooled convective boiling, Podowski et al. [16] proposed a mechanistic model based on one-dimensional transient heat transfer from pipe wall surface to bubble while accounting for dwelling and growth of bubbles. These authors considered a comprehensive set of parameters such as subcooled temperature, bubble departure diameter, transient heat flux, wall characteristics (material properties and cavity radius), making the model one of the most trusted for predicting bubble departure frequency in subcooled convective boiling.

Situ et al. [17] obtained an extensive experimental dataset which indicated fundamentally different bubble behaviour under pool and convective boiling conditions. This provided an important basis for grouping of boiling models according to bubble behaviour. For instance, as boiling regime approaches to a pool boiling case, where bubble detachment process is purely controlled by buoyancy force, bubble shape, density number, and critical diameter are significantly different from convective boiling. On the other hand, in presence of shear flows, in convective boiling, new phenomena should be accounted such as bubble deformation, bubble sliding on the wall and more complicated detachment process.

Advances in boiling knowledge, inclusive of bubble dynamics, phase interaction and phase transition, have warranted the development of many numerical simulation models that would overcome limitations of traditional approaches and extend analytical capabilities into deeper examination of boiling regimes.

Forming the basis for majority of published numerical models, Kurul and Podowski [18] first presented the wall heat partitioning concept as a key technique for developing boiling closure. At the forefront of this is the RPI (Rensselaer Polytechnic Institute) model of which the wall heat flux is taken to be contributed by three components of heat transfer through liquid, that due to quenching and liquid evaporation. The model also assumes thermal equilibrium between the phases of vapour and liquid, therefore the vapour is treated to be fixed at saturation temperature.

For several decades, the RPI model, which is a Eulerian two-phase approach, has been applied for various boiling cases with attempts to modify and validate it. Koncar et al. [19] utilised the simplest form of RPI, with three heat partitioning components and pool boiling bubble departure frequency. This study investigated local subcooled flow boiling at low pressure and demonstrated the model validity for maximum void fraction of 0.3. For analysing a fuel assembly design, Krepper et al. [20] modified the RPI model by introducing a correction for liquid wall temperature to be mesh-independent and based on liquid temperature at a fixed y^+ value. Koncar and Krepper [21] used the RPI model through CFX commercial CFD package for investigating boiling of R-113 in a vertical annulus. These authors reported a good agreement with experimental measurements. For a parametric investigation of subcooled jet impingement boiling, Abishek et al. [22], obtained satisfactory outcomes for isothermal and isoflux jet impingement boiling that were validated against test data.

Review of current literature identifies limitations of published boiling simulation models with possibilities for improvements. In these, the closures for bubble dynamics are typically estimated through pool boiling data ignoring the significance of surface characteristics, varying nature of bulk flow within regimes and the influence of flow shear on vapour bubbles. Addressing these limitations, this book chapter demonstrates feasible enhancements to convective boiling models for better representation of flow intricacies over the entire range of flow boiling regimes, where the void fraction varies from low values in the bubbly region to high magnitudes in the mist flow regime. Based on Eulerian multi-phase framework, the proposed model incorporates a mechanistic description for bubble dynamics accounting for surface properties (wall material properties and temperature) and bulk flow velocity. It also includes an enhanced momentum exchange scheme for accurate estimation of slip velocity between phases that undergo extreme variations with slug and mist flow conditions. For ascertaining the effectiveness of the modified closures, well-established experimental data for subcooled flow boiling within rectangular ducts by Pierre and Bankoff [23] are used as the evaluation benchmark. The model is validated with not one, but both void fraction and phase slip velocity to ensure accuracy and conformity through overall and interfacial mass exchange rates. The paper also provides, as a guide, a description of essential modelling elements to be considered for improved convective boiling simulations. It is viewed that these modelling enhancements will improve and consolidate the current convective boiling design practice, leading to better thermal efficiency, energy saving and resource utilisation in boiling systems, hence contributing to a sustainable energy future.

2 Development of Computational Framework

The model developed and presented in this book chapter paper uses the following governing equations with Eulerian approach where the liquid and vapour phases are separately considered in solving the momentum, mass and energy conservation equations. The turbulence equations is solved for mixture (defined according to mixture velocity and material properties) while pressure is shared between both phases. In writing these equations, liquid and vapour phases are denoted by subscripts "p" and "q", respectively.

(a) Mass, momentum and energy conservation

The continuity equations are written as (only phase "q" is show to prevent repetition),

$$\nabla.(\alpha_q \rho_q \vec{v}_q) = \dot{m}_{qp} - \dot{m}_{pq} \tag{1}$$

The momentum conservation equations, which are coupled by mass transfer, momentum exchange coefficient and other interfacial forces, are defined as (only phase "q" is show to avoid repetition),

$$\begin{aligned}\nabla(\alpha_q \rho_q \vec{v}_q \vec{v}_q) = {} & -\alpha_q \nabla p + \nabla.\bar{\bar{\tau}}_q + \alpha_q \rho_q \vec{g} + K_{pq}(\vec{v}_p - \vec{v}_q) + \dot{m}_{pq}\vec{v}_{pq} - \dot{m}_{qp}\vec{v}_{qp} \\ & + \vec{F}_q^{TD} + \vec{F}_{wl} + \vec{F}_{lift}\end{aligned} \tag{2}$$

Stress tensor of each phase accounting for the effects from molecular and turbulence viscosity is written as (only phase "q" is show to prevent repetition),

$$\bar{\bar{\tau}}_q = \alpha_q \mu_q \left(\nabla \vec{v}_q + \nabla \vec{v}_q^T \right) + \alpha_q \left(\lambda_q - \frac{2}{3} \mu_q \right) \nabla . \vec{v}_q \bar{\bar{I}} \tag{3}$$

Energy conservation equation for liquid/vapour mixture is given by,

$$\nabla (\alpha_q \rho_q \vec{v}_q H_q) = -\alpha_q k_q \nabla T_q + Q_{pq} + \dot{m}_{pq} H_{pq} - \dot{m}_{qp} H_{qp} \tag{4.a}$$

and mixture scalars are calculated as,

$$\rho_m = \alpha_l \rho_l + \alpha_v \rho_v \text{ and } \vec{v}_m = \frac{\alpha_l \rho_l \vec{v}_l + \alpha_v \rho_v \vec{v}_v}{\alpha_l \rho_l + \alpha_v \rho_v}, E = \frac{\alpha_l \rho_l E_l + \alpha_v \rho_v E_v}{\alpha_l \rho_l + \alpha_v \rho_v} \tag{4.b}$$

(b) Turbulence

Using k-ω SST turbulence model, transport equations will be shared for liquid/vapour mixture (i.e. using mixture velocity and material properties for each phase) as follows,

$$\nabla . (\rho_m k_m \vec{v}_m) = \nabla . \left((\mu + \frac{\mu_t}{\sigma_k}) \nabla k_m \right) + G_k - Y_k \tag{5.a}$$

$$\nabla . (\rho_m \omega_m \vec{v}_m) = \nabla . \left((\mu + \frac{\mu_t}{\sigma_\omega}) \nabla \omega_m \right) + G_\omega - Y_\omega \tag{5.b}$$

In this, $G_{k,\omega}$ and $Y_{k,\omega}$, which are generation and dissipation rate term for k and ω, have the common formulation for standard and SST version of k-ω closure [24, 25], whereas σ_k and σ_ω (k and ω Prandtl number) are specifically derived for SST version of k-omega model by Mentor [24]. Turbulence induced by the presence of bubbles/droplets as dispersed phase has to be accounted for in the closure when used in a multi-phase frame. In taking this into consideration, various methodologies propose either explicit source terms to be included in transport turbulence equation [26] or alternatively, suggest modifying turbulence viscosity incorporating random primary phase motion originated by dispersed phase [27]. Following Sato et al. [28], the current model uses turbulent viscosity modification for dispersed phase as,

$$\mu_{td} = C_{\mu d} \rho_m \alpha_v d_{b,d} |\vec{v}_v - \vec{v}_l| \tag{6}$$

where $C_{\mu d}$ is an adjustable coefficient in the range of [0.5, 0.75]. For the analysis, $C_{\mu d}$ is chosen as 0.65, which is an estimation for density ratio $\left(\frac{\rho_l}{\rho_v}\right)$ range of (25, 130). This source term may be applied to various regimes by assuming particle as bubbles (d_b) or droplet (d_p). Then for the liquid/vapour mixture, total turbulence viscosity is formulated as,

$$\mu_{tm} = \frac{\rho k}{\omega} \frac{1}{\max\left[\frac{1}{\alpha^*}, \frac{SF_2}{\alpha_1 \omega}\right]} + \mu_{td} \tag{7}$$

where S is the mean rate of strain tensor while the coefficient α^* dampens turbulent viscosity if low-Reynolds correction is applied with F_2 and α_1, which are part of functions proposed in the SST closure. Since all the interfacial exchange terms are interpreted as a source term in the primary (continuous) phase, in turbulence equations, solving k and ω equations for the mixture or separately in phases is not anticipated to make a noticeable difference. Besides, turbulence dispersion is independently accounted as a source term in the momentum equation while the turbulence agitation effect from dispersed phase (turbulence interaction) is included in RANS equations separately. In implementing the turbulence equation in the mixture domain, the turbulence controlling parameters are monitored with respect to the mixture phase parameters, for example y^+ through $y_m^+ = \frac{\rho_m u_m^* y_{wall\_cell}}{\mu_m}$. This approach is confirmed by comparing results from the trial runs having mixture versus per-phase schemes that give identical results. Therefore, the mixture scheme is applied to enhance numerical stability and reduce computational effort whilst achieving the same level of turbulence resolutions.

(c) Wall heat partitioning

Current flow boiling simulation is developed as a non-equilibrium model alleviating a major modelling drawback in the RPI approach, where thermal equilibrium is assumed between liquid and vapour phases. It considers three wall heat partitioning components, namely the liquid convective, quenching and evaporative heat fluxes as with the RPI approach. In addition, it also includes a fourth heat partitioning component to account for diffusive heat flux within bubble vapour phase, hence removing the RPI assumption of vapour being in equilibrium at saturation temperature. Contribution from the forth heat partitioning term is particularly important for convective boiling with large void fraction and improves the applicability of analysis.

Summing up the four heat flux contributions, the total wall heat flux is obtained to be,

$$\dot{q}_{Total} = (\dot{q}_L + \dot{q}_Q + \dot{q}_E) f(\alpha) + \dot{q}_V (1 - f(\alpha)) \tag{8}$$

where $f(\alpha)$ is a function which is defined according to the phase distribution and the flow regime. Following Lavieville et al. [29], the current model uses the expression given below,

$$f(\alpha) = \begin{cases} 1 - \frac{1}{2}\exp(-20(\alpha_l - 0.2)) & \alpha_l \geq 0.2 \\ \frac{1}{2}\left(\frac{\alpha_l}{0.2}\right)^4 & \alpha_l < 0.2 \end{cases} \tag{9}$$

Known as boiling closure, Eq. (10.a, 10.b) below provides the diffusive heat component for each phase during boiling/condensation processes. In this, $\dot{q}_L$ and $\dot{q}_V$ are convective terms calculated for liquid and vapour phases using temperature gradient at wall, their area of influence and local fluid properties.

$$\dot{q}_L = h_L(T_W - T_L)(1 - A_b) \tag{10.a}$$

$$\dot{q}_V = h_V(T_W - T_V) \tag{10.b}$$

Convective liquid and vapour heat transfer coefficients (h_L, h_V) are computed from the wall function formulations in a RANS framework.

In Eq. (10.a, 10.b), A_b is the area covered by bubble and is calculated based on bubble departure diameter and an empirical constant K ranged between 1.8 and 5. These parameters are estimated following the suggestions by Del Valle and Kenning [30] from,

$$A_b = K\frac{N_w \pi D_w^2}{4} \tag{11.a}$$

$$K = 4.8\ \exp(-\frac{Ja_{sub}}{80}) \tag{11.b}$$

where N_w is nucleation site density. N_w is obtained from the empirical expression of Lemmert and Chawala [31] as,

$$N_w = 210^{1.805}(T_w - T_{Sat})^{1.805} \tag{12}$$

while the Bubble departure diameter (D_w) is computed from the semi-empirical correlation developed by Tolubinski and Kostanchuk [32],

$$D_w = \min(0.0014, 0.0006\ \exp(-\frac{\Delta T_{Sub}}{45}) \tag{13}$$

where local subcooling is defined as $\Delta T_{Sub} = T_{Sat} - T_{bulk}$

In Eq. (8), $\dot{q}_Q$ and $\dot{q}_E$ are cyclic-averaged heat transfer rates for quenching (i.e. heat removal by liquid re-entering the wall region after bubble detachment) and evaporation (i.e. latent heat) processes. These heat transfer rates are evaluated over one cycle period of bubbles defined as the time difference between two consecutive bubble departures. Based on experimental observations, one cycle period is taken to comprise of bubble dwelling (or waiting) phase and bubble growth phase [11, 15, 16]. Following Podowski model, the dwelling and growth times are calculated for the present model to estimate the bubble departure frequency. In this, bubble dwelling time is given by,

$$\tau_D = \left[(-C_2 + \sqrt{C_2^2 - 4C_1C_3})/2C_1 \right]^2 \tag{14.a}$$

where

$$C_1 = \frac{2q''_w}{\Re\pi} \tag{14.b}$$

$$C_2 = (\frac{k_w T_w}{\sqrt{\lambda_w}} + \frac{k_l T_{bulk}}{\sqrt{\lambda_l}})/(\frac{k_w}{\sqrt{\lambda_w}} + \frac{k_l}{\sqrt{\lambda_l}}) - T_{sat} - \frac{q''_w r_c}{\Re\sqrt{\pi\lambda_l}} - \frac{2\sigma}{r_c}\frac{T_{sat}(\alpha_v/\rho_v - \alpha_l/\rho_l)}{h_{lv}} \tag{14.c}$$

$$C_3 = -\left[(\frac{k_w T_w}{\sqrt{\lambda_w}} + \frac{k_l T_{bulk}}{\sqrt{\lambda_l}})/(\frac{k_w}{\sqrt{\lambda_w}} + \frac{k_l}{\sqrt{\lambda_l}}) - T_b \right] \frac{r_c}{\sqrt{\pi\lambda_l}} \tag{14.d}$$

and

$$\Re = \frac{k_w}{\sqrt{\pi\lambda_w}} + \frac{k_l}{\sqrt{\pi\lambda_l}} \tag{15}$$

while the bubble growth time is given by,

$$\tau_G = \left[(-A_2 + \sqrt{A_2^2 - 4A_1A_3})/2A_1 \right]^2 \tag{16.a}$$

where

$$A_1 = \frac{q''_w}{k_w} \tag{16.b}$$

$$A_2 = 2[(\frac{k_w T_w}{\sqrt{\lambda_w}} + \frac{k_l T_b}{\sqrt{\lambda_l}})/(\frac{k_w}{\sqrt{\lambda_w}} + \frac{k_l}{\sqrt{\lambda_l}}) + \frac{2q''_w\sqrt{\tau_D}}{\Re\pi} - T_{sat}]/\sqrt{\pi\lambda_w} \tag{16.c}$$

$$A_3 = \frac{D_w\rho_g h_{lv}}{2k_w} \tag{16.d}$$

Explained formulations for dwelling and growth times are derived by coupling transient heat transfer solutions for the heated wall and for the liquid filling the space vacated by departing bubbles. This approach is later discussed and compared against Cole model, which accounts solely for buoyancy force, in detachment process.

The bubble departure frequency is then obtained as,

$$f_{Departure} = \frac{1}{\tau_{B-Cycle}} = \frac{1}{\tau_D + \tau_G} \tag{17}$$

Hence, quenching and evaporation heat fluxes are computed from,

$$\dot{q}_Q = \frac{2k_l}{\sqrt{\pi \lambda_l \tau_{B-Cycle}}}(T_w - T_l) \tag{18.a}$$

$$\dot{q}_E = \frac{\pi D_w^3}{8} N_w \rho_v h_{lg} \tag{18.b}$$

The formulation explained through Eqs. (8–18.a, 18.b) outlines the scheme used for heat partitioning during flow boiling. The section below describes the processes at the liquid-vapour interface.

(d) Interfacial exchange properties

Eulerian framework is fundamentally based on analysis with designated continuous/dispersed phases within a flow regime. Almost all published Eulerian boiling models [18–21] have been developed with the assumption of bubbly flow to comply with low void fraction for which liquid is designated as continuous phase and vapour as dispersed phase. This assumption breaks down for large void fractions such as flow boiling in the mist regime, wherein liquid (droplets) becomes the dispersed particle in continuous vapour phase. Therefore, all published flow boiling models lack the ability to capture flow characteristics over the entire boiling regime and are limited in applicability. This drawback is effectively overcome in the current model with a smoothing function for interfacial parameters defined by,

$$\Phi = (1 - f(\alpha))\Phi_{Bubbly} + f(\alpha)\Phi_{mist} \tag{19}$$

where Φ is any exchange parameter dependant on local cell-base volume fraction and flow regime.

In applying Eq. (19), bubbly regime is defined as vapour volume fraction $\alpha_v \leq 0.4$ wherein vapour bubble is treated as dispersed phase in continuous liquid phase. For intermediate volume fraction of $0.4 < \alpha_v \leq 0.8$, liquid is taken to be primary phase with vapour as secondary phase. In here, drag force coefficient is modified according to churn turbulent regime formulation. If vapour volume fraction $\alpha_v > 0.8$, mist regime is assigned with liquid droplets as dispersed phase within continuous vapour phase [25, 33]. Based on such a framework, interfacial exchange values and applied closure are briefly explained as follows:

(e) Bubble and droplet diameter

Following a formulation by Unal [34], the bubble diameter is expressed as a function of local subcooling temperature as,

$$D_b = \begin{cases} 0.0015 & \Delta T_{Sub} < 0 \\ 0.00015 - 0.0001 \Delta T_{Sat} & 0 < \Delta T_{Sub} < 13.5 \\ 0.00015 & \Delta T_{Sub} > 13.5 \end{cases} \tag{20}$$

However, in the mist flow regime, where liquid droplets constitute dispersed phase, the droplet diameter is estimated using the correlation suggested by Kotaoka et al. [35] as,

$$D_d = 0.028 \frac{\sigma}{\rho_v V_v^2} \mathrm{Re}_l^{-1/6} \mathrm{Re}_v^{2/3} \left(\frac{\rho_v}{\rho_l} \right)^{-1/3} \left(\frac{\mu_v}{\mu_l} \right)^{2/3} \tag{21}$$

where Re_l and Re_v are liquid and vapour Reynolds number, respectively. These correlations have been proven to be successful in many published boiling investigations.

(f) Momentum exchange coefficient and drag force

As a key parameter in Eulerian approach, it is required to define momentum exchange coefficient (K_{pq}) that determines slip velocity between phases, which in turn significantly influence other interfacial parameters such as mass transfer and void fraction. This exchange coefficient in Eulerian scheme is defined as,

$$K_{pq} = -K_{qp} = \frac{\rho f_D}{6 \tau_p} D_p A_i \tag{22}$$

where, subscript "p" denotes dispersed phase which could be bubble or droplet depending on the flow regime. A_i is the interfacial area, and f_D is drag function defined according to drag coefficient. In the drag function, the particulate relaxation time τ_p is given by,

$$\tau_p = \frac{\rho_p D_p^2}{18 \mu_q} \tag{23}$$

Ishii and Zuber [36] have provided a comprehensive drag coefficient and slip velocity functions for liquid-gas flows. These authors provided different correlations to estimate drag function for a wide range of flow regimes including bubbly, mist and churn turbulent regimes, similar to flow regimes adopted for the current model. Both bubbly and mist flow regions are assumed to have continuous-dispersed drag coefficient where viscosity ratio of phases would decide on the appropriate correlation. From Ishii-Zuber, drag coefficient is expressed for the three regimes as,

$$C_D = \begin{cases} \frac{4}{6} D_b \sqrt{\frac{g\Delta\rho}{\sigma}} (1 - \alpha_v)^{-0.5} & bubbly \quad regime \\ \frac{8}{3} (1 - \alpha_v)^2 & Churn \quad turbulent \quad regime \\ \frac{4}{6} D_d \sqrt{\frac{g\Delta\rho}{\sigma}} (1 - \alpha_l)^{-1.5} & mist \quad regime \end{cases} \tag{24}$$

This use of applicable drag coefficient for each flow regime is one of the key improvements in the current model compared to previously published work, where such models indiscriminately assigned a single drag coefficient and assumed continuous liquid phase and dispersed vapour bubbles over the entire computational domain. The drag function in Eq. (22) is then computed with the drag coefficient applicable for a particular regime from Eq. (23) using,

$$f_d = \frac{C_D \mathrm{Re}_{lv}}{24} \tag{25}$$

Relative Reynolds number $\mathrm{Re}_{lv} = \frac{\rho |v_l - v_v| D_p}{\mu}$ in Eq. (25) is computed by using continuous phase density, viscosity and dispersed phase particle sizes.

(g) Lift force

Correlation developed by Morega et al. [37] has been examined for various flow conditions and geometries. It is reported to have reasonable consistency for estimating interfacial lift force in nucleate boiling regime. This, bubbly flow, is formulated in the current model as,

$$\vec{F}_{lift} = C_{lift} \rho_l \alpha_v (\vec{v}_v - \vec{v}_l) \times (\nabla \times \vec{v}_l) \tag{26.a}$$

and

$$C_{lift} = \begin{cases} 0.0767 & \varphi \le 6 \times 10^3 \\ \left[0.12 - 0.2 \exp(-\frac{\varphi}{36000})\right] \exp(\varphi / 3 \times 10^7) & 6 \times 10^3 \le \varphi \le 1.9 \times 10^5 \\ 0.002 & \varphi \ge 1.9 \times 10^5 \end{cases} \tag{26.b}$$

where $\varphi = \mathrm{Re}_B \mathrm{Re}_\nabla$. For Eq. (26.a, 26.b), the bubble Reynolds number and bubble shear Reynolds numbers are respectively defined as $\mathrm{Re}_B = D_b \rho_l |v_l - v_v| / \mu_l$ and $\mathrm{Re}_\nabla = D_b^2 \rho_l |\nabla \times \vec{v}_l| / \mu_l$.

In the mist flow, however, lift force will be applied from continuous vapour phase to dispersed droplets and lift coefficient (C_{lift}) will be obtained using droplet Reynolds number $(\mathrm{Re}_d = D_d \rho_v |v_l - v_v| / \mu_v)$ and droplet shear Reynolds number $(\mathrm{Re}_\nabla = D_d^2 \rho_v |\nabla \times \vec{v}_v| / \mu_v)$.

(h) Turbulence dissipation force

This interfacial parameter influences the process of vapour transportation from wall to core area of flow. Following a formulation by Lopez de Bertodano [38], turbulence dissipation force is evaluated from,

$$\vec{F}_{pq}^{TD} = C_{TD}\rho_l k \nabla \alpha_v \tag{27}$$

where C_{TD} is taken to be 1 although it could be calibrated for a certain case.

(i) Wall lubrication force

In many boiling experiments, the maximum void fraction is reported to occur near the wall separated by a liquid layer. This imparts a force called wall lubrication force on the secondary (vapour) phase. In the current model, this is accounted for by the formulation,

$$\vec{F}_{wl} = C_{wl}\rho_l\alpha_v\left|(\vec{v}_v - \vec{v}_l)_{||}\right|^2 \vec{n}_w \tag{28}$$

where $\left|(\vec{v}_v - \vec{v}_l)_{||}\right|$ is the relative velocity component tangential to wall and $\vec{n}_w$ is the unit normal vector on the wall. Among various correlations suggested, this analysis uses the correlation for C_{wl} by Hosokawa et al. [39],

$$C_{wl} = \max\left(\frac{7}{\mathrm{Re}_B^{1.9}}, 0.021\frac{g\Delta\rho D_B^2}{\sigma}\right) \tag{29}$$

(j) Interfacial heat transfer and mass transfer

When vapour bubbles depart the wall, the heat transfer from vapour bubble to liquid phase is given by,

$$\dot{q}_{lt} = A_i h_{sl}(T_{sat} - T_l) \tag{30}$$

where the heat transfer coefficient h_{sl} is calculated from the correlation suggested by Ranz–Marshal [40] as,

$$h_{sl} = \frac{k_l}{D_b}\left(2 + 0.6\mathrm{Re}^{0.5}\,\mathrm{Pr}^{0.33}\right) \tag{31}$$

Additionally, the vapour interfacial heat transfer (or heat transfer from superheated liquid to vapour) is given by,

$$\dot{q}_{vt} = \frac{\alpha_v\rho_v c_{pv}}{0.05}(T_{sat} - T_v) \tag{32}$$

The bubble/droplet interfacial mass transfer is calculated from,

$$\dot{m}_i = \dot{m}_{lt} + \dot{m}_{lv} = \frac{\dot{q}_{lt} + \dot{q}_{vt}}{h_{fv}} \tag{33}$$

Interfacial formulation in Eqs. (27–32) is represented considering bubbly flow regime where bubbles are dispersed in continuous liquid phase. When applying these equations for mist flow regime, dispersed and continuous phases are interchanged to account for dispersed liquid droplets in vapour phase.

All the equations, accounting for interaction between particle and continuous phase, in the core area of flow, include particle characteristics (e.g. diameter, velocity). Particle characteristics, in these equations, are defined with reference of disperses phase which could be either bubble or droplet. Nevertheless, boiling closures, which accounts for bubble nucleation process and restricted to nucleation sites on the heated surface [Eqs. (11.a, 11.b–18.a, 18.b) and (27–33)], are applied merely with reference to droplet characteristics even in high volume fraction.

3 Model Evaluation and Application

(i) Numerical validation

For ascertaining the validity of the model developed and its hydro-thermal assumptions, the numerical predictions are compared with the experimental work of Pierre and Bankof [23] using the original ANL report [41]. In this, comparisons were carried out for both transverse and axial variations of void fraction throughout the channel to ascertain the predictive accuracy of phase patterns. The agreement of phases and slip velocity were additionally evaluated, and are presented later in the text where the enhancement of drag scheme is discussed.

Pierre and Bankof [23] experiments investigated flow boiling in vertically mounted straight stainless steel rectangular heated ducts with water flowing upward against gravity. The ducts were 1550 mm long with cross sectional dimensions of 44.45 mm × 11.5 mm and 0.43 mm wall thickness. The ducts were heated over a length of 1257 mm from inlet by passing electrical current through the duct walls. Ducts were provided with 13 test windows where γ-attenuation technique was installed. Using 0.8 mm collimator windows, the authors measured the transverse radially averaged void fraction over 11 mm width of chosen cross sections for volumetric flow rates.

The development of vapour phase is first examined in the axial direction, where the experimental void fraction is available as averaged values on successive cross sections located at dimensionless distance ($Z^* = z/L$) from the duct inlet. As such, numerically predicted void fraction is radially averaged at each plane (A to M) and plotted against the corresponding experimental values [27], as illustrated in Fig. 2.

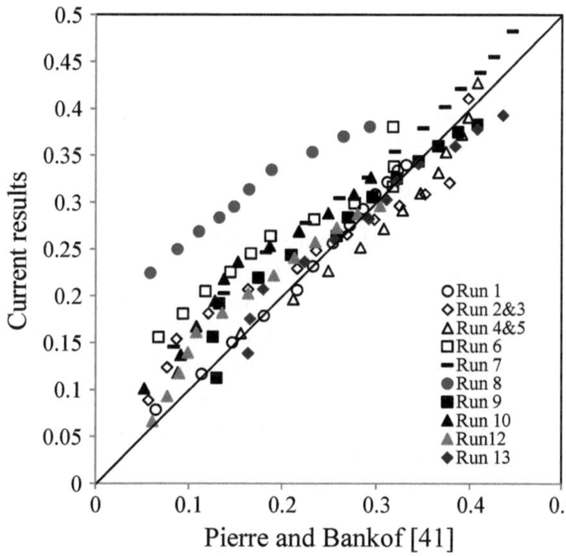

Fig. 2 Development of axial void fraction

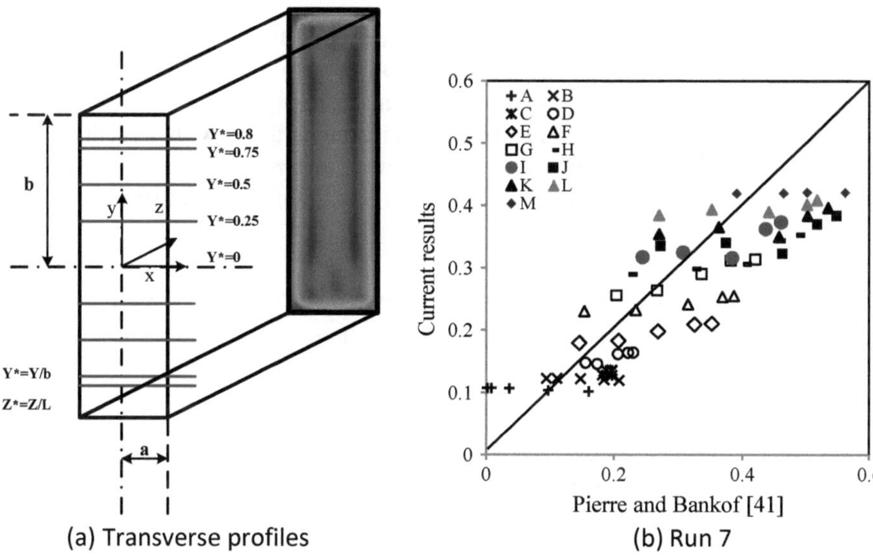

Fig. 3 Comparison of transversely averaged void fraction for different cross sections (A—M)

Considering the narrow scatter band and possible experimental uncertainties, this comparison is regarded as a very good agreement.

As an extended test of validation, transverse measurements of phase patterns are also compared in Fig. 3. Figure 3a shows five lateral profiles ($Y^* = y/b = 0$ to 0.8), where transverse void fraction is experimentally measured for each cross sections at A to M ($Z^* = 0.104$ to 0.831). Figure 3b illustrates the comparison of the numerically predicted transverse profile for these cross sections with the corresponding experimental data [23]. A very good agreement is again clearly evident from this comparison (Table 1).

(ii) Model Setup

Within FLUENT commercial CFD code, full three-dimensional duct geometry is implemented using finite-volume solver with non-equilibrium heat partitioning scheme. Default submodels were modified, as explained in sections describing computational framework, to include User Defined Functions (UDF) to incorporate drag coefficients, bubble departure frequency, quenching corrections and bulk temperature estimation. Saturation properties of water were extracted from the database of National Institute of Standards and Technology [42] for operating pressures, as provided in Table 2. Coupled and modified HRIC (high-resolution interface capturing) schemes were applied, respectively for pressure-velocity coupling and volume fraction discretization method with flow Courant number ranging between 5–10 depending on stability and convergence conditions. In ensuring reliability of convergence, in addition to essential checking of continuity and energy convergence over the entire domain, total mass and heat balance over the computational domain were also monitored.

Dimensions of flow passage geometry are obtained from duct in the Pierre and Bankof experiment (i.e. cross section of 11.5 × 44.45 mm, 1550 mm total length). A fully hexagonal mesh was considered with a sensitivity analysis that was carried out for mixture velocity, volume fraction and temperature with less than 0.5% variation allowance. Accordingly, a minimum mesh size of 0.5 × 0.5 × 0.8 mm was used in core flow area which ends up with maximum cell number of 4566,003.

Table 1 Experimental operating conditions and test results by Pierre and Bankof [23, 41]

	Pressure (atm)	Inlet Velocity (m/s)	Average Heat flux (kW/m^2)	Inlet Subcooling	Saturation Temperature (K)
Run 1	13.61	1.15	71.88	0.28	466.86
Run 2 and 3	20.41	0.77	71.88	1	487.22
Run 4 and 5	27.21	0.77	143.75	2.67	502.37
Run 6	27.21	0.77	143.75	5.28	502.37
Run 7	27.21	1.15	215.63	0.67	502.37
Run 8	27.21	1.15	215.63	4.5	502.37
Run 9	40.82	1.15	287.51	2.33	524.7
Run 10	40.82	1.15	287.51	7	524.7
Run 12	54.42	0.77	143.75	2.72	543.28
Run 13	54.42	1.15	287.51	2.28	543.28

Table 2 Material properties of Liquid and Vapour phases in various pressures

Pressure (atm)	Phase	Density (kg/m3)	Viscosity (Pa.s)	Thermal Conductivity (W/m-K)	C_p (j/kg-K)	Enthalpy (kJ/kg)	Surface Tension (n/m)	T_{sat} (K)
13.6	Liquid	871.9	0.0,001,389	0.6668	4464.6	824	0.038,964	466.86
	Vapour	6.9105	1.55×10^{-5}	0.03,892	2896.4	2787.9		
20.4	Liquid	847.7	0.00,012,505	0.65,411	4576	916.23	0.034,441	487.22
	Vapour	10.375	1.62×10^{-5}	0.042,921	3220.7	2799		
27.2	Liquid	828.17	0.00,011,638	0.642	4681.5	986.5	0.03,092	502.37
	Vapour	13.786	1.67×10^{-5}	0.046,197	3511.9	2802.8		
40.8	Liquid	796.58	0.00,010,541	0.61,942	4886.9	1093.3	0.025,492	524.7
	Vapour	20.512	1.76×10^{-5}	0.051,676	4053.8	2800.4		
54.4	Liquid	767.24	9.74×10^{-5}	0.59,573	5125	1185.9	0.021,306	543.28
	Vapour	28.135	1.83×10^{-5}	0.057,154	4661.3	2789.6		

For achieving y^+ value of less than 5 on the wall (k-ω SST requirement), a mesh inflation coefficient was initially defined on the wall and corrected adaptively for every 100 iterations depending on the flow conditions during computation process. In adopting mesh size required for y^+ limit on the wall, cell aspect ratios were checked to avoid any large magnitudes. The maximum aspect ratio over all cases was recorded to be 4.38. For executing this coupled Eulerian multi-phase model with high mesh refinement and submodels calculating heat partitions, parallel computing having 16 computational cores was deployed.

Boundary conditions are set according to experimental assumptions and common numerical restrictions. Fully developed profiles for velocity and turbulence characteristics, for liquid phase are calculated through sufficient straight length (with the given cross section), and applied on the inlet boundary where vapour fraction is zero and liquid temperature is set with reference of subcooling temperatures. Outlet boundary is adjusted as pressure outlet with zero gauge back pressure and all the external walls are considered as adiabatic which forces all the generated heat to flow through solid-fluid interface. Heat generation rate (heat flux) is distributed in the solid zone with two forms of uniform or non-uniform as it is explained in the validation section to match with experimental assumptions.

(iii) Heat partitioning scheme

In this fully nucleate boiling regime, the evaporation and quenching heat flux components do remain the dominating terms for most areas of the channel. On the other hand at sharp corners of the channel, the vapour heat flux begins to become comparable with the evaporative and quenching terms, indicating the relative importance of vapour diffusion. This effect is much more prominent when the wall heating is increased and the flow boiling approaches dry-out conditions. Therefore this fourth (vapour heat flux) heat partitioning component will be essential for the model to be comprehensive.

(iv) Momentum exchange and Drag scheme

In the Eulerian scheme, the momentum exchange term is the strongest interfacial parameter coupling momentum conservation equations of phases. As indicated by Eqs. (22–25), this interfacial parameter is heavily dependent on the drag force along with particulate relaxation time and bubble diameter. Therefore, the selection of appropriate drag coefficient in accordance with the boiling flow regime is crucial for precise modelling of convective boiling process. In addition, the slip velocity between phases is also a parameter affecting phase momentum exchange while being dependent on the overall mixture velocity and the liquid evaporation rate.

As mentioned earlier, almost all published boiling models assume bubbly flow and incorporates a single drag coefficient to estimate drag from continuous liquid phase on bubbles as dispersed phase. This approach breaks down as the void fraction is increased as the multi-phase regime no longer remains purely bubbly flow. This modelling weakness is overcome in the current work by modifying the drag coefficient depending on the boiling flow regime through Eq. (24). This is demonstrated in Fig. 4 using the test data from Pierre and Bankof [41].

Pierre and Bankof [41] measured the radially averaged velocity of each phase ($\bar{v}_v = \bar{\alpha}_v G_v / A_c$) and calculated the corresponding mixture velocity $\bar{v}_m = \bar{\alpha}_v . \bar{v}_v + (1 - \bar{\alpha}_v) . \bar{v}_l$ to correlate slip velocity for the range of operational conditions given in Table 1. For these identical test conditions, predicted vapour (gas) and mixture velocities were obtained from the current model with the modified drag scheme, which accounts for bubbly ($\alpha_v < 0.4$) or mist ($\alpha_v > 0.8$) or churn turbulent regimes. To further ascertain the effectiveness of the proposed drag scheme, (gas) and mixture velocities were also computed using only the traditional approach of bubbly flow scheme. As a plot of gas velocity against mixture velocity, Fig. 4a compares the predicted velocities using the pure bubbly flow assumption (red symbols) with Pierre and Bankof [23, 41] results (black symbols) given in Table 1. Similarly, Fig. 4b provides a comparison between predicted results using the proposed new drag scheme (blue symbols) and Pierre and Bankof [23, 41] results.

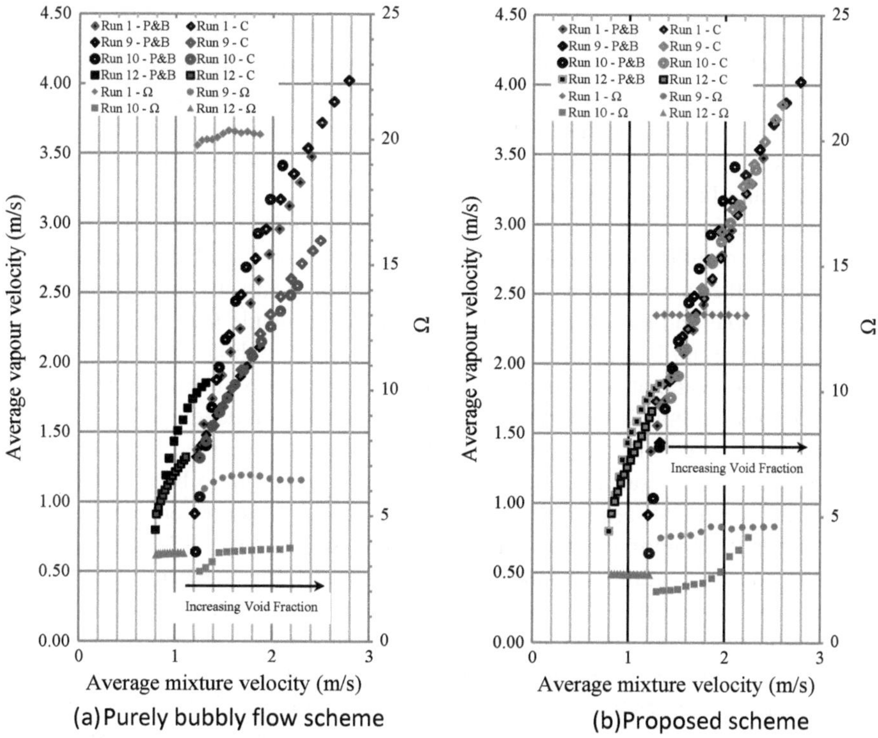

Fig. 4 Comparison of gas velocity and dimensionless wall temperature for bubbly and proposed drag schemes—P&B (Pierre and Bankof), C (Current work) (*Black symbols* represent experimental values from Pierre and Bankof [23, 41])

Figure 4a clearly shows that the predicted results using the bubbly flow scheme deviate significantly from the experimental data with increasing void fraction while continually underestimating the values. To the contrary as illustrated in Fig. 4b, the current scheme highly improves data matching in values and trends with the experimental data consistently over the entire test range. This is a clear indication of the effectiveness in using the proposed multi-tiered drag scheme for accurate representation of convective boiling within all regimes from bubbly to mist flows.

Phase velocity is envisaged to have direct influence on the heat transfer rates within liquid and vapour phases and indirectly affecting the evaporative heat flux. To demonstrate this thermal dependency, a dimensionless temperature term Ω is defined as $\Omega = \frac{T_w - T_{in}}{T_{Sat} - T_{in}}$ and included in Fig. 4 (green symbols) for both drag schemes. Comparison indicates up to 54% difference in Ω (representing up to 7° difference in wall superheat) between schemes and a higher dependency on scheme for large void fractions. Accordingly, the improvement of drag scheme could be interpreted as equally being significant for the accuracy of liquid and vapour convective heat fluxes in the proposed partitioning approach.

(v) Bubble departure frequency and quenching correction

Process of vapour bubble generation is essential for a time-averaged convective boiling model in estimating wall heat flux components due to quenching and evaporation, expressed by Eq. (18.a, 18.b). Between two successive bubble departures at a nucleation site, bubble formation is divided into two stages known as periods of bubble dwelling (waiting) and growth of which the total duration determines the bubble departure frequency.

In the dwelling phase, a bubble resides within a cavity and develops in size to reach cavity mouth. Heat conduction from the wall to liquid during this phase accounts for the quenching heat flux defined by Eq. (18.a). In subsequent growth period, a bubble undergoes much rapid development outside the cavity attached to the wall surface until its detachment. During this phase, bubble removes heat from the surface by evaporation defined as evaporative heat flux in Eq. (18.b).

In determining bubble departure frequency, a correlation developed by Cole [13] is traditionally used in boiling models. This is originally derived for pool boiling and estimates the overall time period between two consecutive bubble departures, hence the detachment frequency. However, this model does not separate individual time periods for dwelling and growth phases. In addition, Cole [13] formulated this departure model treating buoyancy to be the key driving mechanism for bubble detachment. Nonetheless in convective boiling, both buoyancy and inertial (shear-induced) forces are essentially involved in bubble departure mechanism. Consequently, the Cole model is thought to be inadequate for convective boiling.

To evaluate the relative influence from inertial force and surface tension, the current model makes use of Weber number (We) (inertial to surface tension ratio), where We towards 1 indicates inertia dominance while We toward 0 signifies surface tension control. For this, surface Weber number $\left(We_s = \frac{\rho D_h u_*^2}{\sigma}\right)$ is defined with respect to frictional velocity of liquid phase on the heated surface (u_*) and duct hydraulic diameter (D_h). This is then used for appraising the applicability of available bubble detachment models.

Podowski [16] and Basu [15] models accounted for buoyancy and inertial (shear-induced) forces and are potentially considered for convective boiling. However, Basu [15] formulation requires prior knowledge of contact angle values (on the surface) whilst Podowski does not depend on such additional uncertainties. In the current simulation, Podowski model is deployed, supported by its additional analytical advantages.

Podowski model quantifies bubble dwelling and growth times separately whereby it warrants accurate determination of quenching heat flux at the heated wall. Recalling Eq. (18.a), the quenching heat flux is computed by integrating over the bubble cycle period $\tau_{B-Cycle}$. With Cole model, $\tau_{B-Cycle}$ is compelled to assume as the time duration between consecutive bubble detachments. This is inaccurate since wall the quenching process only occurs during the dwelling stage followed by evaporation over the bubble cycle. To account for this in the current scheme, the total duration of bubble cycle $\tau_{B-Cycle}$ is split into dwelling (τ_D) and growth (τ_G) times. Then, a time correction factor defined by $\sqrt{\frac{\tau_D}{\tau_D+\tau_G}}$ is applied for quenching heat flux computed from $\tau_{B-Cycle}$. Similarly, the evaporative heat flux is modified accordingly with $\sqrt{\frac{\tau_G}{\tau_D+\tau_G}}$.

Figure 5a shows, the contours of surface Weber number at the fluid-wall interface of the heated duct wherein red regions having $We \approx 1$ represent the inertia dominated high convective flow while blue areas with $We \approx 0$ where surface tension controls flow dynamics. This large variation of We affirms the fact that Podowski [16] approach for bubble departure is more applicable in convective

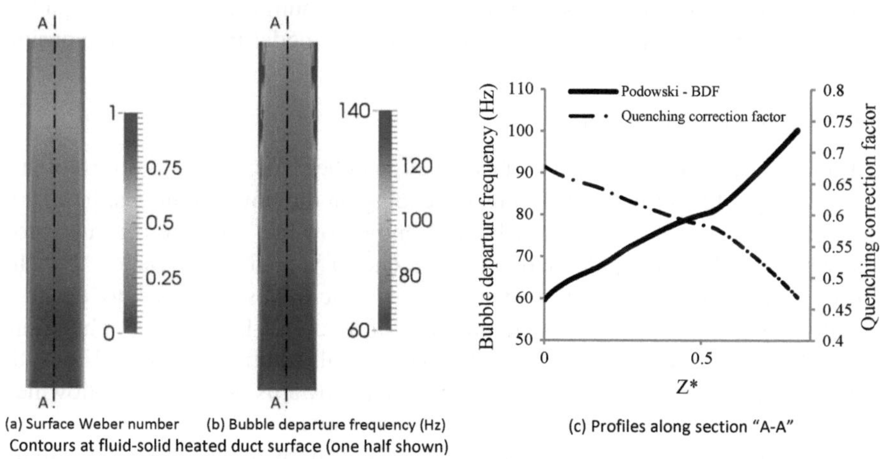

Fig. 5 Surface Weber number, Bubble departure frequency and quenching correction factor using Podowski scheme with Run 10 (Table 3)

boiling than the Cole [13] model, where that latter would only be meaningful in near-stagnant regions of the flow.

Figure 5b illustrates the distribution of bubble departure frequency based on Podowski approach over the heated duct surface. It clearly shows a wide variation to the bubble departure frequency ranging from low values (blue) to high values (red) in convective flow. Additionally, Fig. 5c depicts the local values of correction factor for quenching heat flux along the section A-A. It shows the trend of reducing dwelling time (or such correction) whilst increasing departure frequency due to inertial effects of convective flow. On the contrary, Cole bubble departure model based on pool boiling would predict indiscriminately a single constant frequency of 145.98 Hz throughout the heated section. Accordingly as a major convective boiling model improvement, Podowski scheme is incorporated into the current boiling simulation, accounting for both inertial and buoyancy forces with correction to quenching and evaporative time scales.

Bubble departure frequency invariably affects the void fraction and bubble nucleation rate. Figure 6 compares the predicted void fraction results obtained by using Cole and Podowski models against the Pierre and Bankof [41] experimental data. The figure indicates that the Cole scheme consistently overpredicts the void fraction because of the overestimation in the bubble departure frequency by this scheme, whereas the Podowski approach is much more compatible with the void fraction date of Pierre and Bankof [41]. This signifies the need to assign appropriate of bubble frequency model in convective boiling simulation analysis.

Quenching heat flux along section "A-A" in Fig. 5, are shown on the secondary axis of Fig. 6. It is evident that the quenching heat flux is underestimated near duct inlet compared to Podowski model owing to the overprediction of evaporative heat flux through bubble mechanism captured in Cole scheme. This observation further establishes effectiveness in Podowski model for convective boiling.

For reference and comparison, Table 3 provides a summary of bubble departure frequency and quenching correction factor for each model in terms of dimensionless temperature.

(vi) Wall lubrication model

Bubbles growing at a heated surface undergo either a process of sliding and then lift-off or rapid lift-off without sliding, depending on the flow conditions. In both of these cases following detachment, bubbles move towards the core (or bulk) flow area where bubble may potentially collapse due to vapour condensation. For this reason in convective boiling, maximum void fraction is expected to occur in vicinity of the heated wall with the presence of a contact liquid layer. This liquid layer offers an interfacial shear effect known as Wall Lubrication Force [Eq. (28)] on the vapour phase, whose influence is significant towards predicting the flow field hydrodynamics, void generation and heat transfer rate.

The influence of lubrication force on the phase pattern and location of maximum vapour vicinity is illustrated in Fig. 7a where void fraction contours at a cross section near duct exit are compared for non-lubricated (NL) and lubricated (L) cases.

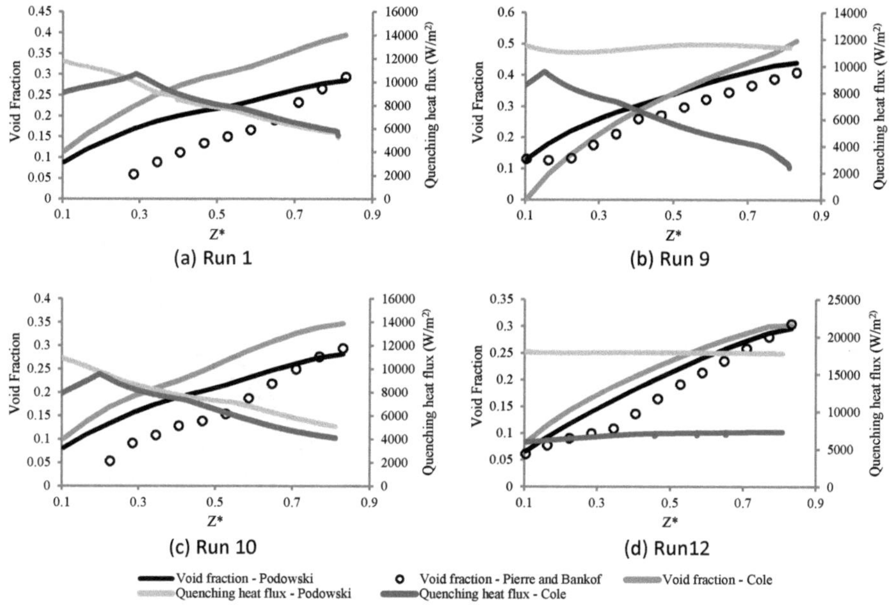

Fig. 6 Variations of void fraction and quenching heat flux along section A-A (Fig. 3)

Table 3 Average bubble departure frequency, quenching correction factor and dimensionless temperature at fluid-solid interface

Run No	Bubble Departure Frequency (Hz)		Quenching Correction Factor		Dimensionless Temperature Ω	
	Cole	Podowski	Cole	Podowski	Cole	Podowski
1	145.46	145.67	–	0.33	9.30	7.58
2 and 3	147.48	62.14	–	0.93	8.34	3.03
4 and 5	146.41	67.91	–	0.91	7.47	2.02
6	143.13	62.71	–	0.88	6.32	1.51
7	146.41	95.45	–	0.91	13.11	5.28
8	147.08	107.86	–	0.25	7.86	2.39
9	145.73	63.46	–	0.44	9.65	2.54
10	145.98	83.35	–	0.32	2.20	1.96
12	144.92	23.96	–	1.00	1.50	1.94
13	145.05	46.37	–	0.93	12.05	2.46

Figure 7b compares radially averaged void fraction and averaged dimensionless temperature over duct cross section with (L) and without (NL) wall lubrication model for Runs 12 and 13. It is observed that the wall temperature is overpredicted in the absence of wall lubrication forces with reduced heat transfer rates through vapour at the wall. This effect is more pronounced for flow with higher void fraction.

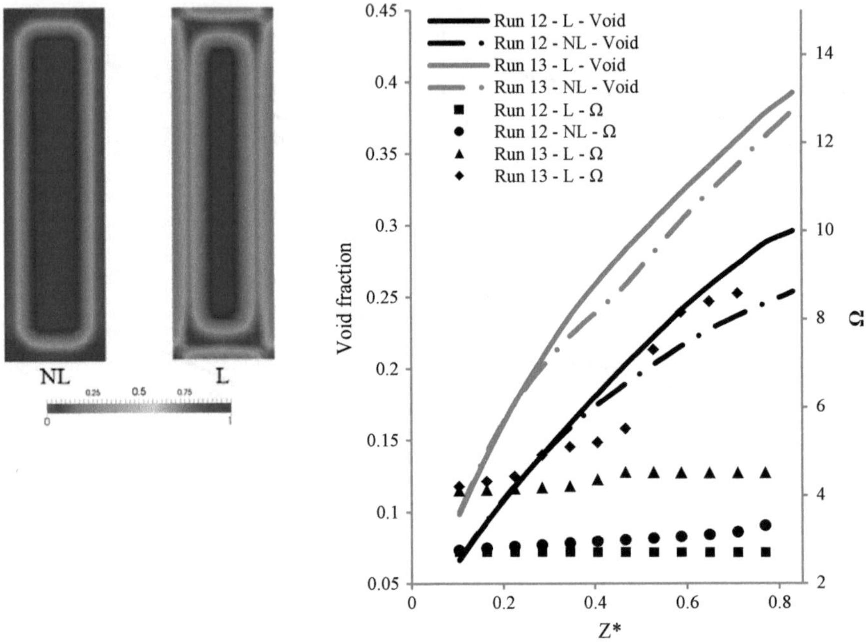

(a) Void fraction at Z*=0.77, Run 13

(b) Axial evaluation of void fraction and normalised temperature

Fig. 7 Influence of lubrication force of phase field and domain parameters

For improved thermal modelling, the current model includes the wall lubrication effect where by wall heat removal occurs as a combination of quenching and evaporation processes. Without this consideration of lubrication forces, the vapour diffusion governs the heat flux component and predicts premature wall liquid layer dry-out, which is unrealistic as the case with existing boiling models.

4 Appraisal of Thermal Benefits and Potential for Energy Saving

Above modelling strategies are proposed to capture the unique characteristics of convective boiling thereby imparting a better physical representation of boiling heat and mass exchange phenomena in the numerical simulation practice. In these, a key contributor is the use of Podowski model [16] in conjunction with a dwelling and growth time modification, instead of the traditional Cole closure [13] approach. The former model is more compliant with thermal and fluid mechanics of convective boiling, for it is built upon wall heat partitioning technique, and accounts for inertia and buoyancy effects imposed by flow fields in the vicinity of bubble interface.

On the other hand, the latter model carries inherent mechanistic weakness as it is specifically developed for pool boiling with no implications from flow fields. Moreover, Cole model does not differentiate the significance of dwelling and growth phases of bubble life in evaluating wall heat flux-it computes both quenching and evaporative heat fluxes by considering the entire bubble life or departure time, hence grossly over-estimating these heat components. Such predictive inaccuracies are translated into and reflected as non-optimal system design, energy and resource wastage, and poor overall plant thermal efficiency. Also, inadequately designed systems may experience operating instabilities and load fluctuations.

The proposed modelling strategy uses the Podowski model and integrates appropriately weighted dwelling time (quenching) and growth time (evaporative), to refine quenching and evaporative wall heat flow components. Furthermore, it incorporates wall lubrication and turbulence models, which are not considered in the current boiling simulation methodologies. Consequently, these modelling refinements permit precise capturing of convective boiling characteristics beyond the realms of current practice limited to low void fraction boiling and extending the applicability from bubbly flow towards annular flow, where void fraction is much larger. Resulting degree of modelling enhancements and energy-saving potential are clearly evident from Fig. 8, where local boiling wall heat fluxes along the pipe are illustrated and compared for identical simulations using Podowski and Cole models, separately. In this, Run 1 and Run 12 test conditions given in Tables 1 and 2 are used for the comparison, covering both low and high void fraction flow conditions in convective boiling.

In Fig. 8, the results for Cole model clearly show a consistent overestimation of wall heat flux by about 19% over the Podowski model for Run 1 and 35% for Run 2, that arise from the reasons explained above. This is indicative of the

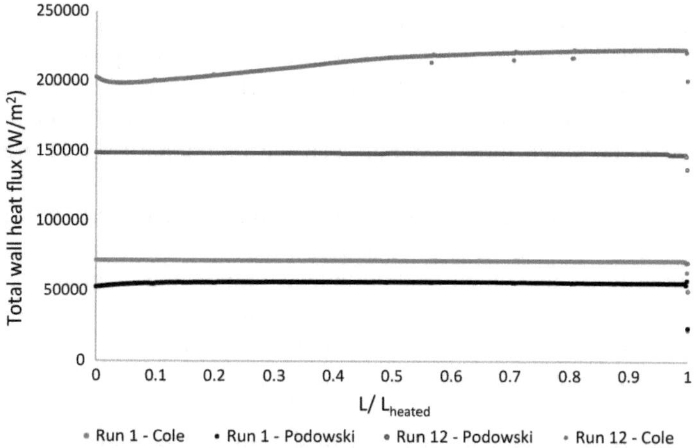

Fig. 8 Local wall heat flux estimated by Cole and Posowski bubble departure frequency models

energy-saving potential and resource utilisation efficiency brought about by the use of proposed enhancement strategies, allowing the application of much stringent parametric and thermal energy estimations leading to improved overall boiling system design.

5 Conclusions

Review of reported literature indicates significant weaknesses in the published CFD simulation models for convective boiling. Contributing to model enhancement, this book chapter presents a numerical study introducing modifications to bubble dynamics and momentum exchange closures while adopting enhanced modelling elements that were not previously considered or included. The suggested framework could be referenced as a comprehensive modelling approach for internal convective boiling within pipe systems.

These contributions are summarised below:

- Based on Eulerian framework, the study developed a modified methodology for capturing momentum transfer between phases undergoing vapour–liquid phase change. In this, framework of primary and secondary phases was modified and a tiered approach was included for the local drag coefficient to improve the applicability of the analysis from bubbly regime (low void fraction) to mist regime (high void fraction). These improvements were very tangibly noticed in reducing results over prediction, particularly in void fraction, that consistently prevalent in previous convective boiling models. Consequently, the study identified that the slip velocity between phases play, although indirectly, a pivotal role in interfacial properties exchange and convective boiling heat and mass transfer.
- Current model incorporates Podowski [16] approach for accurate determination of bubble departure frequency by accounting for dwelling and growth periods. This methodology significantly improved estimations of quenching and evaporative heat fluxes associated with bubble growth. The predicted results showed improved agreement with experimental data over all boiling regimes and were observed to be much more accurate than the traditional Cole [11] correlation extensively used in previous studies.
- The study identifies critical modelling elements that were not previously included in numerical analysis of convective boiling. These include wall lubrication force and additional turbulence modelling compliances both of which were shown to improve validation and enhanced overall simulation approach for convective boiling with high void fraction.
- The proposed simulation enhancement strategies are shown to be very proficient towards accurate estimation of convective boiling design parameters leading to much improved overall thermal system design and resource utilisation, with significant technical contributions towards a sustainable energy future.

References

1. Collier, J. G. (1972). Convective Boiling and Condensation. McGraw-Hill book Company (UK). Second edition.
2. Lockhart, R. W., & Martinelli, R. C. (1949). Proposed correlation of data for isothermal two-phase two-component flow in pipes. Chem eng prog. 45.
3. Martinelli, R. C., & Nelson, D. B. (1948). Prediction of pressure drop during forced circulation boiling of water. *Tran Journal ASME, 70,* 695.
4. Zuber, N., & Findaly, J. (1965). Average Volumetric Concentration in two-phase flow systems. *Trans ASME Journal of Heat Transfer, 87,* 453.
5. Bangkof, S. G. (1960). A variable density single-fluid model of two-phase model with particular reference to water-vapour flow, Trans ASME. J. Heat Trans. Series C.82.
6. Marchaterre, J. F., & Hoglund, B. W. (1962). Correlation for two-phase flow. *Nucleonic, 142.*
7. Griffith, P. (1963). The slug-annular flow regime transition at elevated pressure. ANL-6796.
8. Bergles, A. E., & Rohesnow, W. M. (1964). The determination of forced-convection surface-boiling heat transfer. *ASME. Journal Heat Transfer, 86,* 365–372.
9. Kandlikar, S. G. (1990). A general correlation for saturated two-phase flow boiling heat transfer inside horizontal and vertical tubes. *ASME. Journal Heat Transfer, 112,* 219–228.
10. Boyd, R. D., & Meng, X. (1995). Boiling curve correlation for subcooled flow boiling. *International Journal Heat and Mass Transfer, 38–4,* 758–760.
11. Cole, R. (1960). A photographic study of pool boiling in the region of the critical heat flux. *AIChE Journal, 6–4,* 533–538.
12. Thorncroft, G. E., Klausnera, J. F., & Mei, R. (1998). An experimental investigation of bubble growth and detachment in vertical up-flow and down-flow boiling. *International Journal Heat and Mass Transfer, 41,* 3854–3871.
13. Situ, R., Mi, Y., Ishii, M., & Mori, M. (2004). Photographic study of bubble behaviours in forced convection subcooled boiling. *International Journal Heat and Mass Transfer, 47,* 3659–3667.
14. Situ, R., Hibiki, T., Ishii, M., & Mori, M. (2005). Bubble lift-off size in forced convective subcooled boiling flow. *International Journal Heat and Mass Transfer, 48,* 5536–5548.
15. Basu, N., Warrier, G. R., & Dhir, V. K. (2005). Wall heat flux partitioning during subcooled flow foiling: part 1—model development. *ASME Joural Heat Transfer, 127,* 131–140.
16. Podowski, R. M., Drew, D. A., Lahey, R. T., Podowski, J. R., & M.Z. (1997). A mechanistic model of ebullition cycle in forced convection sub-cooled boiling. In: *Proceeding of the 8th International topical meeting on nuclear reactor thermal hydraulic.* Kyoto, Japan. vol. 3.
17. Situ, R., Ishii, M., Hibiki, T., Tu, J. Y., Yeoh, G. H., & Mori, M. (2008). Bubble departure frequency in forced convective subcooled boiling flow. *International Journal Heat and Mass Transfer, 51,* 6268–6282.
18. Kurul, N., & Podowski, M. Z. (1990). Multidimensional effects in forced convection Subcooled Boiling. In: *9th International Heat Trans Conference.* Jerusalem. Israel, Proceedings.
19. Koncar, B., Kljenak, I., & Mavko, B. (2004). Modelling of local two-phase flow parameters in upward subcooled flow boiling at low pressure. *International Journal Heat and Mass Transfer, 47,* 1499–1513.
20. Krepper, E., Koncar, B., & Egorov, Y. (2007). CFD modelling of subcooled boiling-Concept, validation and application to fuel assembly design. *Nuclear Engineering and Design, 237,* 716–731.
21. Koncar, B., & Krepper, E. (2008). CFD simulation of convective flow boiling of refrigerant in a vertical annulus. *Nuclear Engineering and Design, 238,* 693–706.
22. Abishek, S., Narayanaswamy, R., & Narayanan, V. (2013). Effect of heater size and Reynolds number on the partitioning of surface heat flux in subcooled jet impingement boiling. *International Journal Heat and Mass Transfer, 59,* 247–261.

23. Pierre, C. C. S. T., & Bankoff, S. G. (1967). Vapour volume profiles in developing two-phase flow. *International Journal Heat and Mass Transfer, 10,* 237–249.
24. Menter, F. R. (2009). Review of the SST turbulence model experience from an industrial perspective, Int. *Journal Computational Fluid Dynamics, 23–4,* 305–316.
25. Inc, A. N. S. Y. S. (2012). ANSYS FLUENT theory guide. *Release, 14,* 5.
26. Troshko, A. A., & Hassan, Y. A. (2001). A two-equation turbulence model of turbulent bubbly flows. *International Journal Multi-phase Flow, 27,* 1965–2000.
27. Sato, Y., & Sadatomi, M. (1981). Momentum and heat transfer in two-phase bubble flow—I. Theory. *International Journal Multiphase Flow, 7–2,* 167–177.
28. Sato, Y., & Sekoguch, K. (1975). Liquid velocity distribution in two-phase bubble flow. *International Journal Multiphase Flow, 2–1,* 79–95.
29. Lavieville, J., Quemerais, E., Mimouni, S., Boucker, M., & Mechitoua, N. (2005). NEPTUNE CFD V1.0 Theory Manual, EDF.
30. Del Valle M, V. H., & Kenning, D. B. R. (1985). Subcooled flow boiling at high heat flux. International Journal Heat and Mass Transfer *28,* 1907–1920.
31. Lemmert, M., & Chawla, L. M. (1977). *Influence of flow velocity on surface boiling heat transfer coefficient in Heat Transfer in boiling*. NY, USA: Academic Press and Hemisphere.
32. Tolubinski, V. I., & Kostanchuk, D. M. (1970). Vapour bubbles growth rate and heat transfer intensity at subcooled water boiling. In: *4th International Heat Transfer Conference*, Paris, France. Proceeding.
33. Apte, S. V., Gorokhovski, M., & Moin, P. (2003). LES of atomizing spray with stochastic modelling of secondary breakup. *International Journal Multiphase Flow, 29,* 1503–1522.
34. Unal, H. C. (1979). Maximum Bubble diameter, maximum bubble growth time and bubble growth rate during subcooled nucleate flow boiling of water up to 17.7MN/m2. *International Journal Heat and Mass Transfer, 19,* 643–649.
35. Kataoka, I., Ishii, M., & Mishima, K. (1983). Generation and size distribution of droplet in annular two-phase flow. *ASME Journal Fluid Engineering, 105,* 230–238.
36. Ishii, M., & Zuber, N. (1979). Drag coefficient and relative velocity in bubbly, droplet or particulate flows. *AlChE Journal, 25–5,* 843–855.
37. Moraga, F. J., Bonetto, F. J., & Lahey, R. T. (1999). Lateral forces on spheres in turbulent uniform shear flow. *International Journal Multiphase Flow, 25,* 1321–1372.
38. De Bertodano, M. L. (1991). Turbulent bubbly flow in a triangular duct. Ph.D. Thesis, Rensselaer Polytechnic Institute, New York.
39. Hosokawa, S., Tomiyama, A., Misaki, S., & Hamada, T. (2002). Lateral migration of single bubbles due to the presence of wall. In: *ASME Joint U.S.-European Fluids Engineering Division Conference*, Montreal, Canada.
40. Ranz, W. E., & Marshall, W. R. (1952). Evaporation from drops. *Parts I & II. Chemical Engineering, 48*(141–6), 173–180.
41. Pierre, C. C. S. T. (1965). Frequency-Response analysis of steam voids to sinusoidal power modulation in a thin-walled boiling water coolant channel, Arogan National Labratoary, ANL-7041.
42. Lemmon, E. W., McLinden, M. O., & Friend, D. G., (2013). Thermophysical Properties of Fluid Systems, NIST Chemistry WebBook, NIST Standard Reference Database Number 69, National Institute of Standards and Technology, Gaithersburg MD, 20899. http://webbook.nist.gov.

A Method of Three-Dimensional Thermo-Fluid Simulation of the Receiver of a Standard Parabolic Trough Collector

M. Islam, Suvash C. Saha, M.A. Karim and Prasad K.D.V. Yarlagadda

Abstract A parabolic trough collector (PTC) is the most proven concentrating collector system for indirect steam generation in solar thermal power plants. The receiver of the collector is fabricated enveloping a metal absorber tube using an evacuated glass tube. Depending on the level of evacuation, the glass envelope reduces the convection heat loss from the tube almost to zero. However, sometimes the envelopes are broken, damaged or removed that causes potential convection loss from an open-to-air receiver tube. On the other hand, the solar irradiance profile around the receiver tube is likely to be highly nonuniform. In order to study the heat transfer mechanism of an exposed receiver tube of a standard PTC under the actual optical and environmental conditions, a 3-dimensional Computational Conjugate Heat Transfer (CCHT) model of the receiver tube was developed. The CCHT model was developed applying finite volume technique of computational fluid dynamics integrating with a verified Monte Carlo ray tracing optical model. The CCHT model was verified extensively, and different heat loss models were developed to investigate the heat loss characteristics. The convection heat loss from the outer surface of the receiver tube was observed very high as it was exposed to a high-temperature gradient with its surroundings. Therefore, it is obvious that the receiver tube should be enveloped and evacuated properly. A well-managed and efficiently operated PTC solar energy field could be the best candidate for sustainable energy management for a sustainable future.

Keywords LS2 collector · Parabolic trough collector · Computational fluid dynamics · Conjugate heat transfer modelling · Monte Carlo ray tracing · Fluent

M. Islam (✉) · S.C. Saha (✉) · M.A. Karim · P.K.D.V. Yarlagadda
Faculty of Science and Engineering, School of Chemistry, Physics and Mechanical Engineering, Queensland University of Technology, Brisbane, QLD 4001, Australia
e-mail: murad99me@yahoo.com

S.C. Saha
e-mail: suvash.saha@qut.edu.au

M. Islam
Department of Mechanical Engineering, Chittagong University of Engineering and Technology, Chittagong 4349, Bangladesh

M.M.K. Khan et al. (eds.), *Application of Thermo-fluid Processes in Energy Systems*, Green Energy and Technology, https://doi.org/10.1007/978-981-10-0697-5_9

1 Introduction

Parabolic trough collector (PTC) is a mature and widely used concentrating collector technology in the solar energy field [1, 2]. A PTC, as shown in Fig. 1, consists of a single axis North–South tracking (parallel to YZ plane) parabolic trough (PT) mirror that focuses solar radiation on to a receiver. The receiver, which, ideally, placed axially along the focal line (along X axis) of the mirror, consists of a receiver tube (RT) and an evacuated glass tube (GT) envelop. The receiver is the main part of the trough collector system that is used to convert incident solar radiation to thermal energy.

The spectral energy harnessing process of the collector involves: incidence of solar radiation on mirror aperture, reflection and concentration of the incident energy onto the outer surface of the receiver tube, and absorption of the radiant energy as thermal energy. Most of the radiant energy is conducted to the inner surface of the receiver tube as thermal energy due to temperature gradient developed by inducing forced convection heat transfer phenomenon [3]. Some suitable Heat Transfer Fluid (HTF) is used to induce the phenomenon. This is an example of a coupled heat transfer problem with complex geometry condition; this, and fairly similar concepts have been studied theoretically as well as experimentally by many researchers.

A buoyancy-driven flow and heat transfer in a narrow annular gap between co-axial, horizontal cylinders have been investigated experimentally and computationally by Vafai et al. [4], Hamad and Khan [5] and Dyko et al. [6] on different occasions.

Borjini and his research group numerically studied the effect of radiation on steady two-dimensional (2D) laminar natural convection in a participating medium between two horizontal con-focal elliptical cylinders by using an elliptic-cylinder coordinates system [7], and unsteady natural convection in a two-dimensional

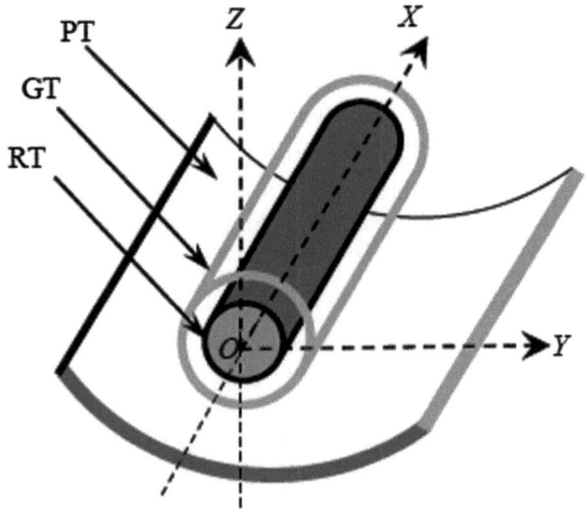

Fig. 1 Isometric view of a typical PTC (In the figure, PT, GT and RT refer the parabolic trough mirror, glass tube envelop and the receiver tube, respectively)

participating medium between two horizontal concentric and vertically eccentric cylinders by using a bi-cylindrical coordinates system [7]. The laminar natural convection in air inside a differentially heated horizontal bare and finned rhombic annulus was studannulus was studied computationallyied computationally by Farinas et al. [8]. Investigations specific to the trough solar collector were also conducted remarkably in recent years.

Dudley et al. [9] investigated the energy performance of the Luz Solar 2 (LS2) PTC performing an extensive experimental procedure. Referring the results for verifications, Forristall [10] developed a detailed one-dimensional and a two-dimensional mathematical models for a PTC to investigate its heat transfer characteristics implementing in an Engineering Equation Solver (EES). An isothermal condition for the glass envelop and a sinusoidal local heat flux distribution for the receiver tube were simulated by Kassem [11], and the free convection heat transfer in the annular space between the receiver tube and the glass envelope was studied. Computational study by Reddy and his research team [12–14] showed that the thermal performance of a PTC with a porous receiver is better than that with a traditional nonporous receiver. Neglecting nonlinearity effect on heat loss, and assuming constant solar radiation, Odeh et al. [15] established a detailed thermal model to calculate the heat loss of trough collector. Later the model was used by Hou et al. [16] to analyse thermal efficiency of a trough system. Effects of the realistic nonuniform heat flux distribution boundary condition on the conjugate heat transfer phenomenon of the collector system have been studied theoretically by different researchers. He et al. [17] simulated the heat transfer process inside the absorber tube and analysed the characteristics under the influence of the nonuniform heat flux distribution around the absorber tube. They coupled a Monte Carlo Ray Tracing (MCRT) optical model with a finite volume Computational Fluid Dynamics (CFD) model to obtain 3D fluid flow fields and temperature distribution for the coupled heat transfer problem in the absorber tube. Tao and He [18] developed a unified computational model for the coupled heat transfer process inside the receiver tube and the annuli space between the receiver tube and the glass envelop of the collector. Islam et al. [19, 20] on different occasions computationally studied the heat transfer mechanism of the receiver tube of a standard PTC.

The foregoing literature review confirms that a plenty of investigations were performed to reveal energy performance and conjugate heat transfer phenomenon of the receiver with evacuated glass envelop, and heat transfer phenomenon inside the annuli space between the receiver and the glass envelop. However, indeed the annuli space is not always evacuated, and many a times the glass envelops are either broken or removed. Again, usually there is no evacuated glass cover around the Concentrating Photovoltaic/Thermal (CPV/T) receiver. Therefore, the receiver is exposed to its surroundings and experiences both free convection and radiation heat losses. This article details a method of 3D Computational Conjugate Heat Transfer (CCHT) model of a bare receiver tube of a standard parabolic trough collector applying Finite Volume (FV) discretization technique of CFD.

The receiver tube of the LS2 PTC from Dudley et al. [9] was modelled and simulated. The irradiance distribution around the receiver tube was calculated applying a verified MCRT optical model [21, 22] and was integrated with the FV model applying a special MCRT-FV coupling technique. Several test conditions of the bare receiver of the LS2 collector were simulated. As the receiver was bare to the environment, special care was taken in calculating the radiation heat loss and the natural convection heat loss from the receiver walls. Moreover, the flow condition at the inlet was considered a fully developed flow. A comprehensive approach was adopted to verify the accuracy of the FV model. Typical results of the simulation are presented. A finite volume based software package, ANSYS Fluent 15.0 [23], was used for this modelling and simulation.

2 Physical Model

The LS2 PTC was used in the Solar Energy-Generating System (SEGS) III–VII 150 MW plants, Kramer Junction, California, USA, which is a proven solar collector for solar thermal electricity generation [9]. The geometric configuration of the collector and the receiver are shown in Figs. 2 and 3, respectively. The mirror aperture is $5 \times 7.8\ m^2$, and the receiver tube is 8 m long with 66 mm inner diameter and 2 mm wall thickness. In order to increase the flow velocity of the HTF, Dudley et al. [9] inserted a 50.8 mm diameter closed-end plug inside the receiver tube. The HTF was Syltherm800 silicone based oil. The LS2 collector module was tested on the AZTRAK rotating platform at Sandia National Laboratory (SNL) by Dudley et al. [9]. Thermal energy performances of the collector under three physical conditions relating to the glass envelope of the receiver: (i) the glass envelope was evacuated, (ii) lost vacuum or air inside the envelope and (iii) removed glass cover or bare receiver was investigated. As has been mentioned earlier, since, the glass cover has direct contact with the Photovoltaic (PV) solar cells of a CPV/T collector; the convection loss, along with the radiation loss, must

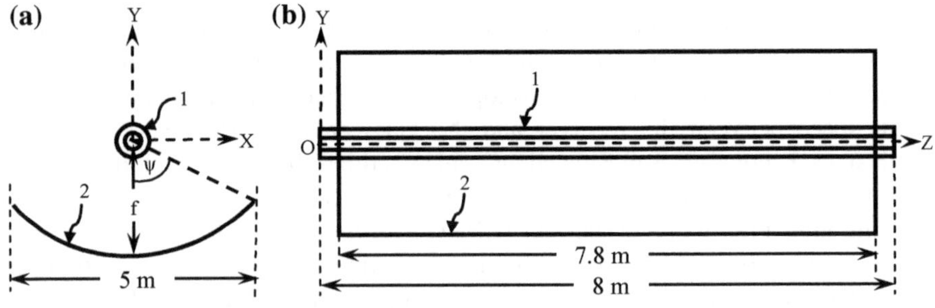

Fig. 2 Schematic of the LS2 collector on Cartesian coordinate system: **a** cross-sectional view on XY plane, and **b** top view on XZ plane (In the figure, the numerals 1 and 2 refer the evacuated receiver and the parabolic mirror, respectively, f is the focal length (=1.84 m), and ψ is the rim angle of the mirror ($\approx$70°))

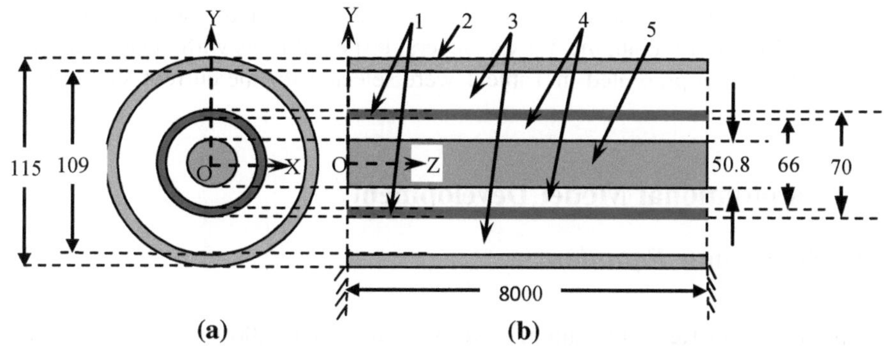

Fig. 3 Geometry of the LS2 collector receiver: **a** Cross-sectional view on XY plane, and **b** longitudinal section of the receiver on YZ plane (In the figure, the numerals 1, 2, 3, 4 and 5 refer to the absorber tube, glass tube, evacuated passage between the glass tube and the absorber tube, heat transfer fluid passage and the flow restriction device, respectively. All dimensions are in mm)

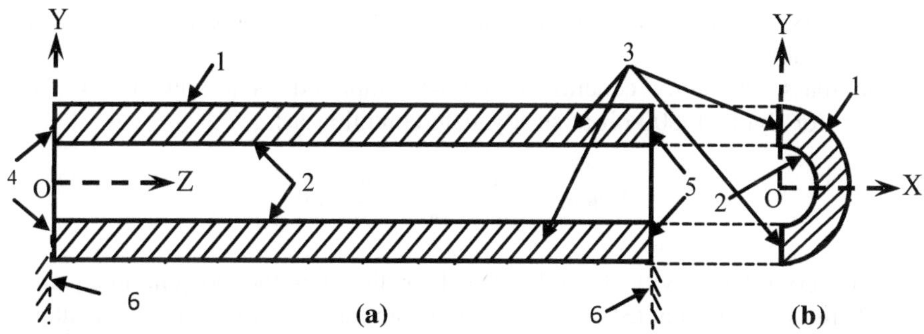

Fig. 4 Heat transfer fluid (HTF) domain of the receiver for the finite volume (FV) model on the OXYZ coordinate system, and the boundary conditions: **a** longitudinal section of the domain on YZ plane, and **b** the angular cross-section on XY plane (In the figure, the numerals 2, 3, 4, 5 and 6 refer to no-slip heat flux wall, no-slip adiabatic inside wall, symmetry, fully developed velocity inlet, fully developed pressure outlet, and adiabatic ends of the receiver, respectively)

take place from the PV surface to the environment. Therefore, to account for this environmental effect on the CPV/T collector receiver, the bare receiver of the LS2 collector was modelled for the current finite volume simulation so that the heat losses could be simulated and verified.

The irradiance distribution along the periphery of the receiver tube of the LS2 collector under ideal conditions, applying the MCRT optical simulation technique was found perfectly symmetric to the plane of geometric symmetry of the collector (YZ plane in Fig. 2). Therefore, the HTF domain of right-hand-side half of YZ plane was modelled as shown in Fig. 4 so as to minimize the computational expense utilizing of the characteristics of physical symmetry. The flow restriction device and the metal tube were not modelled for this conjugate heat transfer analysis as the flow properties were assumed steady state. Therefore, the shape of the computational domain was semi-cylindrical and horizontal. Gravity was acting vertically downward along the Y-axis. Several test conditions with the bare receiver of the collector as presented in Table 1 were selected for the current simulation.

3 Computational Model Development

3.1 Governing Equations

As the minimum Reynolds number (Re) was more than 6000, the flow criteria of the selected test conditions as shown in Table 1 were turbulent considering the critical Reynolds number, Re_c 4000; incompressible; and steady state [9]. The turbulent energy production, κ, and the turbulent energy dissipation, ε, were calculated using the RNG $\kappa-\varepsilon$ transport equations. The governing equations and the transport equations were as given below:

The mass conservation equation or the continuity equation can be expressed in differential form as.

The mass continuity equation could be simplified as a volume continuity equation for Syltherm800 oil at steady state condition as:

$$\nabla \bullet u_i = \frac{\partial u}{\partial x} + \frac{\partial v}{\partial y} + \frac{\partial w}{\partial z} = 0 \tag{1}$$

Considering the gravity along the global $-y$ direction, the momentum equations for a Three-Dimensional (3D) steady state incompressible flow can be rewritten as

$$\rho(u\frac{\partial u}{\partial x} + v\frac{\partial u}{\partial y} + w\frac{\partial u}{\partial z}) = -\frac{\partial p}{\partial x} + (\mu + \mu_t)\left\{\frac{\partial^2 u}{\partial x^2} + \frac{\partial^2 u}{\partial y^2} + \frac{\partial^2 u}{\partial z^2}\right\} \tag{2}$$

$$\rho(u\frac{\partial v}{\partial x} + v\frac{\partial v}{\partial y} + w\frac{\partial v}{\partial z}) = -\frac{\partial p}{\partial y} + (\mu + \mu_t)\left\{\frac{\partial^2 v}{\partial x^2} + \frac{\partial^2 v}{\partial y^2} + \frac{\partial^2 v}{\partial z^2}\right\} - \rho g_y \tag{3}$$

Table 1 Selected test conditions of the bare receiver of the LS2 collector from Dudley et al. [9]

Test conditions	DNI (W/m^2)	T_{amb} (°C)	T_{in} (°C)	V_{HTF} (m/s)	Re	T_{out} (°C)	V_{air} (m/s)
1st	919	22.6	301.4	0.608,157	12,710.1	318	0.1
2nd	867.6	19.8	203.4	0.597,403	6431.098	219.6	0.5
3rd	929.8	21.8	252.2	0.606,962	9218.136	269	1
4th	941.1	13.5	313.1	0.667,897	14,950.33	322	8
5th	961.3	15.1	313.3	0.667,897	14,967.46	320.7	9.3

DNI daily normal insolation, *T* temperature, *V* velocity, *Re* Reynolds number, *HTF* heat transfer fluid, *amb* ambient, *in* inlet, and *out* outlet

$$\rho(u\frac{\partial w}{\partial x}+v\frac{\partial w}{\partial y}+w\frac{\partial w}{\partial z})=-\frac{\partial p}{\partial z}+(\mu+\mu_t)\left\{\frac{\partial^2 w}{\partial x^2}+\frac{\partial^2 w}{\partial y^2}+\frac{\partial^2 w}{\partial z^2}\right\} \tag{4}$$

where, μ and μ_t were the molecular viscosity and turbulent (eddy) viscosity, respectively.

The conservation of energy equation was based on the energy balance that can be computed by the formula

$$\rho c_p\frac{\partial T}{\partial t}+\rho c_p\nabla.(u_iT)=-\nabla p+\left[k+\frac{c_p\mu_t}{\mathrm{Pr}_t}\right]\nabla^2 T+(\mu+\mu_t)\frac{\partial u_i}{\partial x_j}\left\{\frac{\partial u_i}{\partial x_j}+\frac{\partial u_j}{\partial x_i}-\frac{2}{3}\frac{\partial u_k}{\partial x_k}\delta_{ij}\right\}+S_r \tag{5}$$

where, S_r is solar radiation source term.

Two transport equations for RNG κ-ε model to calculate the turbulent energy production, κ and the turbulent energy dissipation, ε were

$$\frac{\partial}{\partial t}(\rho\kappa)+\nabla.(\rho\kappa u_i)=\alpha_\kappa(\mu+\mu_t)\nabla^2\kappa+G_\kappa+G_b-\rho\varepsilon \tag{6}$$

$$\frac{\partial}{\partial t}(\rho\varepsilon)+\nabla.(\rho\varepsilon u_i)=\alpha_\varepsilon(\mu+\mu_t)\nabla^2\varepsilon+C_{1\varepsilon}\frac{\varepsilon}{\kappa}(G_\kappa+C_{3\varepsilon}G_b)-\rho C^*_{2\varepsilon}\frac{\varepsilon^2}{\kappa} \tag{7}$$

,respectively.
Where, $C_{1\varepsilon}$ and $C_{3\varepsilon}$ were two model constants equal to 1.42 and 1.68, respectively, G_κ and G_b were turbulent kinetic energy generation due to mean velocity gradient and buoyancy effect, respectively, and α_κ and α_ε were the inverse effective Prandtl numbers for κ and ε, respectively.

Eddy viscosity was modified for swirl generation in the RNG model as

$$\mu_t=\rho C_\mu\frac{\kappa}{\varepsilon}f(\alpha_s,\Omega,\frac{\kappa}{\varepsilon})-\rho C^*_{2\varepsilon}\frac{\varepsilon^2}{\kappa} \tag{8}$$

Table 2 Temperature (K)-dependent correlations for the HTF physical properties

Properties	$=a+bT+cT^2+dT^3+eT^4+$ ———				Temperature range (K)
	a	*b*	*c*	*d*	*e*	
ρ(kg/m^3)	1139.2	−0.546	−4.87e−04			233 to 673
c_p(J/kg K)	1108.2	1.7073				233 to 673
k (W/m K)	0.1901	−1.88e−04				233 to 673
μ (Pa S)	0.2591	−8.93e−04				233 to 273
	0.157	−8.00e−04	1.03e-06			273 to 340
	0.0848	−5.54e−04	1.39e−06	−1.57e−09	6.67e−13	340 to 673
k_{Steel} = 15.906 + 0.0025T^2 + 5e-05T^3 W/mK						

ρ Density, c_p isobaric heat capacitance, k thermal conductivity, μ dynamic viscosity

where, C_μ = *0.0845*, α_s was the swirl constant roughly set to *0.07* for mild swirl flows and even higher value can be set for a strong swirl flow, and **Ω** was a characteristics swirl number that was calculated inside the ANSYS Fluent.

3.2 Physical Properties of the Computational Domain

The constituent of the computational domain as shown in Fig. 4 was the Syltherm 800 fluid. The physical properties including the density, isobaric heat capacitance, thermal conductivity and the dynamic viscosity of the HTF were correlated with its instantaneous absolute temperature. The polynomial correlations were developed as given in Table 2. The tube material, which was not modelled, was steel, and the thermal conductivity of steel, k_{Steel}, was also correlated with its absolute temperature as included in the same table.

3.3 Assumptions and the Boundary Conditions of the Computational Domain

As Fig. 4 shows, the boundary conditions in the computational domain were: **1:** no-slip heat flux wall, **2:** no-slip adiabatic wall, **3:** symmetry, **4:** fully developed velocity inlet and pressure outlet, and **5:** adiabatic ends of the receiver. They are explained in details below.

1. **The no-slip heat flux wall**: The inside surface and the outside surface of the absorber tube formed this no-slip heat flux boundary wall of the computational domain. The outside surface of the tube was exposed directly to the concentrated solar radiation and the environment; and the inside one was in direct contact with the viscous, pressurized and incompressible HTF. Shell conduction in the wall was active.

The outside surface of the tube was absorbing the incident solar irradiation energy as heat flux, and the inside surface was losing the heat energy to the flowing HTF. Simultaneously, because of the environmental effect, the outside surface was also losing some of the heat energy due to the radiation and convection heat losses. The rate of the concentrated solar energy and its density distribution around the receiver tube was calculated by applying the MCRT technique, which was coupled with the current FV model applying a special coupling technique as discussed in Sect. 3.4. The theoretical framework and the calculation technique of the heat loss from the outside surface of the tube are described in Sect. 3.5.

On the other hand, the convective and the conductive energy transfer from the inside surface of the tube to the HTF were calculated applying the built-in finite volume technique of the ANSYS Fluent software package. Moreover, the inside surface of the tube was assumed sufficiently frictional to form a no-slip wall to the viscous HTF.

2. **The no-slip adiabatic wall**: The circumferential surface of the cylindrical flow restriction device that was immersed into the viscous, pressurized and incompressible HTF as shown in Fig. 3 formed this wall boundary condition. The surface of the devise was assumed sufficiently frictional to the HTF to form a no-slip boundary. On the other hand, at steady state condition, no energy exchange between the device and the HTF was assumed.

As the RNG κ-ε turbulence model was not that efficient in calculation of the flow parameters near the wall, standard wall functions developed by Launder and Spalding [24] were used for near wall treatment. The logarithmic law-of-the-wall for mean velocity is given by the following formula:

$$U^* = \frac{\ln(Ey^*)}{\kappa} \tag{9}$$

where, $U^* \equiv \left(U_P C_\mu^{0.25} K_P^{0.25}\right) / (\tau_w/\rho)$, $y^* \equiv \left(\rho C_\mu^{0.25} K_P^{0.25} y_P\right)/\mu$, κ = von Kármán constant (= 0.4187), E = empirical constant (= 9.793), U_P = mean velocity of the fluid at point P, K_P = turbulence kinetic energy at point P, y_P = distance from point P to the wall, and μ = dynamic viscosity of the fluid.

This law is only valid for y^* between 30 and 300. In Fluent, the log-law is employed when y^* of the mesh adjacent to the wall is larger than 11.225. But the laminar stress-strain relationship, $U^* = y^*$ is used for the values smaller than 11.225. The laws-of-the-wall for mean velocity and temperature are based on the wall unit, y^*, rather than y^+ ($\equiv \rho u_\tau y/\mu$). However, these quantities are approximately equal in equilibrium turbulent boundary layers.

3. **Symmetry**: The shape of the HTF domain in between the absorber tube and the flow restriction device was annular or hollow cylindrical. The thickness of the annulus was assumed to be perfectly uniform. Moreover, the incident concentrated light around the external surface of the annulus was found symmetric either side of the YZ plane (see Fig. 3) along the periphery. The HTF flow inside the annulus was assumed symmetric with respect to the YZ plane. Therefore, the flow criteria and the conjugate heat transfer characteristics were assumed symmetric to the YZ plane.
4. **Fully developed velocity inlet and pressure outlet**: As the HTF inlet of the computational domain of the modelled receiver was downstream of a sufficiently long close-channel flow line, the flow of Syltherm800 viscous oil inside the no-slip annulus at the inlet must be fully developed turbulent flow. The flow was simulated as fully developed flow instead of assuming simplistic average fluid velocity such that [25]:

$$\frac{\partial u_i}{\partial z} = \frac{\partial k}{\partial z} = \frac{\partial \varepsilon}{\partial z} = \frac{\partial p}{\partial z} = \frac{\partial T}{\partial z} = 0 \tag{10}$$

5. **Adiabatic edges**: The rest of the HTF flow line except the computational domain enclosed within the HTF inlet and outlet was assumed perfectly insulated, therefore, no or negligible heat losses were assumed from inlet and outlet edges of the domain. Moreover, the bracket loss at the steady state condition was also assumed negligible.

3.4 Coupling Between MCRT and FV Models

As explained in Sect. 3.3, the no-slip heat flux boundary wall was absorbing the concentrated light that was reflected from the parabolic trough mirror. The distribution of the irradiance along the periphery of the computational domains of the both the receiver elements was calculated using a well verified MCRT model. The irradiance distribution around the receiver was calculated as Local Concentration Ratio (LCR). The integration between the MCRT and the FV models in the literature could be found sharing the same local irradiance data table by the models [17, 25, 26] provided that the grid at the outer surfaces of the physical model was the same, and the analysis of the optical and thermal behaviour of the collector were simultaneous and dependent. On the contrary, a unique approach for the current FV model was adopted in which both models shared the same irradiance profile yet they were completely independent in their grid system and analysis.

Predicting the LCR profiles of the LS2 receiver for a large number of test conditions, two sets of polynomial correlations were developed applying the curve fitting technique as a function of angular location, β (°) of the receiver. Using the correlations, the local irradiance around the receiver with evacuated glass envelope and without glass envelope could be calculated. The correlations with their coefficients of determination, R^2, were given by the following set of equations for the receiver with glass envelope:

$$\begin{aligned}
I_{0^\circ \le \beta \le 15^\circ} &= C_1 \times (2.8e-3 \times \beta^3 - 1.29e-2 \times \beta^2 + 0.1333 \times \beta + 43.333), (R^2 = 0.9973) \\
I_{15^\circ < \beta \le 48^\circ} &= C_1 \times (6.98e-3 \times \beta^2 - 0.108 \times \beta + 52.414), (R^2 = 0.9988) \\
I_{48^\circ < \beta \le 90^\circ} &= C_1 \times (1.2849e-3 \times \beta^3 - 0.2622 \times \beta^2 + 15.74 \times \beta - 229.49), (R^2 = 0.9999) \\
I_{90^\circ < \beta \le 180^\circ} &= C_1 \times (-1.06e-4 \times \beta^2 + 4.24e-2 \times \beta - 2.9507), (R^2 = 0.9973)
\end{aligned} \tag{11}$$

and without glass envelope:

$$\begin{aligned}
I_{0^\circ \le \beta \le 15^\circ} &= C_2 \times (1.98e-4 \times \beta^4 - 3e-3 \times \beta^3 + 2.37e-2 \times \beta^2 + 3.59e-2 \times \beta + 46.453), (R^2 = 0.9857) \\
I_{15^\circ < \beta \le 48^\circ} &= C_2 \times (1.19e-4 \times \beta^3 - 4.1e-3 \times \beta^2 + 0.2074 \times \beta + 49.602), (R^2 = 0.997) \\
I_{48^\circ < \beta \le 90^\circ} &= C_2 \times (1.230434e-3 \times \beta^3 - 0.2511 \times \beta^2 + 15.012 \times \beta - 214.19), (R^2 = 0.9999) \\
I_{90^\circ < \beta \le 180^\circ} &= C_2 \times (-1.055e-4 \times \beta^2 + 4.19e-2 \times \beta - 2.9121), (R^2 = 0.9954)
\end{aligned} \tag{12}$$

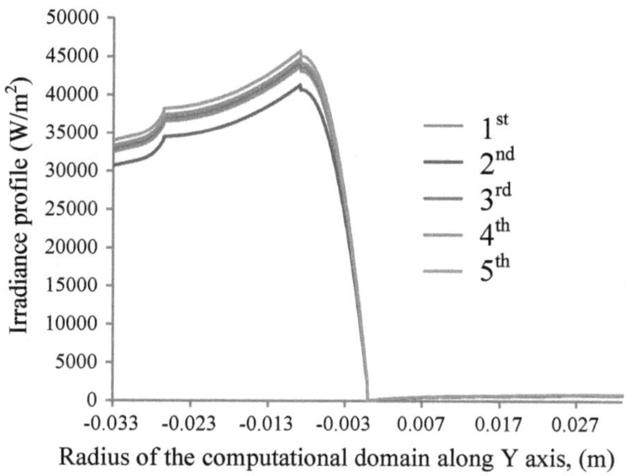

Fig. 5 Calculation of the irradiance distribution along the circumference of the computational domain using the set of correlations of Eq. (12)

where, $C_1 = DNI \times \rho_{PT} \times \tau_{GT} \times \alpha_{RT}$, $C_2 = DNI \times \rho_{PT} \times \alpha_{RT}$, R^2 = Coefficient of determination and DNI = Daily normal insolation (W/m^2).

The irradiance distributions along the circumference of the computational domain for the selected test conditions as presented in Table 1 were calculated using the set of correlations (12) as shown in Fig. 5. As the DNI for the most of the test conditions were very close, the irradiance profiles were also found very close to each other.

Incorporating the correlations, an in-house subroutine was developed by using the ANSYS macros, which was interpreted in the FV model. The solar energy in the macro was treated as heat flux and surface phenomenon of the domain. The irradiance profile was assumed uniform along the tube length. At the inlet and outlet, each edges of the receiver tube were larger than the mirror by 100 mm, therefore, was shaded (see Fig. 2). The macro enabled a grid independent coupling between the MCRT and the FV model. The heat loss from the outside wall of the domain to the environment by means of radiation and convection was also incorporated in the same macro applying the following theories.

3.5 *The Radiation and Convection Heat Losses from the Outside Surface*

As the absorber tube was bare, both the radiation and convection heat losses from the wall to the ambient were considered as surface phenomena. The following algorithm for the heat loss calculation was incorporated into the same MCRT-FV coupling macro.

The radiation heat loss was calculated by the formula

$$q_{rad} = \varepsilon\sigma[F_{gr}(T_w^4 - T_{gr}^4) + F_{sky}(T_w^4 - T_{sky}^4)] \tag{13}$$

where, ε was the emissivity of the cermet coated absorber tube that was given by, $\varepsilon = 0.000327T_w - 0.065971$ [10], σ was the Stefan–Boltzmann constant (= 5.670373 × 10^{-8} W/m^{2}K^4), F was the radiation view factor, and T was temperature in Kelvin. The subscripts 'gr', 'w' and 'sky' referred to the ground, wall and the sky, respectively.

The view factor between the receiver and the ground, F_{gr}, and between the receiver and the sky, F_{sky}, were calculated respectively, as:

$$\begin{aligned} F_{gr} &= 0.5(1 + \cos\beta) \\ F_{sky} &= 0.5(1 - \cos\beta) \end{aligned} \tag{14}$$

where, β was the angular location on the receiver in OXY plane such that β = 0° along −OY axis and 180° along +OY axis (see Fig. 4).

The ground temperature, $T_{gr} = T_a$, whereas, the sky temperature, $T_{sky} = T_a - 8$ [10]. Here, the subscript 'a' refers to the air or ambient.

On the other hand, as the selected test conditions in Table 1 shows wind speed during the data collection, forced convection heat loss from the receiver surface was assumed. The loss per unit surface area was given by:

$$q_{convec} = h(T_w - T_a) \tag{15}$$

where, h was the forced convection coefficient. As the exact event was not known, h was calculated as average of leeward coefficient, h_{lee} and wind-ward coefficient, h_{wind}. According to Sharples and Charlesworth [27], these coefficients were calculated by,

$$\begin{aligned} h_{lee} &= 2.2V_a + 8.3 \\ h_{wind} &= 3.3V_a + 6.5 \end{aligned} \tag{16}$$

where, V was the velocity of air.

3.6 Grid Generation Technique for the Computational Domain of LS2 Receiver

The accuracy and the stability of numerical simulation strongly rely on the grid resolution and distribution inside the computational domain. The mesh resolution should be fine enough, and the distribution should be reasonable to some regions of the domain in order to capture the flow physics properly; otherwise, an exaggerated

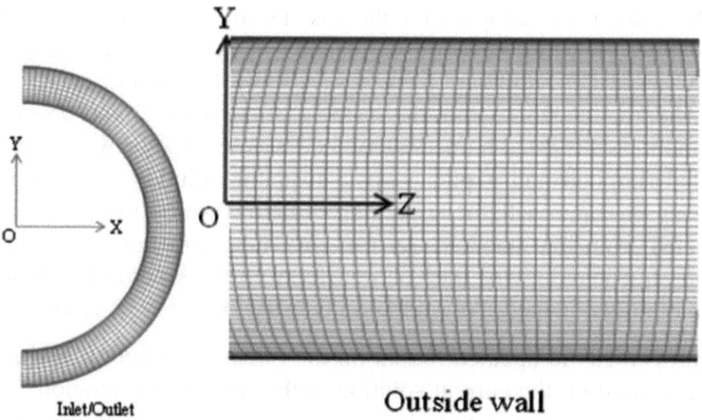

Fig. 6 Grid structure distribution for the computational domain

result would be produced from the simulation. As the current computational simulation involves turbulent flow along the Z axis, and high-temperature heat transfer across the XY plane from the wall to the viscous HTF, the grid resolution near the wall must be fine enough in order to capture the flow physics, that is, the viscous and thermal boundary layers near the frictional wall; and the grid must be distributed such a way that the effect of viscosity and heat transfer by means of convection (diffusion and advection) across the flow is resolved. Therefore, a structured and hexahedral grid system was generated for the current computational domain in such a way that the grids were uniformly distributed along the HTF flow (see the grid system on the YZ plane at the Outside wall, and along the Z axis at the symmetry in Fig. 6) and inflated by 10% across the flow from the wall to the centre of the bulk flow (see the grid system on the XY plane at the Inlet/Outlet face, and along the Y-axis at the Symmetry in Fig. 6). However, the optimum grid resolution was decided by performing the grid independence test as explained in Sect. 4.1.

3.7 Solution Method

The governing equations describing the current conjugate heat transfer fluid flow model were highly nonlinear and coupled in nature that made them almost impossible to obtain an analytical solution applying by available knowledge. Instead, an approximate numerical solution would be an efficient technique for this kind of real world problem. However, the accuracy and/or the acceptability of the solution to the experts simply rely on the adopted numerical scheme and the solution method.

The FV method was adopted for the current simulation to discretise the governing equations. The discretization was facilitated by employing one of the highly reliable FV based CFD software packages, ANSYS Fluent 14.5.

As the shell conduction within the material of the receiver was accounted for in the simulation, a segregated numerical method was selected between the segregated and coupled (implicit and explicit)—two available solution methods in ANSYS Fluent. Using this approach, the governing equations were solved sequentially, that is, segregated from one another. As the governing equations are nonlinear and coupled, several iterations of the solution loop must be performed before a converged solution is obtained. Any iteration consists of the steps as outlined below:

1. Fluid properties are updated, based on the initialized solution at the first iteration, and based on the current solution at the subsequent iteration.
2. In order to update the velocity field, the u, v, and w momentum equations are each solved in turn using current values for pressure and face mass fluxes.
3. Since the velocities obtained in Step 2 may not satisfy the continuity equation locally, a 'Poisson-type' equation for the pressure correction is derived from the continuity equation and the linearized momentum equations. This pressure correction equation is then solved to obtain the necessary corrections to the pressure and velocity fields and the face mass fluxes such that continuity is satisfied.
4. Where appropriate, equations for scalars such as turbulence, energy, and species are solved using the previously updated values of the other variables.
5. When interphase coupling is to be included, the source terms in the appropriate continuous phase equations may be updated with a discrete phase trajectory calculation.
6. A check for convergence of the equation set is made.
7. These steps are continued until the convergence criteria are met.

As the grid system of the current computational domain was hexahedron, and the upstream and the downstream of the domain were fully specified; in order to achieve a better accuracy in the simulation, a higher order Quadratic upwind differencing scheme: the QUICK scheme [28, 29] was adopted to discretise 3D convection-diffusion problem. Quick type schemes are based on a weighted average of second-order upwind and central interpolations of the variable. However, undertaking calculations with a higher order differencing scheme like QUICK sometimes would experience an instability problem in the solution because of some reason like poor initial guess. In order to minimize this chance of instability, after the initialization of the solution, the calculation was accomplished in three different steps: (1) first 50 to 500 iterations with first order upwind scheme, (2) next 50 to 500 more iterations with the second-order upwind scheme, and (3) finally resume the calculation until convergence of the solution activating the QUICK scheme. Because of the inherent limitations of the standard, linear, second-order and body-force-weighted pressure interpolating schemes, the PRESTO! scheme was adopted for the current simulation as this particular scheme is applicable with all types of grid systems. In FLUENT, SIMPLE is the default pressure-velocity coupling technique. Therefore, the SIMPLE algorithm was selected for the current steady state and turbulent HTF flow model. In the present study, the under-relaxation factors were set as 0.3 for the pressure and momentum, 0.8 for turbulent dissipation rate, and 0.75 for the rest of the parameters. The maximum allowable residuals were set at 10^{-6} for the energy equation, and 10^{-4} for the rest of the parameters. However, at the end of the computations of the present FV model, the residuals of the most of the values were found to be a couple of order less than the set values.

4 Verification of the Model

4.1 Grid Independence Test to Decide the Optimum Grid Resolution

In Sect. 3.6, a grid generation technique for the computational domain of the LS2 receiver element was illustrated. In this section, how the optimum grid system for the receiver element was adopted is explained.

The grid independence test is an approximate measure of whether the grid system in a computational domain is optimally fine; otherwise, the simulation result might be misleading. As the present FV model was a conjugate heat transfer simulation of steady state turbulent flow, the HTF temperature near the outlet of the domain was presumed to be the highest within the domain. Therefore, the optimum grid resolution was decided based on the effect of grid resolution on the static temperatures at two different locations on the heat flux wall (see Sect. 3.3) and the adiabatic wall (see Sect. 3.3) close to the HTF outlet. Nine different grid systems were generated, and the static temperature near the HTF outlet at two points on the inside and outside walls of the domain were calculated as shown in Fig. 7. As the figure shows, analysing the variations in the recorded temperatures, the grid system $20_r \times 60_\beta \times 1600_z$ was found satisfactory for the current computational domain. Here, suffix r, β and z refer to the radial, angular/circumferential and axial direction of the domain. However, for further accuracy and to get maximum benefit of high-performance computing (HPC) facilities at the Queensland University of Technology (QUT) the grid system $20_r \times 80_\beta \times 3200_z$ was chosen for the current model that produced 5,120,000 hexahedral cells.

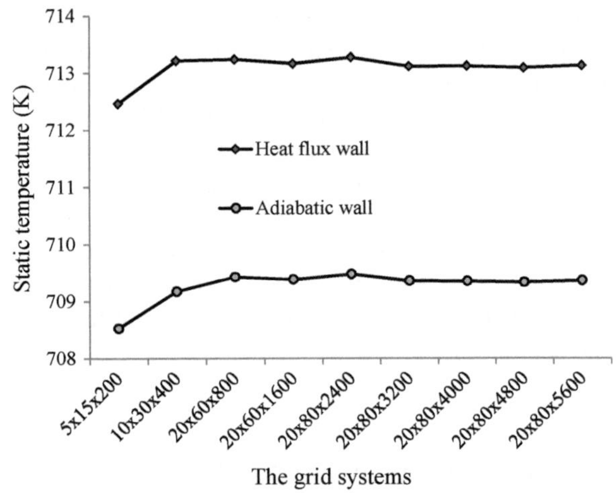

Fig. 7 Grid independence test

4.2 Checking the Near Wall Grid Resolution

The value of wall Y* and/or Y+ is an indication of whether the grid resolutions near the wall are fine enough, or the first cell adjacent to the wall is within the boundary layer so that the flow physics inside the computational domain near the no-slip wall is resolved properly. Lower values of Y* and the Y+ correspond to a fine mesh near the wall. Therefore, Y* and Y+ of both of the walls (the heat flux wall and the adiabatic wall) of the computational domain were calculated from the FV model as shown in Figs. 8 and 9, respectively. The figures show that the maximum Y* value was 4.16, and the maximum Y + value was 4.19 on the outside wall near the HTF outlet where the temperature was maximum. As the maximum Y* value was lower than 11.225, the laminar stress-strain relationship was employed in calculation of this value (see the Sect. 3.3). The minimum values of Y* and Y+ were calculated 2.76 and 2.78, respectively, which were also could be found on the outside wall near the inlet edge where the temperature was minimum relative to the entire domain. That implies that the near wall grid resolution was fine enough.

4.3 Verification of the FV Model and the MCRT-FV Integration

The current FV model was used to calculate the HTF outlet temperature for five different test conditions for the bare receiver of LS2 collector. The outlet temperatures HTF of the simulated test conditions were compared with the experimental results as presented in Table 3. The maximum absolute error between the simulated

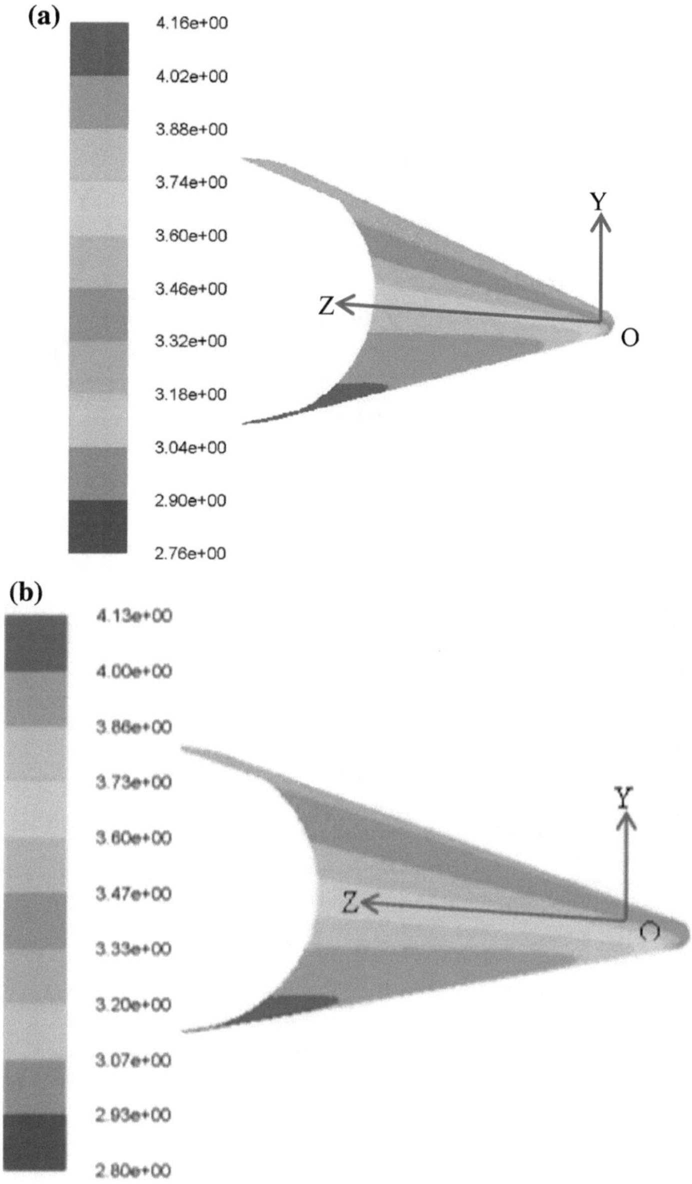

Fig. 8 Wall Y* values: **a** at outside wall, and **b** at inside wall

results and the experimental results were calculated to be 2.92% for the second test condition, and the minimum absolute error was calculated to be 0.23% for the fourth test condition. The average absolute error was calculated to be 1.11%. This

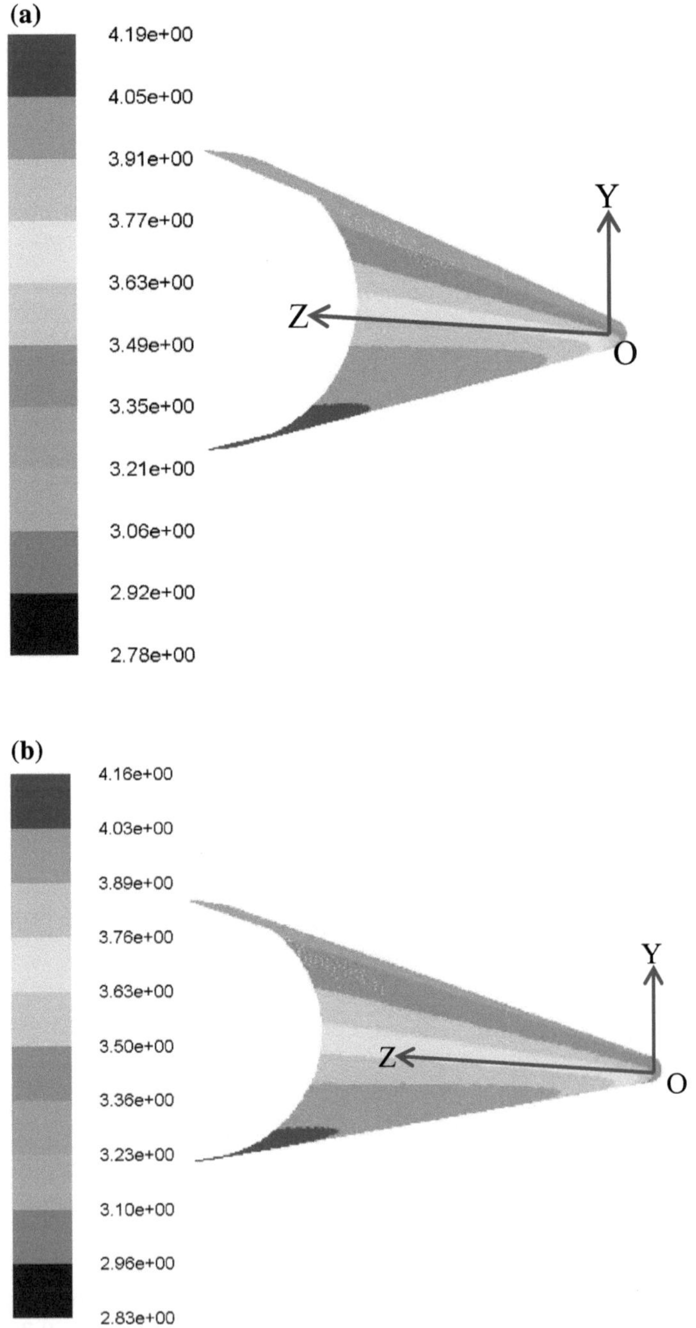

Fig. 9 Wall Y+ values: **a** at outside wall, and **b** at inside wall

Table 3 Comparison of model calculated results with experimental data

TCs	DNI (W/m^2)	V_a (m/s)	T_a (°C)	T_{in} (°C)	T_{o_expt} (°C)	T_{o_model} (°C)	E_{abs} (%)	E_{av} (%)
1st	919	0.1	22.6	301.4	318	320.98	0.94	1.11
2nd	867.6	0.5	19.8	203.4	219.6	213.19	2.92	
3rd	929.8	1	21.8	252.2	269	271.68	1.04	
4th	941.1	8	13.5	313.1	322	322.73	0.23	
5th	961.3	9.3	15.1	313.3	320.7	322.08	0.43	

Acronyms: *TC* Test Conditions, *DNI* Daily Normal Insolation, *V* Velocity, *T* Temperature, *E* Error. Suffixes: *a* air or ambient, *in* inlet, *o_expt* experimental outlet data, *o_model* model calculated outlet data, *abs* absolute, *av* average

Table 4 Comparison of outlet temperature at different heat loss conditions

Test conditions	V_a (m/s)	T_i (°C)	No loss model	Total loss model		Convection loss model		Radiation loss model	
			T_{o_Max} (°C)	T_{o_Min} (°C)	q_{loss_Max} (%)	T_{o_Convec} (°C)	q_{loss_Convec} (%)	T_{o_Rad} (°C)	q_{loss_Rad} (%)
1st	0.1	301.4	325.45	320.98	18.60	322.20	13.52	324.18	5.30
2nd	0.5	203.4	215.90	213.19	21.64	213.57	18.65	215.51	3.11
3rd	1	252.2	275.94	271.68	17.91	272.21	15.69	274.94	4.21
4th	8	313.1	335.74	322.73	57.47	323.85	52.54	334.43	5.80
5th	9.3	313.3	336.43	322.08	62.05	323.18	57.28	335.11	5.71

In the table, *V* is velocity, *T* is temperature and *q* is heat energy per second. Suffix *a, i, o, Max, Min, Convec* and *Rad* stand for air, inlet, outlet, maximum, minimum, convection and radiation, respectively

good agreement between the simulated results and the experimental results was validated the accuracy of the current FV model and the MCRT-FV integration macro.

4.4 Verification of the Heat Loss Calculation Algorithm

As the receiver was bare, ambient conditions strongly affected the convection and radiation heat losses from the surface of the computational domain. The loss was calculated using an in-house subroutine coupled with the MCRT-FV integration macro. The test conditions were simulated assuming four different heat loss conditions including, (i) no loss model, (ii) total loss model, (iii) convection loss model, and (iv) radiation loss model. The estimated HTF outlet temperature and the percentage of heat loss data were arranged in Table 4. One might understand from the general knowledge of heat transfer that the HTF outlet temperature would be the maximum without heat loss, the minimum with total loss (both convection and radiation heat losses), and in between these maximum and minimum temperatures with the convection heat loss and the radiation heat loss. The calculated outlet temperatures of the HTF under all four heat loss conditions as presented in the table support the general knowledge completely. This particular investigation confirmed the reliability of the heat loss algorithm that used in the in-house macro with the MCRT-FV integration macro.

4.5 Checking the Fully Developed Flow Condition at the HTF Inlet

As discussed in Sect. 3.3, the HTF flow at the inlet was considered fully developed instead of average bulk flow. Firstly, in case of the average velocity assumption, the longitudinal velocity (the Z velocity for the current models) across the flow (from the wall to the bulk flow centre) is uniform and becoming fully developed along the downstream gradually or rapidly depending on the turbulence. Whereas, in case of the fully developed flow, the longitudinal velocity is zero adjacent to the wall due to a viscous share effect and developed fully at the bulk flow centre of the HTF where all of the flow properties are in steady state condition according to Eq. (10). Secondly, the velocity components perpendicular to the bulk flow (the X and Y velocities for the current models) are zero in case of average or uniform velocity assumption, whereas, the components are nonzero at fully developed flow. Finally, the longitudinal or bulk flow velocity contours of a steady state flow at the inlet and outlet of a solved computational domain would be exactly similar to each other if the flow is isothermal and adiabatic; otherwise, would be fairly similar. The velocity contours at the HTF inlet, outlet and symmetry boundaries of the computational domains of both of the receiver elements were visually investigated and compared against these three criteria for the flow to be fully developed at the inlet.

The velocity contours at the HTF inlet and outlet boundaries of the computational domain of the LS2 receiver are shown in Figs. 10 and 11, respectively. The longitudinal velocity profile and the Z velocity contour in Fig. 10c match perfectly with the first criterion as the velocity near the wall is zero, and maximum at the centre. Moreover, the velocity components perpendicular to the axial velocity, the X velocity and the Y velocity in Fig. 10a and b, respectively, match perfectly with the second criteria as none of these components are nonzero. Finally, the axial velocity contour at the inlet (see Fig. 10c) could be seen to match perfectly with that at the outlet (see Fig. 11c). However, the dissimilarities, if any, could be seen by visual investigation between the velocity contours of X, Y and Z velocity components at the inlet (see Fig. 10) and those components at the outlet

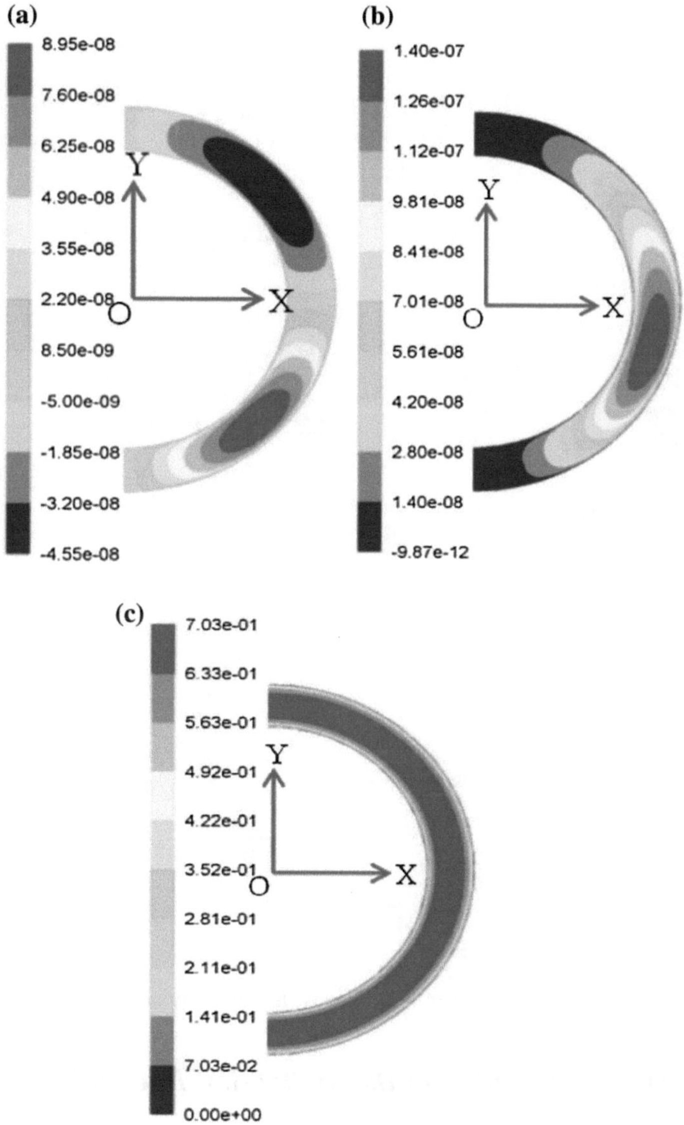

Fig. 10 Velocity contour at the inlet: **a** X velocity, **b** Y velocity and **c** Z velocity

(see Fig. 11) were might be because of conjugate heat transfer effect across the flow. Nonetheless, it could be claimed that the simulation of the fully developed flow at the HTF inlet of the computational domain of the LS2 receiver was appropriate.

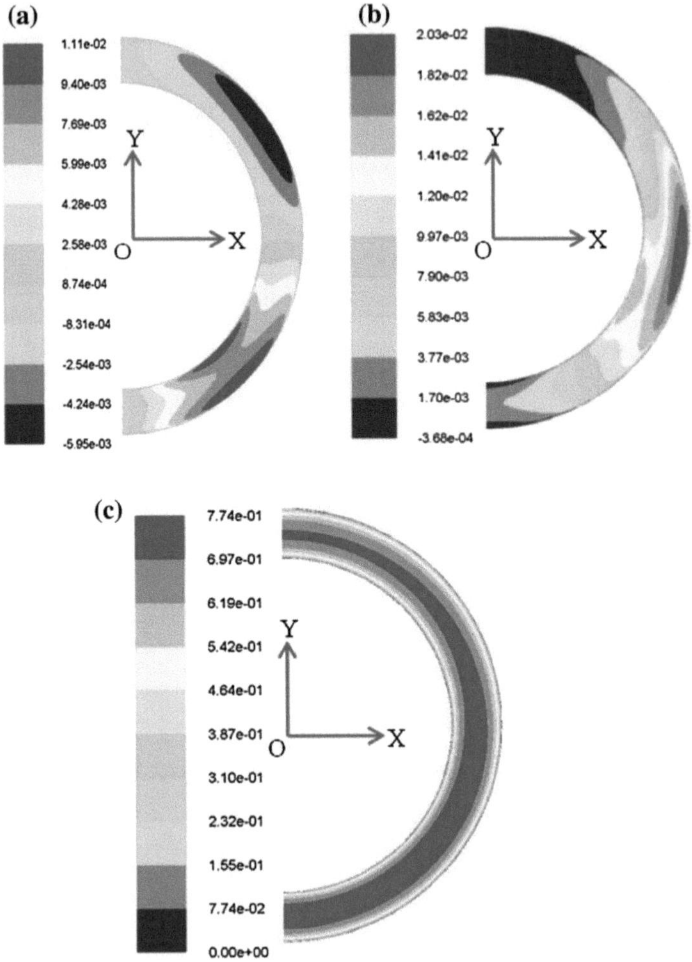

Fig. 11 Velocity contour at the outlet: **a** X velocity, **b** Y velocity and **c** Z velocity

4.6 Further Verification of the MCRT-FV Integration

The accuracy of MCRT-FV integration was already justified in Sect. 4.3. However, the integration is further verified comparing the residual heat flux profile around the circumference of the computational domain as shown in Fig. 12a. The purpose of the integration was to reproduce a realistic and appropriate concentrated solar energy flux profile around the circumference of the computational domain as shown in Fig. 5. Whether the algorithm and the employed macro for the integration worked properly could be justified comparing the similarity between the input concentrated solar flux profile as shown in Fig. 5 with the residual wall heat flux profile in Fig. 12a. However, the magnitude of the solar flux in Fig. 5 could be seen

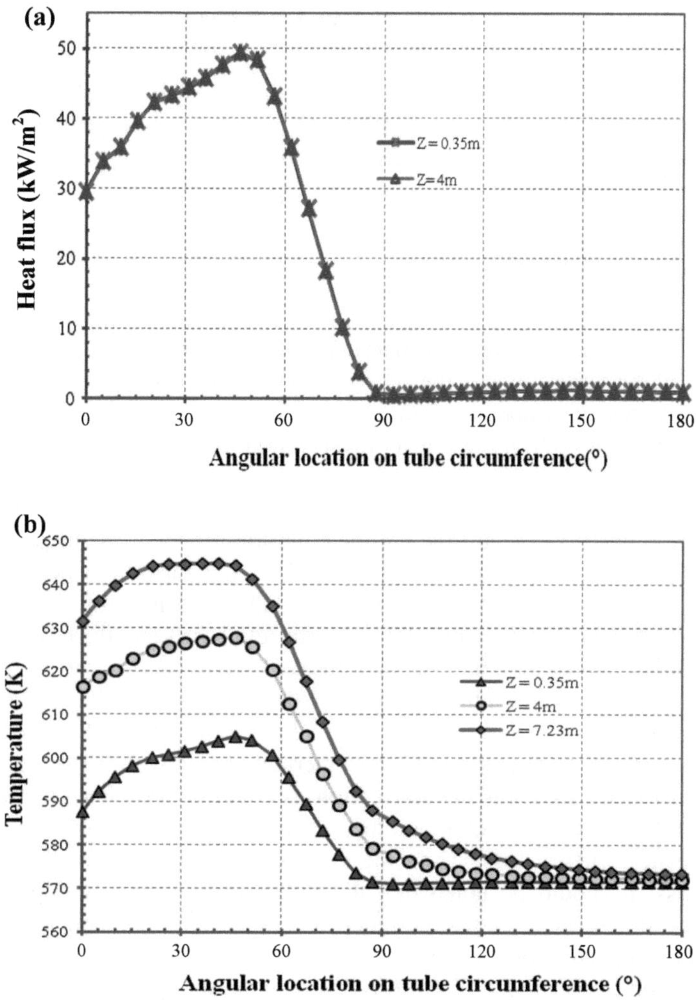

Fig. 12 Thermal characteristics along the circumference of the computational domain: **a** resultant heat flux profile, and **b** the temperature profiles (In the figure, 'Z' denotes the longitudinal location on the tube from the inlet end.)

much lower than the residual heat flux in Fig. 12a, which was because of the residual heat flux was calculated for about 1000 W/m^2 DNI and at ideal conditions without considering any heat loss from the surface in contrast to the realistic test conditions as shown in Fig. 5 for validation of the present FV model. Ignoring the discrepancy in the flux magnitude, the similarity of the both profiles confirmed the accuracy of the coupling and the in-house algorithm.

5 Thermal Characteristics of the LS2 Receiver

Using the present FV model, the heat loss from the receiver surface was investigated, and the resultant heat flux profile and the temperature profiles around the receiver surface were calculated as described below.

5.1 Heat Loss Phenomena

As the receiver was bare, and the operating temperature was significantly higher than that of the environment, the receiver must lose heat energy from its outer surface by virtue of temperature difference due to convection and radiation. The individual effect of both of these heat transfer modes was investigated as presented in Table 4.

In the table, the no loss model represents the maximum possible rise of HTF temperature without any heat loss. On the contrary, the total loss model shows the minimum possible temperature rise of the HTF and maximum heat loss at the respective ambient conditions. On the other hand, the convection loss model and the radiation loss model show the effect of convection loss and the radiation loss on the HTF outlet temperature and the heat loss.

As the table shows, from the total loss model, the maximum heat loss increases moderately with the increase of ambient air velocity from 18% of heat energy with 0.1 m/s air velocity to as much as 62% of thermal energy with 9.3 m/s air velocity. The reason for this loss was the convection as could be seen from the convection loss model. Because, in the total loss, the radiation was found to account for around 5% of the heat energy loss, while the convection accounted for the rest of the thermal loss. It should be noted that the radiation loss directly depends on the temperature difference between the receiver and the environment, while the convection loss depends not only on the temperature difference but also on the air velocity of the environment. Therefore, maintaining an optimum evacuation level between the receiver tube and the glass envelop of a conventional PTC is important to proof the convection loss. However, as a higher working temperature would decrease the electric performance of PV solar cells, heat loss from the PV aperture of a CPV/T collector might be beneficial for the cells' efficient operation.

5.2 Resultant Heat Flux and Temperature Profiles of the Receiver Outside Wall

The current model was adapted to calculate the heat flux profile and the temperature profile around the receiver wall at ideal condition. The DNI for this particular

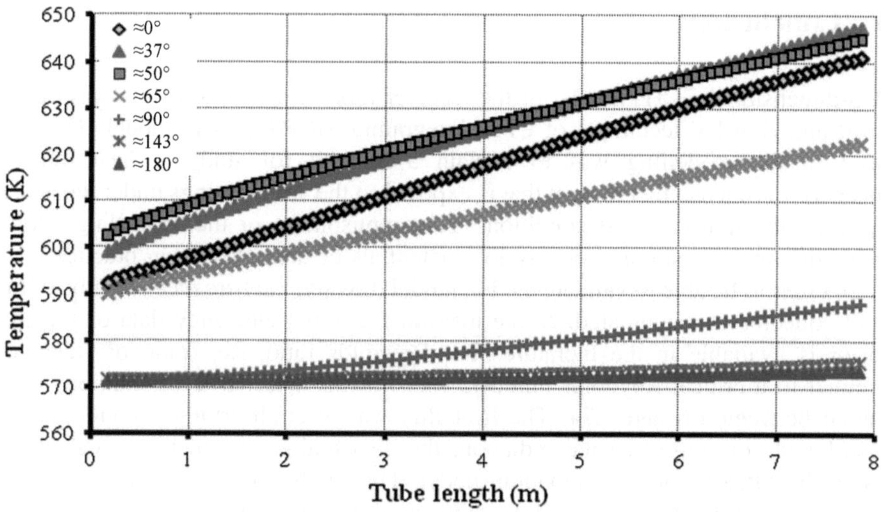

Fig. 13 Longitudinal temperature profile along the length of the tube at different angular locations

investigation was considered about 1 sun. While the circumferential profiles are illustrated in Fig. 12, the longitudinal profiles are presented in Fig. 13.

As the Fig. 12a shows, the heat flux profiles at 0.35 m length location and at 4 m length location of the receiver along the circumference were calculated that were found highly nonuniform and almost similar to that of the solar irradiance distribution. Since both the profiles were coincident, there was no or little variations in heat flux could be seen along the receiver length, which was reasonable as the incident solar flux along the length was assumed uniform.

Again Fig. 12b shows the temperature profiles along the circumference of the receiver at 0.35, 4 and 7.23 m length locations. The temperature profiles were found significantly nonuniform as was found for the heat flux profiles. Unlike the heat flux profile, the temperature was found to increase from the inlet to the outlet along the length of the tube, which was further supported from the longitudinal temperature profile of the receiver as shown in Fig. 13. The increase in temperature, as the figure shows, from inlet to outlet edges of the receiver was found fairly linear.

For a standard PTC, the solar flux distribution along the receiver was found almost fixed. Therefore, it would be possible to decrease the temperature variations along the receiver circumference by increasing the HTF turbulence or HTF mixing inside the domain. The mixing of the HTF inside the receiver could be achieved by using some type of swirl generator such as twisted tape insert into the domain. The insert would induce a swirl or rotation effect into the fluid, so that the HTF may rotate at least 90° while flowing from the inlet to the outlet. By this way, the HTF inside the receiver tube would be heated more evenly, and the resultant temperature profile of the tube wall along the periphery might be gentler, which would not only increase the heat gain of the HTF but also would decrease the heat loss from the receiver surface.

6 Conclusions

A 3-dimensional CCHT model of the receiver tube of a standard PTC was developed applying FV technique of CFD, integrating MCRT optical model. The purpose of the modelling was to explain a method for studying heat transfer mechanism of a receiver system that is exposed to the surroundings under the actual optical and environmental conditions. Accomplishment of the modelling allows predicting energy gain and energy loss potentials of the collector in practice.

In order to be able to validate the FV model directly, the bare receiver tube of the LS2 collector was simulated, as the measured thermal efficiency data of the collector is available in the literature. The Reynolds ratio, Re_r (ratio of Reynolds number to a critical Reynolds number, 4000) of the selected test conditions were varied between 1.6 and 3.74. The heat flux wall was absorbing a nonuniformly distributed concentrated solar radiation, the distribution of which was calculated using the MCRT optical simulation model. The irradiance profile around the circumference of the computational domain was simulated applying MCRT technique and integrated with the FV model using an in-house macro. Radiation and convection loss from the outer surface of the receiver were incorporated in the macro.

The FV model was verified by comparing the calculated and measured HTF outlet temperature data. The absolute deviation between the numerical results and the measured data was estimated would be maximum 2.92%, minimum 0.23% and average 1.11%. This very good agreement between the simulation results and the experimental data confirmed the reliability of the FV model and the accuracy of the MCRT-FV integration.

The FV model was further used to investigate the combined and separate effects of convection heat loss and radiation heat loss. The combined heat loss was found increasing from 19 to 62% of maximum heat gain with the increase of air velocity from 0.1 m/s to 9.3 m/s. Among these losses, radiation loss was accountable for at most 6% of total loss, whereas the convection loss was accountable for the rest of the heat losses that refer to the importance of evacuated glass envelope around the receiver tube of a PTC. On the other hand, the resultant heat flux profiles and the temperature profiles along the periphery of the computational domain were found mostly similar to that of the incident irradiance distribution around the same receiver. While the resultant heat flux was found to be uniform along the length of the tube due to the uniform incident irradiance along the same direction, the temperature was found to gradually increase from the inlet to the outlet.

The convection heat loss from the outer surface of the receiver tube of a PTC with damaged or removed glass envelop was observed very high, because, it is potentially exposed to very high-temperature gradient with respect to the surroundings. Therefore, it is obvious that the receiver tube of a PTC should be enveloped properly with an evacuated glass tube. A well-managed and efficiently operated parabolic trough collector solar energy field could be the best candidate for sustainable energy management of a sustainable future.

Acknowledgements This article is a part of a PhD project that is supported by a QUT post graduate research award and by a CSIRO Flagship collaboration fund PhD top-up scholarship through the Energy Transformed Flagship.

References

1. Grena, R. (2009). Optical simulation of a parabolic solar trough collector. *International Journal of Sustainable Energy, 29*(1), 19–36.
2. Yang, B., Zhao, J., Xu, T., & Zhu, Q. (2010). Calculation of the concentrated flux density distribution in parabolic trough solar concentrators by monte carlo ray-trace method. In: *Photonics and Optoelectronic (SOPO) Symposium.*
3. Molla, M. M., Saha, S. C., & Hossain, M. A. Radiation effect on free convection laminar flow along a vertical flat plate with streamwise sinusoidal surface temperature. *Mathematical and Computer Modelling, 53*(5–6), 1310–1319.
4. Vafai, K., Desai, C. P., Iyer, S. V., & Dyko, M. P. (1997). Buoyancy induced convection in a narrow open-ended annulus. *Journal of Heat Transfer, 119*(3), 483–494.
5. Hamad, F. A., & Khan, M. K. (1998). Natural convection heat transfer in horizontal and inclined annuli of different diameter ratios. *Energy Conversion and Management, 39*(8), 797–807.
6. Dyko, M. P., Vafai, K., & Mojtabi, A. K. (1999). A numerical and experimental investigation of stability of natural convective flows within a horizontal annulus. *Journal of Fluid Mechanics, 381,* 27–61.
7. Borjini, M. N., Mbow, C., & Daguenet, M. (1999). Numerical analysis of combined radiation and unsteady natural convection within a horizontal annular space. *International Journal of Numerical Methods for Heat and Fluid Flow, 9*(7), 742–764.
8. Farinas, M. I., Garon, A., St-Louis, K., & Lacroix, M. (1999). Study of heat transfer in horizontal bare and finned annuli. *International Journal of Heat and Mass Transfer, 42*(21), 3905–3917.
9. Dudley, V. E., Kolb, G. J., Mahoney, A. R., Mancini, T. R., Matthews, C. W., Sloan, M., & Kearney, D. (1994). *Test results: Segs ls-2 solar collector*, 139.
10. Forristall, R. (2003). *Heat transfer analysis and modeling of a parabolic trough solar receiver implemented in engineering equation solver.* National Renewable Energy Laboratory, 1617 Cole Boulevard, Golden, Colorado 80401–3393, Technical report No. NERL/TP-550-34169.
11. Kassem, T. (2007). Numerical study of the natural convection process in the parabolic-cylindrical solar collector. *Desalination, 209*(1–3), 144–150.
12. Reddy, K., Kumar, K. R., & Satyanarayana, G. (2008). Numerical investigation of energy-efficient receiver for solar parabolic trough concentrator. *Heat Transfer Engineering, 29*(11), 961–972.
13. Reddy, K., & Satyanarayana, G. (2008). Numerical study of porous finned receiver for solar parabolic trough concentrator. *Engineering applications of computational fluid mechanics, 2* (2), 172–184.
14. Reddy, K., & Kumar, N. S. (2009). An improved model for natural convection heat loss from modified cavity receiver of solar dish concentrator. *Solar Energy, 83*(10), 1884–1892.
15. Odeh, S. D., Morrison, G. L., & Behnia, M. (1998). Modelling of parabolic trough direct steam generation solar collectors. *Solar Energy, 62*(6), 395–406.
16. Hou, Z., Zheng, D., Jin, H., & Sui, J. (2007). Performance analysis of non-isothermal solar reactors for methanol decomposition. *Solar Energy, 81*(3), 415–423.
17. He, Y.-L., Xiao, J., Cheng, Z.-D., & Tao, Y.-B. (2011). A mcrt and fvm coupled simulation method for energy conversion process in parabolic trough solar collector. *Renewable Energy, 36*(3), 976–985.

18. Cheng, Z., He, Y., & Cui, F. (2013). A new modelling method and unified code with mcrt for concentrating solar collectors and its applications. *Applied Energy, 101,* 686–698.
19. Islam, M., Karim, A., Saha, S. C., Miller, S., & Yarlagadda, P. K. (2012). Three dimensional simulation of a parabolic trough concentrator thermal collector. In: *The Proceedings of the 50th annual conference, Australian Solar Energy Society (AuSES), 6–7 December*. Swinburne University of Technology, Melbourne, Australia.
20. Islam, M., Karim, A., Saha, S. C., Yarlagadda, P. K., Miller, S., & Ullah, I. (2012). Visualization of thermal characteristics around the absorber tube of a standard parabolic trough thermal collector by 3d simulation. In: *the Proceedings of the 4th International Conference on Computational Methods (ICCM2012), November 25–27*. Gold Coast, Australia.
21. Islam, M., Karim, A., Saha, S. C., Miller, S., & Yarlagadda, P. K. (2013). Development of optical ray tracing model of a standard parabolic trough collector. In: *The Proceedings of the Renewable Energy for Sustainable Development & Decarbonisation, World Renewable Energy Congress—Australia 2013, 14–18 July*. Murdoch University, Perth, Western Australia, Australia.
22. Islam, M., Karim, M. A., Saha, S. C., Miller, S., & Yarlagadda, P. K. D. V. (2014). Development of empirical equations for irradiance profile of a standard parabolic trough collector using monte carlo ray tracing technique. *Advanced Materials Research*. Energy Development 860–863, 180–190.
23. ANSYS, I. (2011). *Ansys fluent theory guide*. SAS IP, Inc.: U.S.A.
24. Launder, B. E., & Spalding, D. B. (1974). The numerical computation of turbulent flows. *Computer Methods in Applied Mechanics and Engineering, 3*(2), 269–289.
25. Cheng, Z. D., He, Y. L., Cui, F. Q., Xu, R. J., & Tao, Y. B. (2012). Numerical simulation of a parabolic trough solar collector with nonuniform solar flux conditions by coupling fvm and mcrt method. *Solar Energy, 86*(6), 1770–1784.
26. Cheng, Z. D., He, Y. L., Xiao, J., Tao, Y. B., & Xu, R. J. (2010). Three-dimensional numerical study of heat transfer characteristics in the receiver tube of parabolic trough solar collector. *International Communications in Heat and Mass Transfer, 37*(7), 782–787.
27. Sharples, S., & Charlesworth, P. S. (1998). Full-scale measurements of wind-induced convective heat transfer from a roof-mounted flat plate solar collector. *Solar Energy, 62*(2), 69–77.
28. Leonard, B. P., & Mokhtari, S. (1990). Beyond first-order upwinding: The ultra-sharp alternative for non-oscillatory steady-state simulation of convection. *International Journal for Numerical Methods in Engineering, 30*(4), 729–766.
29. Saha, S. C., Patterson, J. C., Lei, C. Scaling of natural convection of an inclined flat plate: Sudden cooling condition. *Journal of Heat Transfer, 133*(4), 041503.

Enhancement of Confined Air Jet Impingement Heat Transfer Using Perforated Pin-Fin Heat Sinks

Md. Farhad Ismail and Suvash C. Saha

Abstract The development of semiconductor fabrication process and electronic packaging technology causes the size and weight of electronic components to decrease consistently. Along with the increasing operating power, the heat generation rate of the electronic products apparently gets higher. For the sustainable future, we must limit using natural resources and also reduce the greenhouse gas emission. Efficient removal of heat from the electronic products in a limited space becomes a major task in electronics cooling. Air impingement cooling with a heat sink is an attractive option for electronic cooling, because it is inexpensive, robust, and localized. Rapid heat transfer from heated surfaces and reducing material weight is also becoming a major task for the design of heat exchanger equipment for electronic cooling. Rectangular plate fins as extended surfaces are good heat transfer equipment which are widely used for various industrial applications. Heat transfer rate can be improved by introducing perforations, porosity, or slots. Moreover, due to restrictions in setup space and economic reasons, heat transfer equipment has been required to be much more compact in size and lighter in weight. Studies on three-dimensional plate and pin-fin heat sinks are extensive. But no focus has been yet given on air jet impingement heat transfer with perforated pin-fin heat sink. Thermal-fluid characteristics of solid and perforated pin-fin heat sinks cooled by confined air jet impingement are investigated numerically in this study. The SST k-ω turbulence model is used to predict the turbulence flow parameters. The numerical model is verified with previously published experimental data. Flow and heat transfer characteristics are presented for the impinging Reynolds number, Re = 5000−25000 having constant impingement distance (Y/D = 8), fin width (W/L = 0.1), and height (H/L = 0.5). The main objective of this study is to examine the effects of fin perforations on the thermal performance of

Md. Farhad Ismail
School of Mechatronic Systems Engineering, Simon Fraser University, Burnaby, BC V5A1S6, Canada

S.C. Saha (✉)
School of Chemistry, Physics & Mechanical Engineering, Queensland University of Technology, Brisbane, QLD 4001, Australia
e-mail: suvash.saha@qut.edu.au; s_c_saha@yahoo.com

M.M.K. Khan et al. (eds.), *Application of Thermo-fluid Processes in Energy Systems*, Green Energy and Technology, https://doi.org/10.1007/978-981-10-0697-5_10

pin-fin heat sinks. Results show that thermal resistance decreases and fin efficiency increases with the increase of Reynolds number due to perforation. Thus, this kind of heat sink equipment would reduce the cooling power consumption rate.

Keywords Thermal-fluid characteristics · Thermal resistance · Perforated fins · Confined jet impingement

1 Introduction

Heat sinks are utilized to dissipate thermal energy generated by electronic components to maintain a safe temperature. Browne et al. [1] investigated the performance of two in-line micro-jet arrays with air and water in the Reynolds number range $150 < Re_d < 4900$. Area-averaged Nusselt numbers were measured in the range $6 < Nu_d < 75$. Moreover, an area-averaged heat flux of 1100 W/m^2 with an inlet fluid to surface temperature difference of less than 3 °C was achieved with the higher area ratio array. From the data collected using air as the working fluid, the effect of Mach number on the heat transfer performance was determined to be negligible. Brignoni and Garimella [2] found experimentally that for a fixed airflow rate under confined jet impingement condition, a nozzle with a smaller diameter reduced the thermal resistance. El-Sheikh and Garimella [3] investigated the heat transfer enhancement for pin-fin heat sinks under confined air jet impingement and proposed some correlations of the average heat transfer coefficient in terms of Reynolds number, fluid properties, and the geometric parameters of the heat sinks and the nozzles. Fabbri and Dhir [4] studied that single-phase heat transfer of impinging jet arrays with free jets of both water and FC-40 was studied; Reynolds numbers ranged from 70 to 3800. The heat transfer coefficients reported were 6,000 −60,000 W/m^2-K. Recently, Foo et al. [5] found from experimental and numerical studies that number of horizontal perforations is more important for heat transfer than an increase to the perforation diameter. The steady-state forced-convection cooling of a horizontally based pin-fin assembly has been investigated experimentally by Haq et al. [6]. The overall pressure drop and the effect of the shroud clearance were examined.

Kim et al. [7] conducted experimental study to determine the thermal performance of different types of heat sinks under impinging flow condition. The results showed that thermal resistance of pin fins is lower than that of the plate fins for small pumping power. For confined impingement cooling, the inlet geometry affects the heat transfer. Yang and Peng [8] studied numerically the impinging thermal performance of heat sink with non-uniform fin width for high Reynolds number. Sparrow et al. [9] found that the heat transfer and pressure drag for staggered arrays are higher than those for the in-line arrangement. Yang et al. [10] studied numerically forced convective heat transfer in staggered aluminum porous pin-fin arrays. They found higher heat transfer rate in porous pin fins than in solid pins. The pressure drag for staggered pin-fin array can be reduced by introducing either perforated pin fins or pin-fin-dimple arrays.

Maveety et al. [11] investigated the influence of the impingement distance on thermal performance of heat sink. Kim and Kuznetsov [12] suggested that a thin pin-fin heat sink should be designed to have high porosity. Li et al. [13–18] investigated the effects of fin width and height on the performance of the pin-fin heat sinks with air impingement cooling by the numerical method and the experimental technique. Lou et al. [19] studied the effects of geometric parameters of heat sink under jet impingement cooling.

Wang et al. [20] studied three different micro-jet arrays with varying jet diameters, spacing, and configurations using water. They concluded that pool boiling was the dominant heat transfer mechanism. The area ratio (AR), which is the ratio of total jet orifice area to heater surface area, ranged from 0.00016 to 0.0002 in the experiment. The maximum power dissipated was 90 W. Yangming et al. [21] and Ismail et al. [22, 23] studied the application of carbon nanotube-based heat sink as well as perforated heat sink for heat transfer enhancement under different fluid flow conditions. Ledezma et al. [24] conducted a numerical and experimental study of the heat transfer on a pin-finned plate and proposed the correlation equations for an optimal fin-to-fin spacing and a maximum thermal conductance. Maveety and Jung [25] compared between numerical and experimental results for the cooling performance of pin-fin heat sink with air jet impinging flow. The numerical results illustrated a complex pressure gradient inside the fin array and a greater pressure gradient improved mixing and heat transfer. Hwang and Lui [26, 27] studied the heat transfer and pressure drop characteristics between pin-fin trapezoidal ducts with straight and lateral outlet flows. The effect of pin arrangement for the ducts of different direction outlet flows was also examined. Rao et al. [28] and Sahin et al. [29] studied heat transfer enhancement of miro-fins due to perforations. Eid et al. [30] studied the jet impingement cooling for thin-strip pin fins. Very recently, Ismail et al. [31] numerically investigated the fluid flow under the laminar condition for solid and different types of lateral perforated fins. The perforations considered in their studies were of rectangular (square), circular, triangular, and hexagonal cross sections. Fins having three perforations were used in their study.

From the above literature reviews, the present study seems to be the first research that considers perforated fins for heat transfer enhancement under jet impingement condition. Most of the works were directed toward the air jet impingement cooling of solid flat plates or curved plates as target surfaces. The main objective of this study was to examine the effects of fin perforations on the thermal performance of pin-fin heat sinks. Numerical simulations demonstrate the effects of the Reynolds number on the thermal resistance, Nusselt number, and fin efficiency of different types of heat sinks.

2 Computational Method and Numerical Model

2.1 Geometry and Boundary Conditions

The numerical simulation herein is performed using the finite element method using COMSOL Multiphysics, version 5.1. Figure 1 presents the computational domain showing different dimensions and the boundary conditions used in the numerical simulation, which is similar to the domain used by Li et al. [16]. The heat sink has a 6 × 6 array of fins. The length (L) and the thickness (B) of the base plate are 80 mm and 8 mm, respectively. The perimeter and the bottom of the base are thermally insulated. The top of the base and the fins are in contact with the impinging jet. The diameter (D) of the nozzle and the distance (Y) between the exit of the nozzle and the tip of the heat sink are 8 mm and 64 mm, respectively. The dimensions of the confining plates are 200 mm × 200 mm [16]. Different types of heat sink configurations are shown in Fig. 3.

In this work, the flow is assumed to be steady, incompressible, and turbulent. The fluid and the solid properties are regarded as constant. Non-uniform computational meshes that are finer near the heat sink and the nozzle, where the velocities or the temperatures have higher gradient values and are used to increase the accuracy of the results (Fig. 2).

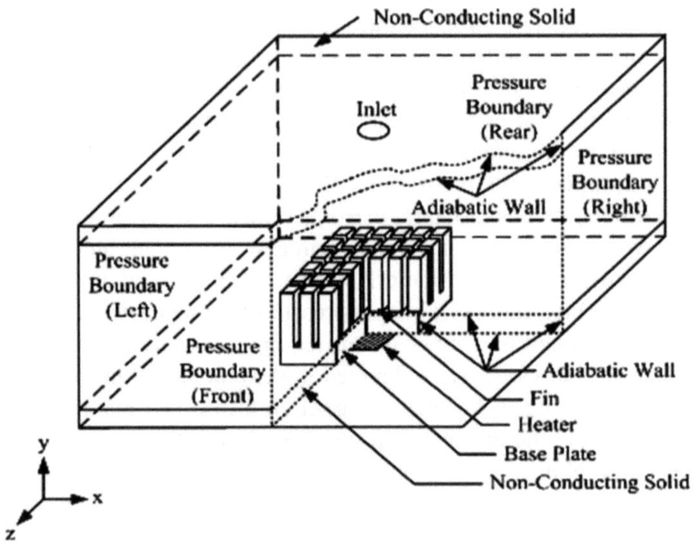

Fig. 1 Computational domain with the boundary conditions [16]

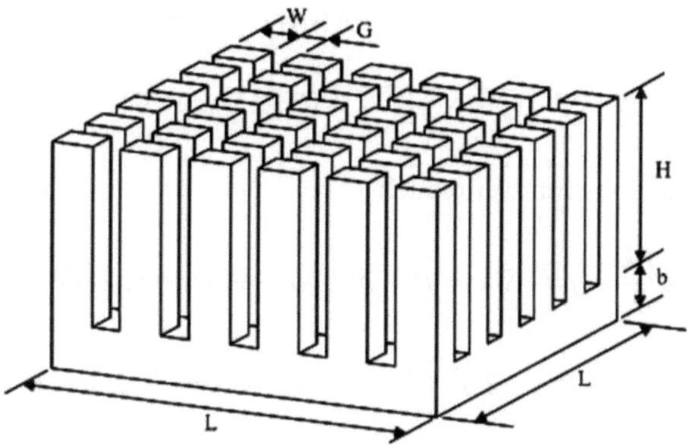

Fig. 2 Geometry of the heat sink showing all the dimensions

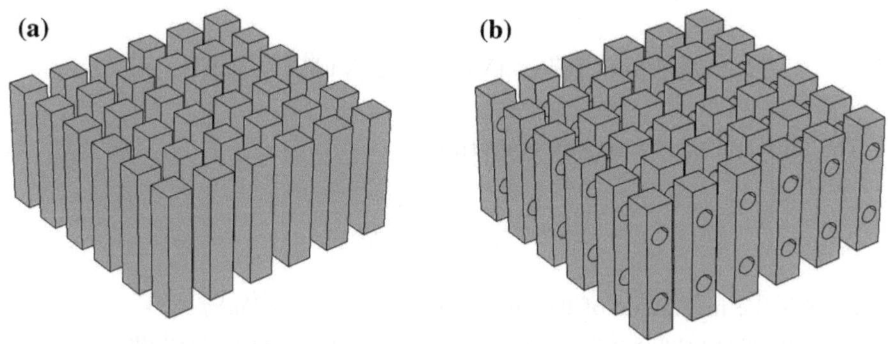

Fig. 3 Geometry of the heat sink: (a) solid fins (b) perforated fins

2.2 Mathematical Model

Three-dimensional governing equations for steady-state incompressible flow and turbulent modeling using the SST-based k-ω model [32–36] are solved using COMSOL Multiphysics 5.0 in this study. For temperature field calculation, the conjugate problem of Fourier's steady-state heat conduction equation with convection in the fluid is solved simultaneously in the present study. The governing equations of mass, momentum, turbulent kinetic energy, turbulent energy dissipation rate, and energy in the steady turbulent main flow using the SST k−ω model are as follows [32–36].

Continuity equation:

$$\nabla \cdot U = 0. \tag{1}$$

Reynolds-averaged Navier–Stokes (RANS) equation:

$$\rho U \cdot \nabla U + \nabla \cdot \left(\overline{\rho \mathrm{u}' \otimes \mathrm{u}'}\right) = -\nabla \mathrm{P} + \nabla \cdot \mu\left(\nabla \mathrm{U} + (\nabla \mathrm{U})^{\mathrm{T}}\right). \tag{2}$$

Transport equations for k-ω model:

$$\rho \frac{\partial k}{\partial t} + \rho u \cdot \nabla k = P - \rho \beta_0^* k\omega + \nabla((\mu + \sigma_k \mu_T)\nabla k). \tag{3}$$

$$\rho \frac{\partial \omega}{\partial t} + \rho u \cdot \nabla \omega = \frac{\rho \gamma}{\mu_T} P - \rho \beta \omega^2 + \nabla((\mu + \sigma_\omega \mu_T)\nabla \omega) + 2(1 - F_1)\frac{\rho \sigma_{\omega 2}}{\omega} \nabla \omega \cdot \nabla k, \tag{4}$$

where $\mathrm{P} = \min\left(\mathrm{P_k}, 10\beta_0^* \mathrm{k}\omega\right)$
and

$$P_k = \mu_{\mathrm{T}}\left(\nabla \mathrm{u} : \left(\nabla \mathrm{u} + (\nabla \mathrm{u})^{\mathrm{T}}\right) - \frac{2}{3}(\nabla \mathrm{u})^2\right) - \frac{2}{3}\rho \mathrm{k} \nabla \cdot \mathrm{u}.$$

The turbulent eddy viscosity is given by

$$\mu_T = \frac{\rho a_1 k}{\max(a_1 \omega, SF_2)},$$

where S is the magnitude of the strain-rate tensor $S = \sqrt{2S_{ij}S_{ij}}$.

Each of the constants is a blend of the corresponding constants of the k-ε and the k-ω model.

$$\emptyset = F_1 \emptyset_1 + (1 - F_1)\emptyset_2.$$

The interpolation functions $\mathrm{F_1}$ and $\mathrm{F_2}$ are defined as

$$F_1 = \tanh\left(\theta_1^4\right).$$

$$\theta_1 = \min\left[\max\left(\frac{\sqrt{k}}{\beta_0^* \omega l_w}, \frac{500\mu}{\rho \omega l_w^2}\right), \frac{4\rho \sigma_{\omega_2} k}{CD_{k\omega} l_w^2}\right].$$

$$CD_{k\omega} = \max\left(\frac{2\rho \sigma_{\omega 2}}{\omega} \nabla \omega \cdot \nabla k, 10^{-10}\right).$$

and

$$F_2 = \tanh\left(\theta_2^2\right).$$

$$\theta_2 = \max\left(\frac{2\sqrt{k}}{\beta_0^* \omega l_w}, \frac{500\mu}{\rho \omega l_w^2}\right),$$

where l_w is the distance to the closest wall. Realizability constraints are applied to the SST model. The model constants are given by [32–36]

$$\begin{aligned}
&\beta_1 = 0.075, \gamma_1 = 5/9, \sigma_{k1} = 0.85, \sigma_{\omega 1} = 0.5, \\
&\beta_2 = 0.0828, \gamma_2 = 0.44, \sigma_{k2} = 1, \sigma_{\omega 2} = 0.856, \\
&\beta_0^* = 0.09, \mathrm{a}_1 = 0.31.
\end{aligned}$$

2.3 *Model Verification*

A comparison between previously published data [13, 16] and present numerical results is shown in Fig. 4. To validate the present simulation, same geometry and boundary conditions are applied as the experimental one [13, 16]. After performing the mesh sensitivity test, in this study, minimum mesh element 70,000 is used for all the simulations. Mostly within all the range of parameters, it is observed that CFD results are in good agreement with experimental results (~10% error).

The thermal resistance of heat sink is defined as

$$R_{\mathrm{th}} = \frac{T_{\mathrm{ave}} - T_{\infty}}{Q}, \tag{5}$$

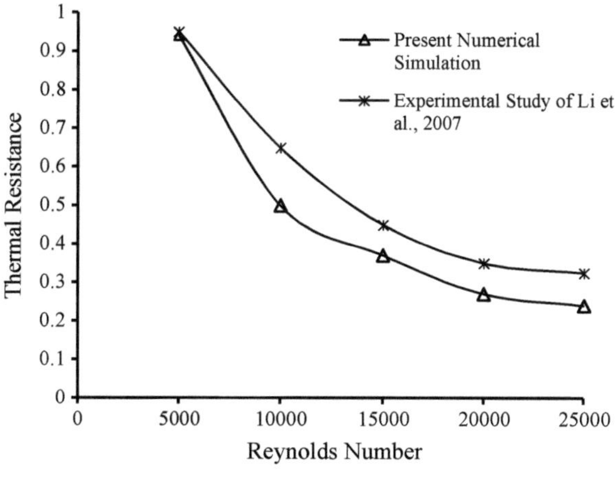

Fig. 4 Comparison between previously published and present numerical results

where T_{ave} is the average base temperature of heat sink, T_{∞} is the inlet temperature (22 °C) of impinging jet and Q is heating power (W). The equations are terminated and considered to be converged when the normalized residual of the algebraic equation is less than 10^{-3}, but for the energy equation, the value is 10^{-7} [16].

3 Results and Discussion

The present analysis of flow field and convection heat transfer for the conjugate problem is carried out for $Re_D = 5 \times 10^3$ to 25×10^3. Thermal field in the solid domain is determined by solving conduction equation. The effects of various parameters with the variation of the Reynolds number are investigated numerically. Three-dimensional finite element analyses are employed to solve the present numerical problem. The properties are set to be $\rho = 1.1614$ kg/m, $C_p = 1005$ J/kg-K, $k = 0.0261$ W/m-K for air and $\rho = 2800$ kg/m, $C_p = 900$ J/kg-K, $K = 168$ W/m-K for aluminum. The heat flux Q is assumed 25 W.

Figures 5 and 6 show the temperature distribution (in °C) for two types of fin geometry at XZ plane (top view near the base) with H/L = 0.5, W/L = 0.1, Y/D = 8, Q = 25 W and Re = 10,000. From the figures, it can be seen that the middle section of the heat sink base is comparatively cooler than the peripheral sections. It is because of the stronger momentum of flow at the middle section due to impinging jet. As the chip is generally attached with the middle section of the heat sink; from Figs. 5 and 6, it can be concluded that perforated fins have better heat removal rate.

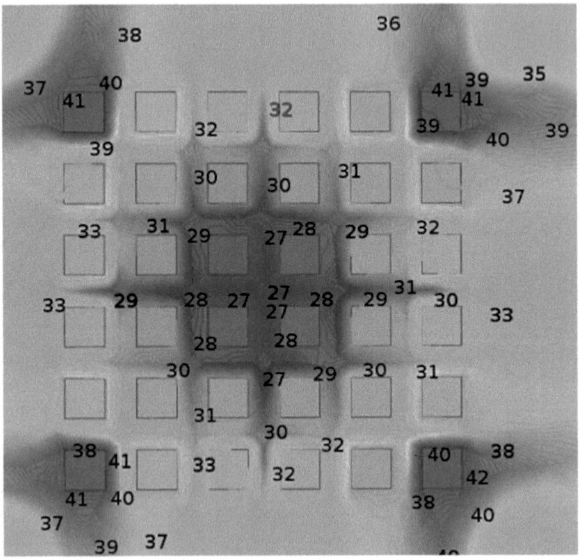

Fig. 5 Temperature distribution near the base (XZ plane) of heat sink having solid pin fins for Re = 10000

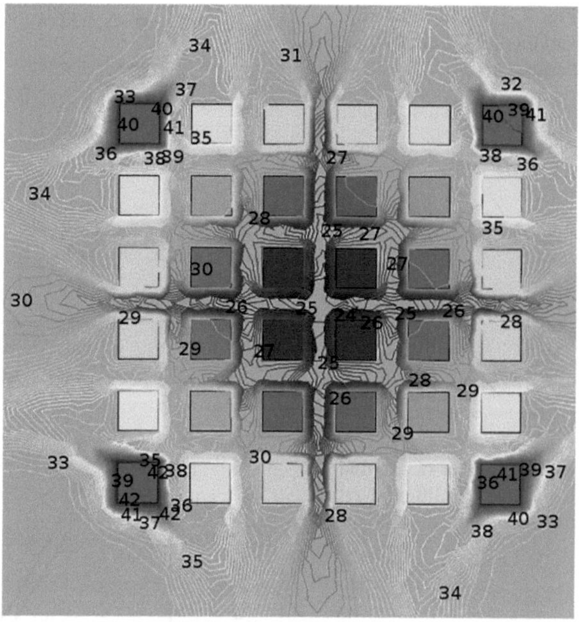

Fig. 6 Temperature distribution near the base (XZ plane) of heat sink having perforated pin fins for Re = 10000

In this study, the Nusselt number can be defined as follows:

$$Nu = \frac{Q}{A_h(T_b - T_{in})k_a}. \tag{6}$$

If d = perforation diameter (5 mm), n = number of fins, and m = number of perforation, then the total heat transfer surface area for the solid (A_{hs}) and perforated (A_{pf}) fins are as follows, respectively:

$$A_{hs} = 4 \times n \times w \times H, \tag{7}$$

$$A_{hp} = 4 \times n \times w \times H + m \times n \times \frac{\pi}{4} d^2. \tag{8}$$

The coefficient of enhancement (COE) can be defined as the improvement in heat transfer rates due to the different types of the heat sink fins.

$$COE = \frac{Nu_{perforated}}{Nu_{solid}}. \tag{9}$$

From Figs. 7 and 8, it can be noted that perforated fins yield the larger convective coefficient and so have the greater Nusselt number and smaller thermal resistance. The results indicate that Nusselt number increases with the increase of

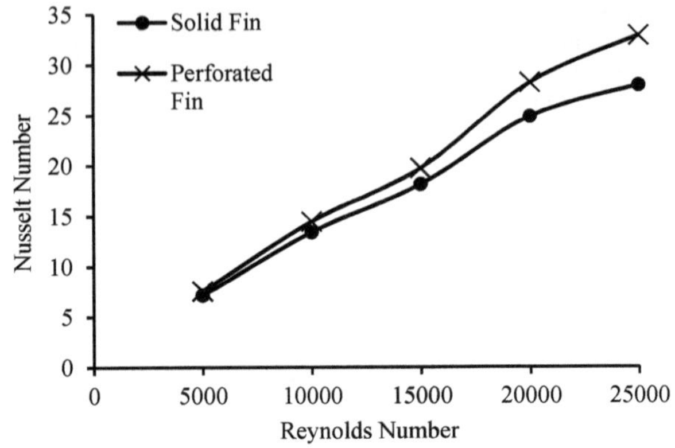

Fig. 7 Nusselt number variation at different Reynolds numbers

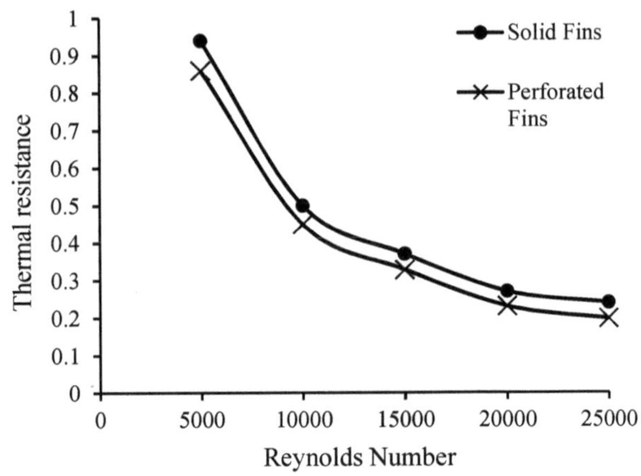

Fig. 8 Effects of thermal resistance for different types of fin configurations

impinging Reynolds number. Due to perforations, heat transfer surface area is increased and thus increased the heat removal rate. It is found that when the inlet air temperature and the heat flux are kept constant, the thermal resistance decreases with increasing the Reynolds number. This is because the heat transfer rate increases as the airflow rate increases. It is necessary to choose appropriate fin dimensions and fin arrangements at a particular Reynolds number for better thermal performance. The heat transfer performance has been increased with the increase of Reynolds number. From Fig. 9, one can find that perforated fins have higher heat transfer performance and can exchange more heat between primary surface and the ambient air.

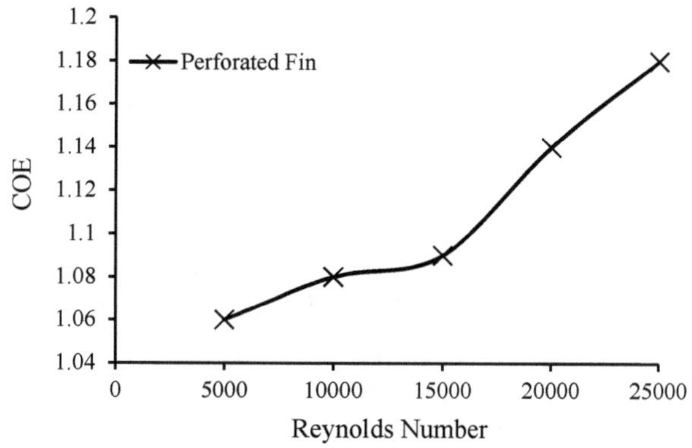

Fig. 9 Coefficient of heat transfer enhancement at different Reynolds numbers for perforated fin configurations

4 Conclusion

A finite element numerical simulation was employed to investigate the thermal performance of perforated heat sinks at various impinging Reynolds numbers. The purpose of this study was to evaluate the possibility of improving the thermal performance by introducing perforated fins. The impinging distance, fin width, and fin heights were kept constant to measure and compare the thermal performance of the heat sinks. From the study, it can be concluded that:

- The impinging Reynolds number plays an important role in measuring the thermal resistance and fin efficiency of the heat sink. Increasing the Reynolds number consistently diminishes the thermal resistance.
- Heat transfer performance enhancement rate increases with the increase of Reynolds number.
- It is found that perforations significantly increase the convective heat transfer coefficient as well as the Nusselt number.
- For all the cases, perforated fins have higher heat removal rate and COE values are >1.
- From the study, it can also be concluded that Nusselt number and fin efficiency may be increased with increasing of fin number and the number perforations. But there should be an optimum value of the perforation number and the perforation diameter.

References

1. Browne, E. A., Michna, G. J., Jensen, M. K., & Peles, Y. (2010). Experimental investigation of single-phase microjet array heat transfer. *Journal of Heat Transfer, 132,* 041013.
2. Brignoni, L. A., & Garimella, S. V. (1999). Experimental optimizations of confined air jet impingement on a pin fin heat sink. *IEEE Transactions on Components and Packaging Technologies, 22*(3), 399–404.

3. El-Sheikh, H. A., & Garimella, S. V. (2000). Enhancement of air jet impingement heat transfer using pin-fin heat sinks. *IEEE Transactions on Components and Packaging Technologies, 23*(2), 300–308.
4. Fabbri, M., & Dhir, V. K. (2005). Optimized heat transfer for high power electronic cooling using arrays of microjets. *Journal of Heat Transfer, 127*(7), 760–769.
5. Ji-Jinn, F., Pui, S.-Y., Lai, Y.-L., & Chin, S.-B. (2012). Forced convective heat transfer enhancement with perforated pin fins subject to an impinging flow. *SEGi Review, 5*(1), 29–40.
6. Haq, R. F. B., Akintunde, K., & Probert, S. D. (1995). Thermal performance of a pin-fin assembly. *International Journal of Heat and Fluid Flow, 16,* 50–55.
7. Kim, D. K., Kim, S. J., & Bae, J. J. (2009). Comparison of thermal performances of plate-fin and pin-fin heat sinks subject to an impinging flow. *International Journal of Heat and Mass Transfer, 52,* 3510–3517.
8. Yang, Y. T., & Peng, H. S. (2009). Numerical study of the heat sink with un-uniform fin width designs. *International Journal of Heat and Mass Trasnfer, 52,* 3473–3480.
9. Sparrow, E. M., Ramsey, J. W., & Altermani, C. A. C. (1980). Experiments on in-line pin-fin array and performance: Comparisons with staggered array. *ASME Journal of Heat Transfer, 102,* 44–50.
10. Yang, J., Zeng, M., Wang, Q., & Nakayama, A. (2010). Forced convection heat transfer enhancement by porous pin fins in rectangular channels. *ASME Journal of Heat Transfer. 132*, 051702-1–051702-8.
11. Maveety, J. G., & Hendricks, J. F. (1999). A heat sink performance study considering material, geometry, nozzle placement, and Reynolds number with air impingement. *ASME Journal of Electronic Packaging, 121,* 156–161.
12. Kim, S. Y., & Kuznetsov, A. V. (2003). Optimization of pin-fin heat sinks using anisotropic local thermal non-equilibrium porous model in a jet impinging channel. *Numerical Heat Transfer (Part A), 44,* 771–787.
13. Li, H. Y., Chao, S. M., & Tsai, G. L. (2005). Thermal performance measurement of heat sinks with confined impinging jet by infrared thermography. *International Journal of Heat and Mass Transfer, 48,* 5386–5394.
14. Li, H. Y., & Chen, K. Y. (2005). Thermal-fluid characteristics of pin-fin heat sinks cooled by impinging jet. *Journal of Enhanced Heat Transfer, 12*(2), 189–201.
15. Li. H. Y., Chao, S. M., & Tsai, G. L. (2005). *Thermal Performance Measurement of Heat Sinks with Confined Impinging Jet by Infrared Thermography, 16th International Symposium on Transport Phenomena*, ISTP-16, Prague.
16. Li, H. Y., Chiang, M. H., & Chen, K. Y. (2007). Performance analysis of pin-fin heat sinks with confined impingement cooling. *IEEE Transactions on Components and Packaging Technologies, 30*(3), 383–389.
17. Li, H. Y., & Chen, K. Y. (2007). Thermal performance of plate-fin heat sinks under confined impinging jet conditions. *International Journal of Heat and Mass Transfer, 50,* 1963–1970.
18. Li, H. Y., & Chen, K. Y. (2009). Thermal-fluid characteristics of pin-fin heat sinks cooled by impinging jet. *Energy Conversion and Management, 50,* 2738–2746.
19. Lou, Z. Q., Mujumdar, A. S., & Yap, C. (2005). Effects of geometric parameters on confined impinging jet heat transfer. *Applied Thermal Engineering, 25*(17–18), 2687–2697.
20. Wang, E. N., Zhang, L., Jiang, L., Koo, J.-M., Maveety, J. G., Sanchez, E. A., et al. (2004). Micro-machined jets for liquid impingement cooling of VLSI chips. *Journal of Microelectromechanical Systems, 13*(5), 833–842.
21. Yangming, L. Y., Zhang, S., & Wang, J. L. (2010). *Numerical investigation on the thermal properties of the micro-cooler 11th International conference on electronic packaging technology and high density packaging* (ICEPT-HDP), China, (pp. 634−638).
22. Farhad, I. M. (2013). Effects of perforations on the thermal and fluid dynamic performance of a heat exchanger. *IEEE Transactions on Components, Packaging and Manufacturing Technology, 3*(7), 1178–1185.
23. Ismail, M. F., Hasan, M. N., & Saha, S. C. (2014). Numerical study of turbulent fluid flow and heat transfer in lateral perforated extended surfaces. *Energy, 64,* 632–639.

24. Ledezma, G., Morega, A. M., & Bejan, A. (1996). Optimal spacing between pin fins with impinging flow. *Transactions of the ASME Journal of Heat Transfer, 118,* 570–577.
25. Maveety, J. G., & Jung, H. H. (2002). Heat transfer from square pin-fin heat sinks using air impingement cooling. *IEEE Transactions on Components and Packaging Technologies, 25* (3), 459–469.
26. Hwang, J. J., & Lui, C. C. (1999). Detailed heat transfer characteristic comparison in straight and 90-deg turned trapezoidal ducts with pin-fin arrays. *International Journal of Heat and Mass Transfer, 42,* 4005–4016.
27. Hwang, J. J., & Lui, C. C. (2002). Measurement of end wall heat transfer and pressure drop in a pin-fin wedge duct. *International Journal of Heat and Mass Transfer, 42,* 877–889.
28. Rao, Y., Wan, C., Xu, Y., & Zang, S. S. (2011). Spatially-resolved heat transfer characteristics in channels with pin fin and pin fin-dimple arrays. *International Journal of Thermal Sciences, 50,* 2277–2289.
29. Sahin, B., & Demir, A. (2008). Thermal performance analysis and optimum design parameters of heat exchanger having perforated pin fins. *Energy Conversion and Management, 49,* 1684–1695.
30. Eid, E. I., Gomaa, A. G., & Gomaa, M. E. (2011). Heat transfer characteristics from an array of thin strips pin fins due to their exposures to a single downward jet impingement. *Heat Mass Transfer, 47,* 211–221.
31. Ismail, M. F., Saha, S. C., & Sarkar, M. A. R. (2015). Effects of lateral perforation geometry on heat transfer characteristics on extended surfaces, In M. M. K. Khan, N. M. S. Hassan (Eds.), Thermofluids modeling for energy efficiency applications, (1st ed.) Elsevier, (Chapter 5) (pp. 103−117).
32. Incropera, F. P., & DeWitt, D. P. (1996). *Fundamentals of Heat and Mass Transfer* (4th ed., pp. 126–128). New York: Wiley.
33. Menter, F. R. The menter shear stress transport turbulence model. http://turbmodels.larc.nasa.gov/sst.html.
34. Menter, F. (1993). Zonal two equation k-w turbulence models for aerodynamics flows. AIAA Pap *93* e 2906.
35. Menter, F. R. (1994). Two-equation eddy-viscosity turbulence models for engineering applications. AIAA Journal, *32(8)* 1598 e 605.
36. COMSOL, 4.3 b. (2013) CFD module User's Guide, COMSOL, Inc.1. New England Executive Park, Burlington, Mass, USA.

Multiphase Flow in Porous Media: Cake Formation During Extreme Drilling Processes

M.A. Kabir and Isaac K. Gamwo

Abstract The successful completion of a wellbore requires a low-permeable filter cake to be deposited on the wellbore walls to seal the porous formation exposed by the drill bit. The filtration processes triggered by the differential pressure between the drilling fluid in the borehole and the pore spaces of formation rocks are well known in both drilling operations and, of greater importance, in the subsequent production of oil. The filter cake acts as a barrier to prevent excessive drilling fluid loss into formation, and invasion of formation fluid. The presence of a filter cake also provides wellbore stability and reduces damage to the formation. Understanding cake formation and fluid flow through porous media is necessary for a successful drilling process. This need becomes even more important during extreme drilling, when pressure and temperature may exceed 35,000 psi and 500 °F. This work presents our effort to simulate the fluid flow and cake formation in extreme drilling processes. Earlier investigations were focused on single-phase flow phenomena in porous media; recent studies have emphasized multiphase (thermo-fluid) flow in porous media to closely mimic the actual drilling fluid composed of fine particles and viscous fluid. In the present study, the Eulerian–Eulerian approach for multiphase flow is employed to evaluate the fluid flow and cake formation patterns during ultra-deep drilling at high-temperature, high-pressure conditions. The rheology of the fluid has been published previously [14] and is repeated here for completeness. Two competitive sub-models were considered: the power-law and the Herschel–Bulkley models. The Herschel–Bulkley rheological model appears superior and best describes the non-Newtonian rheological behavior of drilling fluid due to the yield stress term present in this model.

M.A. Kabir · I.K. Gamwo (✉)
National Energy Technology Laboratory, United States Department of Energy,
Research & Innovation Center, Pittsburgh, PA 15236, USA
e-mail: gamwo@netl.doe.gov

Present Address:
M.A. Kabir · I.K. Gamwo
SABIC Innovative Plastics, Polymer Processing Development Center, Pittsfield, MA, USA

M.M.K. Khan et al. (eds.), *Application of Thermo-fluid Processes in Energy Systems*,
Green Energy and Technology, https://doi.org/10.1007/978-981-10-0697-5_11

Keywords Drilling fluid · Multiphase flow · Filter cake formation · Computational fluid dynamics · Deep drilling

1 Introduction

Drilling fluids play a vital role in drilling operations and perform several important tasks: (a) helping to reduce friction and wear on the drilling bit, (b) transporting the drilled solids, (c) maintaining a favorable pressure difference between the wellbore and the rock formation, (d) cooling the cutters to maintain the temperature below critical, (e) generating a filter cake on the wellbore wall to minimize incursion of drilling fluids into the formation [1, 3, 5, 7, 36, 38].

Additionally, loss of drilling fluids during operations through hydrocarbon-bearing formations is expected to be minimized. The drilling fluid filtrates can lead to formation damage because of rock wettability changes, fines migration, drilling fluid solids plugging, and formation water chemistry incompatibilities [4]. The filtration properties are one of the most important characteristics of all drilling fluids. The invasion of filtrate into the formation can substantially reduce the permeability of the region near the wellbore and may occur through several mechanisms: clay swelling, particles pore plugging, particles migration, and water blocking. Moreover, the nature and thickness of the filter cake deposited on the borehole wall may cause a pressure differential that can lead to sticking [2, 4, 14, 39].

Drilling fluids are designed to exhibit non-Newtonian properties. The drilling mud prevents the rock cuttings from settling at the bottom hole under static conditions as the drilling fluid circulation is stopped, (e.g., to replace the drilling pipe or bit). Yet, at high shear rates when drilling operation resumes, the drilling fluid is expected to flow like a viscous paste removing drilled rock cuttings from the well bottom and carry them to the surface. Accurate modeling of the drilling process and filter cake formation is important so that reliable multiphase flow mathematical models may be developed; these models would predict the drilling fluid flow rate necessary to remove cuttings that would otherwise cause severe problems during drilling such as high drag and torque.

Literature review [1, 5, 10, 27] shows that little research has been carried out modeling on deep drilling processes and cake formation to predict multiphase flow behavior. Recently, the extreme drilling process attracted the attention of fluid researchers due to the drilling fluid vital role in rock cutting removal and in filter cake formation.

In this study, we utilized the Eulerian–Eulerian multiphase flow model to evaluate the drilling fluid flow pattern and filter cake formation patterns for two rheological models: the power-law model and the Herschel–Bulkley model. The parameters in the Herschel–Bulkley rheological model best describe the non-Newtonian rheological behavior of drilling fluid. This includes a yield stress that would allow cuttings to float under static conditions. Our simulations of the flow patterns indicate that the Herschel–Bulkley model is a more accurate model of rheological behavior, and it exhibits better cutting removal performance.

This investigation was further extended to study the effect of differential pressure on filter cake thickness. The ability to optimize filter cake characteristics is extremely useful [10]. The presence of filter cake reduces both formation damage and fluid loss. With thicker cake, the effective diameter of the hole is reduced and problems may arise, such as excessive torque when rotating the drill string and excessive drag when pulling it. Here we have demonstrated a correlation between cake thicknesses and pressure drop by varying the pressure drop from 250 psi (1.72 MPa) to 1000 psi (6.89 MPa) with incremental pressure drop of 250 psi (1.72 MPa) under extreme drilling conditions.

Most of the previous studies have been carried out with Newtonian, single-phase and isothermal conditions for the shallow drilling process. Here, we have performed initial research on the computational fluid dynamics (CFD) modeling and simulations of the flow pattern and filter cake formation resulting in multiphase flow in porous media at extremely high-pressure and high- temperature (up to 25,500 psi or 175.8 MPa and 170 °C) drilling process. The drilling fluid was treated as a two-phase system with solid (45 μm) particles suspended in a non-Newtonian fluid where fluid phase was modeled using the power-law model without yield stress and the Herschel–Bulkley model with yield stress. The CFD code ANSYS Fluent was used for solving mass, momentum, and energy equations for fluid and solid phases where a Eulerian–Eulerian multiphase model was employed.

In this chapter, we have used three main interlinked solution aspects to modeling [10] the filter cake formation in multiphase flow in porous media for a deep drilling process (25,000 psi or 172.4 MPa and 170 °C). The solution aspects used are (i) the Eulerian–Eulerian multiphase fluid flow model in the pipe as well as in the annulus between the pipe and the borehole, (ii) the filter cake deposition model by retaining of the annulus fluid particles at the formation cavity during the seepage, and (iii) the drilling fluid seepage into the formation during and after the cake formation.

Our main focus was to numerically study the effect of two rheological models on the fluid flow pattern and filter cake formation at high temperatures and high pressures utilizing CFD tools. Numerical predictions were performed for a range of overbalance differential pressure from 250 psi (or 1.72 MPa) to 1250 psi (or 8.62 MPa) with incremental pressure of 250 psi (or 1.72 MPa) to study the effect of differential pressure over cake thickness.

2 Governing Equations—Multiphase Flow in Porous Media

The dynamics of solids in fluid media have a large effect on various flow phenomena such as density, viscosity, and pressure. Thus, the hydrodynamics of solids must be modeled correctly [6]. The Eulerian approach is preferred over the

Lagrangian due to the large volume fraction of solids in the drilling fluid. In the Eulerian approach, fluid and solid phases are treated as interpenetrating continua, and momentum and continuity equations are defined for each phase [12]. Therefore, the Eulerian–Eulerian multiphase fluid model has been used to simulate fluid flow and filter cake formation for deep drilling process conditions. The Eulerian model solves a set of n momentum and continuity equations for each phase. Coupling is achieved through the pressure and interphase exchange coefficients. The manner in which this coupling is handled depends upon the type of phases involved. For granular flows, properties are obtained by applying kinetic theory. Mass transfer between the phases is negligible and, therefore, ignored here. The momentum equation for the solid phase differs from the equation used for the fluid phase, since the former contains a solids pressure [12, 17, 18, 28]. Lift and virtual mass forces are assumed to be negligible in the momentum equations. Details of the rheological models used in this study can be found elsewhere [14].

2.1 *Modeling Fluid Flow in the Annulus*

Multiphase equations for modeling the flow of steady, laminar, non-isothermal, incompressible fluid are given in the following sections [6, 12, 17, 18, 28].

2.1.1 Conservation of Mass

$$\text{For liquid,} \quad \nabla.(\alpha_l v_l) = 0 \tag{1}$$

$$\text{For solids,} \quad \nabla.(\alpha_s v_s) = 0 \tag{2}$$

where α is the volume fraction and subscripts l and s denote liquid and solid phases, respectively. Moreover, $\alpha_l + \alpha_s = 1$ must be satisfied. v_l and v_s are the velocities of the solid and liquid phases, respectively.

2.1.2 Momentum Balance

Liquid Phase

The momentum equation for the liquid phase in a solid–liquid system [6, 12, 17, 18, 28] is as follows:

$$\underbrace{\nabla.(\alpha_l \rho_l \vec{v}_l \vec{v}_l)}_{\text{Convective}} = -\underbrace{\alpha_l \nabla p}_{\text{Pressure}} + \underbrace{\nabla.\overline{\overline{\tau_l}}}_{\text{Stress}} + \underbrace{\alpha_l \rho_l \vec{g}}_{\text{Body}} - \left\{ \left(\underbrace{K_{sl}(\vec{v}_l - \vec{v}_s)}_{\text{Forces Exchange}} \right) \right\} \tag{3}$$

where ρ_l and ρ_s are the densities of liquid and solid phases, respectively.

To address non-Newtonian behavior of the liquid phase in the multiphase drilling fluid, we have used the power-law model input parameters in the simulation [10, 12].

For the fluid, the stress tensor, $\overline{\tau_l}$, is related to the fluid strain rate tensor, $\bar{\dot{\gamma}}_l = \nabla\vec{v}_l + (\nabla\vec{v}_l)^{tr}$, by:

$$\overline{\tau_l} = \alpha_l \tau \bar{\dot{\gamma}}_l + \alpha_l \left(\lambda_l - \frac{2}{3}\tau \right) \nabla.\vec{v}_l \bar{I} \tag{4}$$

where $\tau = \tau_o + k\left|\bar{\dot{\gamma}}_l\right|^{n-1}$ or $\tau = k\left|\bar{\dot{\gamma}}_l\right|^{n-1}$ and $\left|\bar{\dot{\gamma}}_l\right|$ is the magnitude of the strain rate tensor defined as $\left|\bar{\dot{\gamma}}\right| = \sqrt{\frac{1}{2}\sum_i \sum_j \dot{\gamma}_{ij}\dot{\gamma}_{ji}}$, τ_o is the yield stress, and k and n are consistency factor and power-law exponent, respectively [10, 12, 16].

Solid Phase

The momentum equation for the solid phase in a solid–liquid system [6, 12, 17, 18, 28] is:

$$\nabla.(\alpha_s \rho_s \vec{v}_s \vec{v}_s) = -\alpha_s \nabla p - \underbrace{\nabla p_s}_{\text{Solid Pressure}} + \nabla.\bar{\tau}_s + \alpha_s \rho_s \vec{g} + \{(K_{ls}(\vec{v}_l - \vec{v}_s))\} \tag{5}$$

The solids pressure p_s, stress $\bar{\tau}_s$, and viscosity μ_s are determined by particle fluctuations, the kinetic energy associated with these fluctuations, and the granular temperature Θ.

The stress-strain relationship for the solid phase s is:

$$\overline{\tau_s} = \alpha_s \underbrace{\mu_s \bar{\dot{\gamma}}_s}_{\text{Shear stress}} + \alpha_s \left(\underbrace{\lambda_s}_{\text{Bulk viscosity}} - \frac{2}{3}\mu_s \right) \nabla.\vec{v}_s \underbrace{\bar{I}}_{\text{Unit tensor}} \tag{6}$$

where solid strain rate tensor $\bar{\dot{\gamma}}_s = \nabla\vec{v}_s + (\nabla\vec{v}_s)^{tr}$.

Interaction forces are considered here to account for the effects of other phases and are reduced to zero for single-phase flow [6, 12]. The momentum exchange coefficients are indistinguishable $(K_{ls} = K_{sl})$

$$K_{sl} = \frac{\alpha_s \rho_s f}{\mathrm{T}_s^p} \tag{7}$$

This function and these coefficients are suitable for drilling process modeling where re-circulating multiphase fluids contain high solid fraction.

Here, T_s^p is the particulate relaxation time and f is the model dependent drag function.

The relaxation time is expressed as

$$T_s^p = \frac{\rho_s d_s^2}{18\mu_l} \tag{8}$$

where d_s is the solid particle diameter.

When the Syamlal–O'Brien Drag function f [6, 12] is used:

$$f = \frac{C_D R_{e_s} \alpha_l}{24 v_{r,s}^2} \tag{9}$$

The relative Reynolds number R_{e_s} can be written as follows [6, 12]:

$$R_{e_s} = \frac{\rho_l d_s |\vec{v}_s - \vec{v}_l|}{\mu_l} \tag{10}$$

The drag function f includes a drag coefficient C_D and the relative Reynolds number R_{e_s}; however, the drag function differs among the exchange-coefficient models. For the drilling process, multiphase drilling fluid with a high solid fraction continuously cycles through the drill assembly and carries away debris produced by the drilling process.

In the Syamlal–O'Brien model, the drag function of Dalla Valle is used [6, 12] where $v_{r,s}$ is the terminal velocity correlation:

$$C_D = \left[0.63 + \frac{4.8}{\sqrt{Re_s / v_{r,s}}}\right]^2 \tag{11}$$

The terminal velocity correlation $v_{r,s}$ for solid phase has the following form:

$$v_{r,s} = 0.5\left(A - 0.06\mathrm{Re}_s + \sqrt{(0.06\mathrm{Re}_s)^2 + 0.12\mathrm{Re}_s(2B - A) + A^2}\right) \tag{12}$$

where

$A = \alpha_l^{4.14}$
$B = 0.8\alpha_l^{1.28}, \quad \alpha_l \leq 0.85$
$B = \alpha_l^{2.65}, \quad \alpha_l > 0.85$

This correlation is based on measurements of the terminal velocities of particles in fluidized or settling beds where high solid volume fractions, similar to solid volume fractions in drilling fluids, are encountered.

The solid pressure P_s is composed of a kinetic term (first term), a particle collisions term (second terms) and a friction term (3rd term) [6, 12]:

$$P_s = \alpha_s \rho_s \Theta_s + 2\rho_s (1 + e_{ss}) \alpha_s^2 g_{0,ss} \Theta_s + F_r \frac{(\alpha_s - \alpha_{s,\min})^n}{(\alpha_{s,\max} - \alpha_s)^p} \quad (13)$$

Both kinetic and collision terms are dependent on the granular temperature Θ. The term e_{ss} is the particle–particle coefficient of restitution (taken here to be $e_{ss} = 0.9$; this choice is consistent with values reported in the literature under similar simulation conditions), $g_{0,}$ is the radial distribution function. This is a correction factor (the non-dimensional distance between spheres) that modifies the probability of collisions between particles when the granular phase becomes dense. The friction is included in this study because the solid volume fraction is relatively high, which may give rise to friction. In this work, the friction pressure is modeled using the semi-empirical model proposed by Johnson et al. [19], where $\alpha_{s,\min}$ and $\alpha_{s,\max}$ are the minimum and maximum packing respectively; $\alpha_{s,\min}$, assumed to be 0.5, is the solid concentration when friction stresses becomes important. The values of empirical materials constants F_r, n, and p are taken to be 0.5, 2.0, and 5.0 respectively following other investigators [19].

2.1.3 Energy Equation

To describe the conservation of energy in Eulerian multiphase applications, a separate steady-state enthalpy equation can be written for each phase q (liquid or solid) [6, 12] as follows:

$$\nabla . \left(\underbrace{\alpha_q \rho_q \vec{u}_q h_q} \right) = \underbrace{\overline{\tau_q} : \nabla \vec{u}_q} - \underbrace{\nabla . \vec{q}_q} + \sum_{p=1}^{n} \left(\underbrace{\vec{Q}_{pq}} \right) \quad (14)$$

where h_q is the specific phase enthalpy, $\vec{q_q}$ is the heat flux, and $\vec{Q}_{pq}$ is the intensity of heat exchange between phases.

2.1.4 Granular Temperature

The granular temperature for the solid phase must be specified for particulate viscosities. We used a partial differential equation, which was derived from the transport equation by neglecting convection and diffusion. It takes the following form [6, 12]:

$$0 = (-p_s \bar{I} + \overline{\tau_s}) : \nabla \vec{v_s} - \gamma_{\Theta_s} + \phi_{ls} \quad (15)$$

where $(-p_s\bar{I}+\overline{\tau_s}):\nabla\vec{v}_s$ is the generation of energy by the solid stress tensor, γ_{Θ_s} is the collisional dissipation of energy, and ϕ_{ls} is the energy exchange between the fluid and the solid phase.

The collisional dissipation of energy, γ_{Θ_s}, represents the rate of energy dissipation within the solid phase due to collisions between particles. The term is represented by the following expression derived by Lun [26]:

$$\gamma_{\Theta_s} = \frac{12\left(1-e_{ss}^2\right)g_{0,ss}}{d_s\sqrt{\pi}}\rho_s\alpha_s^2\Theta_s^{3/2} \tag{16}$$

The transfer of the kinetic energy of random fluctuations in particle velocity from the solid phase to the liquid phase is represented by ϕ_{ls}:

$$\phi_{ls} = -3K_{ls}\Theta_s \tag{17}$$

The radial distribution function $g_{0,ss}$ is modeled as follows [6, 8, 12]:

$$g_{0,ss} = \left[1-\left(\frac{\alpha_s}{\alpha_{s,\max}}\right)^{1/3}\right]^{-1} \tag{18}$$

(The symbols are defined in Table 1.)

where $\alpha_{s,\max}$ is the maximum packing, assumed here to be 0.63.

The viscosity for the solids stress tensor is the sum of the collisional, kinetic, and frictional viscosity elements:

$$\mu_s = \mu_{s,col}+\mu_{s,kin}+\mu_{s,fr} \tag{19}$$

The collisional element of viscosity is modeled as follows [6, 8, 12, 15]:

$$\mu_{s,col} = \frac{4}{5}\alpha_s^2\rho_s d_s g_{0,ss}(1+e_{ss})\left(\frac{\Theta_s}{\pi}\right)^{1/2} \tag{20}$$

The kinetic part of viscosity is modeled using the equation of Syamlal [12]:

$$\mu_{s,kin} = \frac{\alpha_s d_s\rho_s\sqrt{\Theta_s\pi}}{6(3-e_{ss})}\left[1+\frac{2}{5}(1+e_{ss})(3e_{ss}-1)\alpha_s g_{o,ss}\right] \tag{21}$$

Shear stress includes bulk viscosity, λ_s, which in granular flows is related to the particles' resistance to compression and expansion. The bulk viscosity expression of Lun et al. [26] was used in this simulation:

$$\lambda_s = \frac{4}{3}\alpha_s\rho_s d_s g_{0,ss}(1+e_{ss})\left(\frac{\Theta_s}{\pi}\right)^{1/2} \tag{22}$$

Table 1 Definition of symbols

Symbol	Description	Units
Alphabetic		
C_D	Drag coefficient	Dimensionless
d_s	Solid particle diameter	m
e	Coefficient of restitution	Dimensionless
g	Gravitational acceleration	m/s^2
g_0	Radial distribution function	Dimensionless
k	Consistency index	Dimensionless
K	Interphase exchange coefficient	Dimensionless
K_p	Porous media permeability	m^2
P	Pressure (fluid)	Pa
R_e	Relative Reynolds number	Dimensionless
t	Time	s
D_p	Porous media mean particle diameter	m
F_r	Materials constant in Eq. 17 (=0.5)	Dimensionless
p	Materials constant in Eq. 17 (=5.0)	Dimensionless
n	Materials constant in Eq. 17 (=2.0)	Dimensionless
n_i	Flow index	Dimensionless
Greek letters		
α	Volume fraction (solid or liquid)	Dimensionless
ρ	Density	kg/m^3
Θ	Granular temperature	m^2/s^2
$\bar{I}$	Unit stress tensor	Dimensionless
γ_{Θ_s}	Collision dissipation of energy	kg/s^3 m
$\bar{I}_{2D}$	Second invariant of deviatoric stress tensor	Dimensionless
λ	Bulk viscosity	Pa s
μ	Shear viscosity	Pa s
$\overrightarrow{v_s}$	Solid velocity vector	m/s
$\overrightarrow{v_l}$	Fluid velocity vector	m/s
v	Seepage velocity	m/s
$\bar{\tau}$	Stress tensor	Pa
ε	Porous media void volume fraction	Dimensionless
Subscripts		
col	Collision	
fr	Friction	
kin	Kinetic	
l	Liquid phase	
Max, min	Maximum, minimum value	
q	Either liquid or solid phase	
s	Solid phase	

When the solids volume fraction is near the packing limit, the friction between particles is important. The friction element of the shear viscosity can be defined using Schaeffer's expression:

$$\mu_{s,fr} = \frac{p_{sfr} \sin \theta}{2\sqrt{I_{2D}}} \tag{23}$$

where θ is the angle of internal friction and I_{2D} is the second invariant of the deviatoric stress tensor [12].

2.2 *Modeling Fluid Flow in Porous Rock Formation*

The multiphase fluid flow through the porous rock is modeled using an extension of Darcy's law for multiphase flow, also referred to as the Ergun equation for laminar flow or the Blake–Kozeny equation. This equation reads

$$\nabla P = -\frac{\mu}{K_p} v \tag{24}$$

where v is the seepage fluid velocity in the formation and μ the fluid dynamic viscosity. The porous media permeability, K_p, is given below in terms of formation porosity (ε) and the porous media mean pore size (D_p). Here, we set a formation void fraction of 0.2 following Parn-anurak [31]:

$$K_p = \frac{D_p^2}{150} \frac{\varepsilon^3}{(1 - \varepsilon)^2} \tag{25}$$

The differential pressure in between the porous media formation and annulus was maintained at 500 psi (3.4 MPa).

2.3 *Mechanism of Filter Cake Formation in the Porous Rock Surface*

Figure 1 displays a simplified diagram of the drilling process model for oil and gas reservoirs. In our simulation, we zoomed in the bottom drilling zone to capture detailed phenomena occurring during the drilling process. The filter cake is shown on the vertical wellbore wall. The particulate multiphase drilling fluid is pumped-down into the drilling zone through a drilling pipe where drilling fluid interacts with rock debris. As particulate-laden drilling fluid flows upward to the surface through the annulus in between the walls of the well and the drill string,

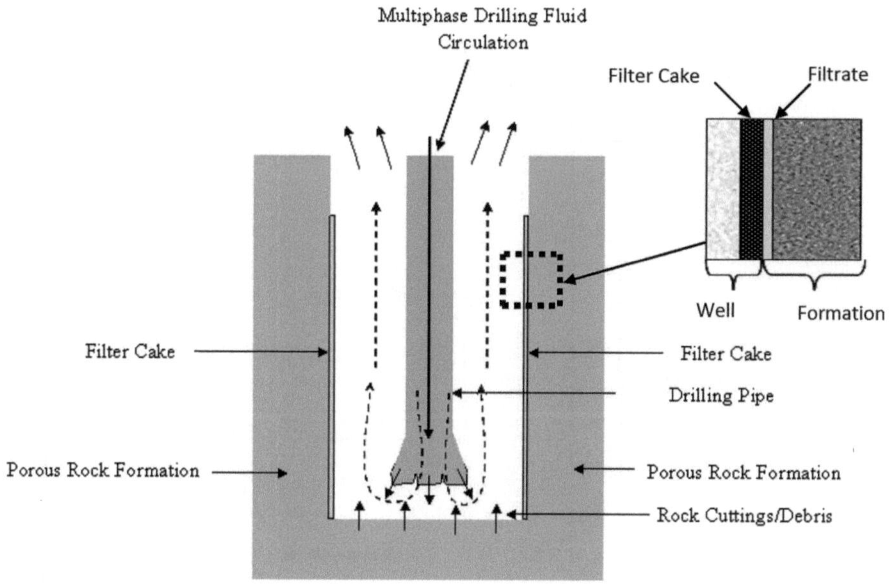

Fig. 1 Schematic diagram of the drilling process model. "Reprinted from *Asia-Pac. Journal of Chemical Engineering, 10, 809, 2015-Gamwo and Kabir, with permission from John Wiley and Sons, September 12, 2016*"

differential pressure causes filter cake to form on the porous rock reservoir surface as shown in Fig. 1.

As the annulus pressure exceeds the formation pressure, the overbalance pressure forces the fluid phase of the drilling fluid through the porous formation and deposits particles on the porous rock surface in the form of filter cake. Fluid seepage in the porous rock surface is related to the rock resistance, fluid viscosity, and differential pressure—this relationship can be described by Darcy's Law [5, 10, 13, 31]. As time progresses, filter cake will grow on the rock surface; therefore, filter cake itself will also resist fluid permeation into porous rock formations and, hence, fluid permeation will decrease. The resistance from filter cake can be related to the concentration of mass loading per unit area (kg/m^2) and specific resistance (m/kg). The filter cake builds up to a maximum thickness, which is determined by particle characteristics and fluid shear [10, 13].

2.4 *Two-Dimensional Wellbore Model*

A two-dimensional (2-D) wellbore model of a vertical well was created and meshed with Gambit as shown in Fig. 2. Symmetry along the central axis was assumed. To simulate the drilling process, multiphase particulate ($\alpha_s = 0.2$) drilling fluid was pumped into the model inlet, and multiphase particulate ($\alpha_s = 0.8$) rock debris was

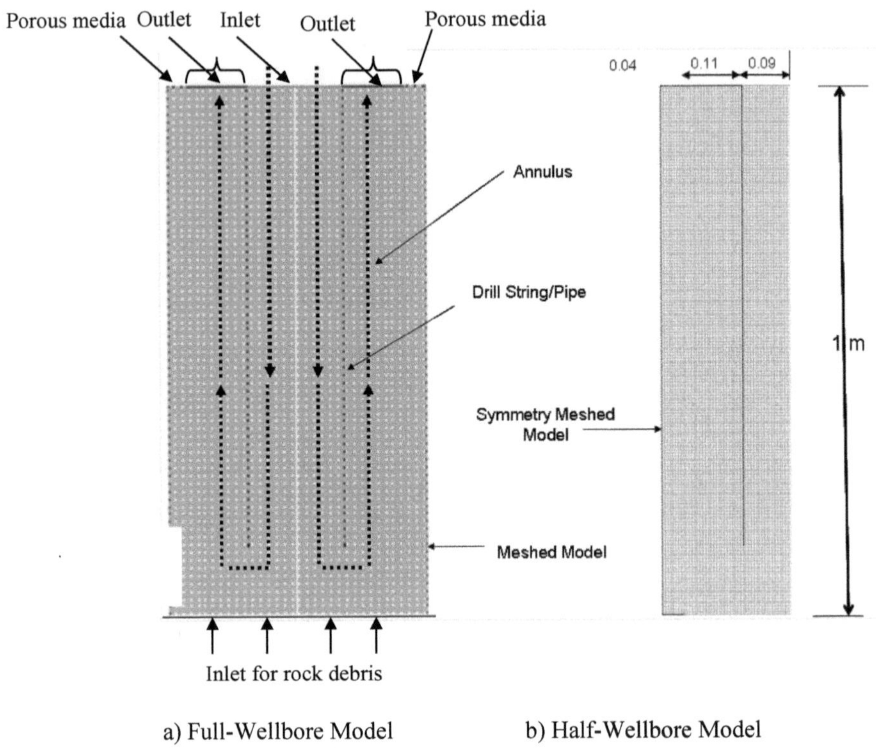

Fig. 2 Meshed vertical wellbore model. "Reprinted from *Asia-Pac. Journal of Chemical Engineering, 10, 809, 2015-Gamwo and Kabir, with permission from John Wiley and Sons, September 12, 2016*"

pushed from the bottom inlet. The model inlet represents drilling fluid pumping in, and the bottom inlet represents rock debris that accumulates during drilling. The solid wall represents the drill string surface. A porous medium with solid volume fraction of 0.8 next to the drill string represents the vertical rock formations on which filter cake builds up. The pressure and temperature are 25,500 psi (175.8 MPa) and 170 °C, respectively, for deep drilling conditions. The formation pressure and temperature were maintained at 25,000 psi (172.4 MPa) and 170 °C to mimic real-world drilling scenarios. Multiphase particulate non-Newtonian drilling fluids were pumped into the drilling zone where the drilling fluids mixed with rock particles. The particle-laden drilling fluid then flowed upwardly, back to the surface, through the annulus between the walls or sides of the wellbore and the drill string. A variety of drilling fluids exist, and, as mentioned earlier, the circulation of the drilling fluid, among others, lubricates the drill bit, removes cuttings from the wellbore as they are produced, exerts hydrostatic pressure on pressurized fluid contained in formations, and seals off the walls of the wellbore so that the fluid is not lost in the permeable subterranean zones [32].

2.4.1 Initial and Boundary Conditions

In our modeling and simulations, the wellbore was initially filled with multiphase particulate drilling fluid or mud while the bottom portion of the drilling zone was filled with rock debris as shown in Fig. 3. The non-Newtonian fluid properties with yield stress were given for the liquid phase, and the granular properties were given for solid particles. The density of the liquid phase was 999 kg/m^3, whereas yield stress (τ_o), consistency (k) index, and power-law (n) index were 3 Pa, 0.1238 $Pa.s^n$, and 0.67 respectively [10, 16]. The solid phase density was set at 2350 kg/m^3. In this study, the particle size in the drilling fluid was 45 μm for wellbore simulation. The domain was discretized with grid where the flow domain was divided into finite surfaces. As mentioned earlier, axisymmetry was assumed for the modeling the drilling process. Several trials were made (from 5500 to 11,000 meshes) to verify grid independent results from CFD simulations. The half-wellbore model consists of 9600 quadrilateral mesh cells with uniform size of 0.5 cm × 0.5 cm for a vertical well. The dimension of the porous media formation in the model was 4 cm × 100 cm, and the porous media formation pressure and temperature were maintained at 25,000 psi (172.4 MPa) and 170 °C for deep drilling conditions.

Our extensive literature review revealed very few experimental and numerical studies have been carried out on the multiphase flow pattern and filter cake formation in deep drilling processes; we therefore had few studies with which to

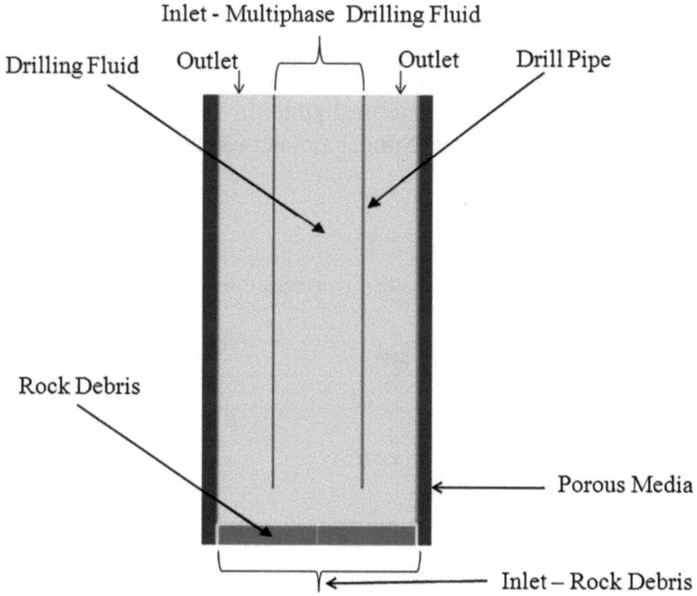

Fig. 3 Initial solid distribution in the vertical well. "Reprinted from *Asia-Pac. Journal of Chemical Engineering, 10, 809, 2015-Gamwo and Kabir, with permission from John Wiley and Sons, September 12, 2016*"

validate our CFD modeling results [5, 10, 16, 31, 32, 34, 35, 37]. To compensate, a single pressure vertical-linear multiphase filtration process was modeled to validate our CFD model. To experimentally validate the filter cake thickness, we previously compared experimental filter cake data of iron ore suspension with CFD simulation results. The analytical, experimental, and numerical results of filter cake heights in multiphase flow porous media compared reasonably well, as described in detail elsewhere [24].

3 Results and Discussion

The simulated initial and boundary conditions described in the previous section (Sect. 2.4.1) are similar to conditions found in field drilling operations [1, 3, 5, 7, 10, 16, 27, 31, 32, 34–38, 40]. The details of these conditions for deep drilling simulations are provided in Table 2.

The deep drilling process was simulated by setting high-pressure (25,500 psi or 175.8 MPa) and high-temperature (170 °C) conditions at the inlet and bottom portion of the well. The differential pressure between wellbore and porous media was maintained at 500 psi (or 3.45 MPa); however, simulations were performed on a range of overbalance differential pressures at 250 psi (or 1.72 MPa), 500 psi (or 3.45 MPa), 750 psi (or 5.17 MPa), 1000 psi (or 6.89 MPa), and 1250 psi (or 8.62 MPa) to study the effect of differential pressure on fluid flow pattern and cake formation.

The bottom portion of the simulated well was maintained at the same pressure and temperature as that of inlet. It was assumed that the pressure and temperature variations over a 1 m long model are negligible. In this study, reservoir formation porosity was assumed to be 0.2. Similar porosity value has been used by other researchers [31].

Table 2 Initial conditions and fluid/formation properties for a vertical well

	Well depth
Inlet pressure (drilling fluid/top) (psi or MPa)	25,500 or 175.8
Pressure (bottom) (psi or MPa)	25,500 or 175.8
Outlet pressure (psi or MPa)	25,000 or 172.4
Formation pressure (porous media) (psi or MPa)	25,000 or 172.4
Particle size (μm)	45
Formation porosity (dimensionless)	0.2
Temperature (°C)	170
Solid fraction (drilling fluid/top) (dimensionless)	0.2
Solid fraction (rock/bottom) (dimensionless)	0.8
Particle density (kg/m^3)	2350
Fluid density (kg/m^3)	999

The wellbore was maintained at a higher pressure than the surrounding porous media to mimic actual overbalance drilling conditions in the well. The differential pressure in the annulus forces fluid through the porous media and deposits solid particles in the form of filter cake on the porous rock surface, as shown in Fig. 4a; filter cake is indicated by a higher volume of solids at the wall. The filter cake grows on the wall in a process similar to simple soil consolidation where differential pressure initially forces some drilling fluid into the formation, and the solids present in the drilling fluid clog the pores of the formation and accumulate against the wall under appropriate conditions. As the pressure difference between the wellbore and the formation forces the filter cake to consolidate, the fluid phase (filtrate) invades the formation. The solid particles become more tightly packed, reducing the permeability of the growing cake and, hence, the fluid moves into the formation [5].

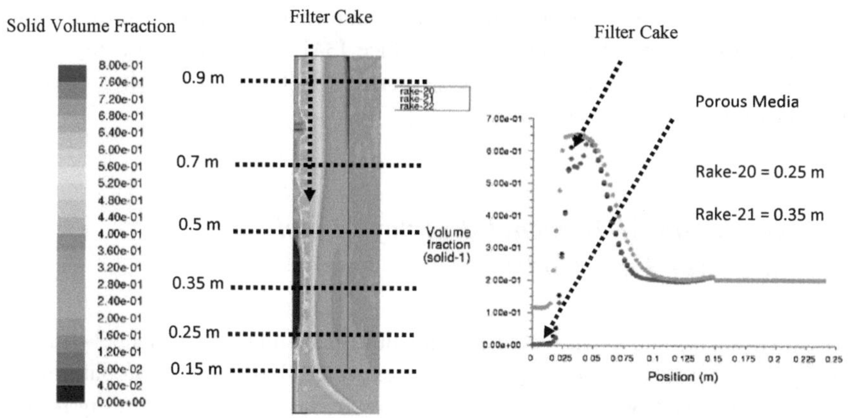

a) Qualitative solid fraction of well (half). b) Solid fraction at different well heights of 0.25 m, 0.35 m and 0.5 m.

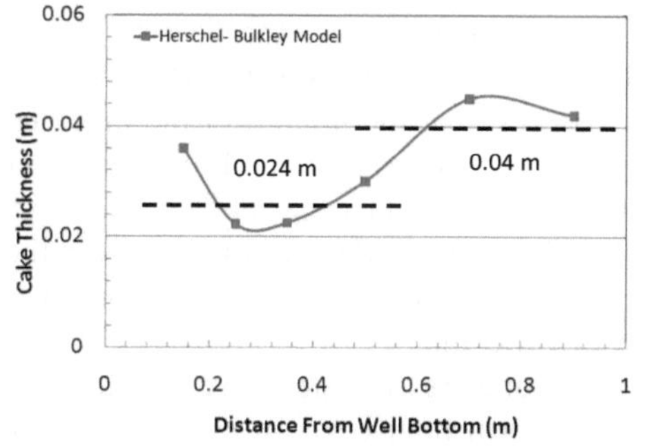

c) Filter cake thickness over well heights.

Fig. 4 a–c Deep drilling—filter cake thickness at different heights of wellbore from bottom. "Reprinted from *Asia-Pac. Journal of Chemical Engineering, 10, 809, 2015-Gamwo and Kabir, with permission from John Wiley and Sons, September 12, 2016*"

3.1 Deep Drilling Modeling and Simulation with Herschel–Bulkley Model

Figure 4a–c shows the simulated cake in the deep vertical wellbore wall for 45 μm particle drilling fluid. In this study, we have defined cake as solid volume fraction above 0.4 at the well wall. Figure 4a qualitatively shows filter cake thickness in half of the well with relatively thinner cake in the lower bottom part followed by thicker cake at the upper part of the well. Figure 4b shows solid volume fractions at different well heights from 0.25 to 0.5 m. Figure 4c displays the filter cake thickness extracted from the solid volume fraction graph, Fig. 4a. Figure 4c shows the cake thickness versus well heights. The average cake thickness varies from 0.024 m near the bottom well to 0.04 m near the top portion (Fig. 5c). This clearly implies that the simulated filter cake formed on the wellbore wall was non-uniform. These results agree qualitatively with the literature, which reports that non-uniform filter cake forms on the vertical porous media surface [34]. The non-uniformity of the filter cake is explained by the presence of vortices in the well annulus, which is discussed in the following section.

3.2 Comparison of Herschel–Bulkley Model with Power-Law Model

Figure 5a compares the steady-state drilling fluid path-lines for Herschel–Bulkley and power-law models. For both models, the expected fluid circulation pattern and intensity are observed with descending fluid flow in the pipe and ascending flow in the annulus. The down fluid flow magnitude is lower because the pressure drop is set to zero. The upflow magnitude is higher because the pressure drop is set to 500 psi (3.45 MPa). Compared to the power-law model, the Herschel–Bulkley model results exhibit higher magnitude of the ascending fluid velocity with significantly higher magnitude at the bottom to propel the debris. This will more efficiently transport cuttings to the surface. In addition, the Herschel–Bulkley model shows only two symmetrical vortices near the bottom of the pipe. The power-law model exhibits six vortices along the annulus section. These numerous vortices along the path of the ascending cuttings will hinder the debris removal process as the cuttings trapped in the vortices will tend to settle near the bottom rather than ascending to the surface. The accumulation of cuttings near the bottom hole may disrupt or prevent the rotation of the drill bit.

Figure 5b qualitatively compares the steady-state solid volume fraction deposited in the form of non-uniform filter cake on the wall for both models. It shows more uniform filter cake thickness for the Herschel–Bulkley model compared to the power-law model. This non-uniformity is probably induced by the numerous vortices along the annulus for the power-law model.

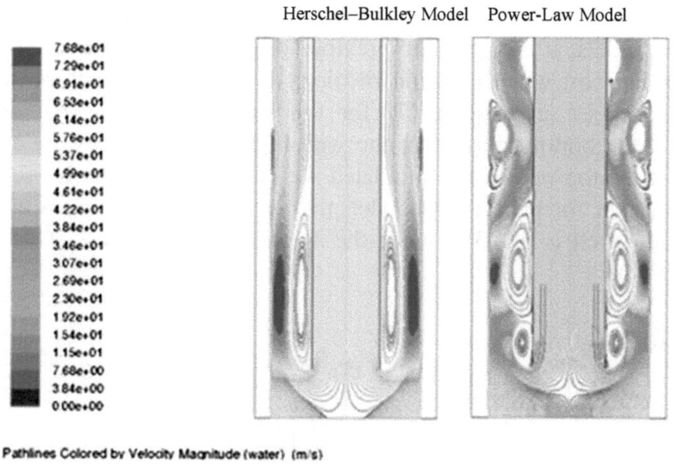

a) Pathlines Colored by Velocity Magnitude (m/s)

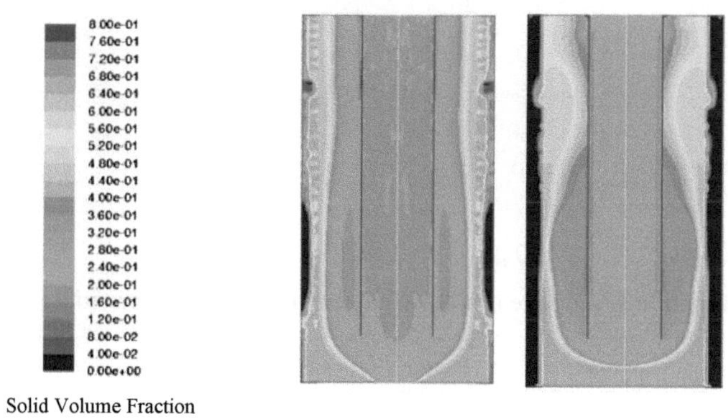

b) Qualitative solid fraction of vertical well (full).

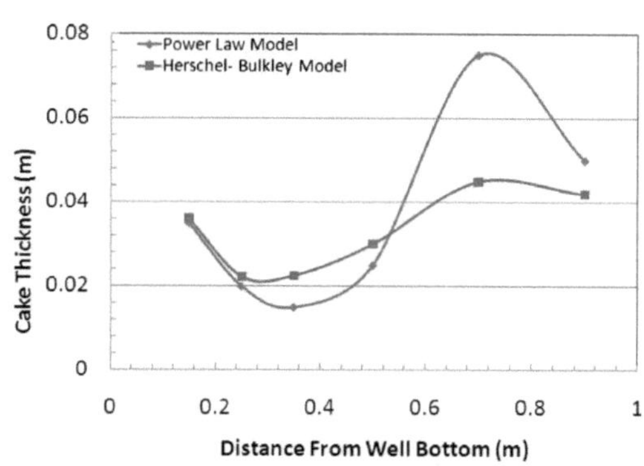

c) Filter cake thickness over well heights from bottom part.

Fig. 5 **a–c** Comparison of non-Newtonian fluid models on filter cake thickness. "Reprinted from *Asia-Pac. Journal of Chemical Engineering, 10, 809, 2015-Gamwo and Kabir, with permission from John Wiley and Sons, September 12, 2016*"

Figure 5c quantitatively displays filter cake thickness for Herschel–Bulkley and power-law models. It confirms the previous observation that the cake thickness is highly non-uniform when the fluid rheology is described by the power law. The numerous vortices are responsible for this non-uniformity. The average cake thickness at the bottom portion of the well is 0.024 m whereas the average cake thickness at the top part of the simulated well is 0.04 m for the Herschel–Bulkley model. However for power-law model, the average cake thickness at the bottom section of the well is 0.023 m while the average cake thickness at the top part of well is 0.05 m.

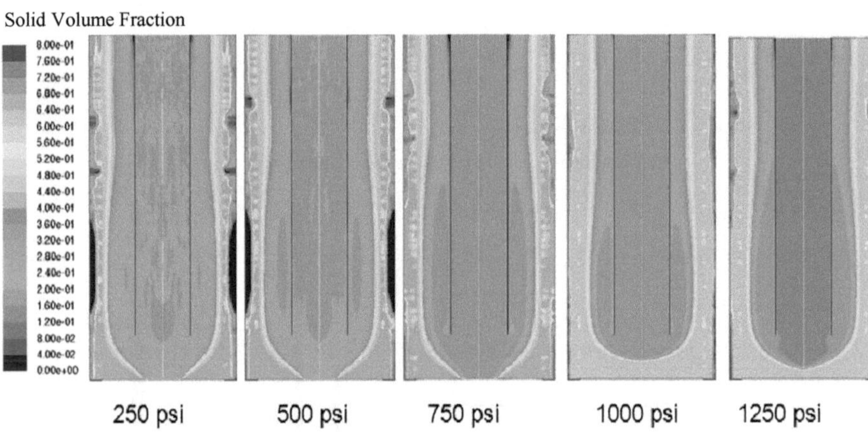

a) Qualitative filter cake and flow pattern at various differential pressures

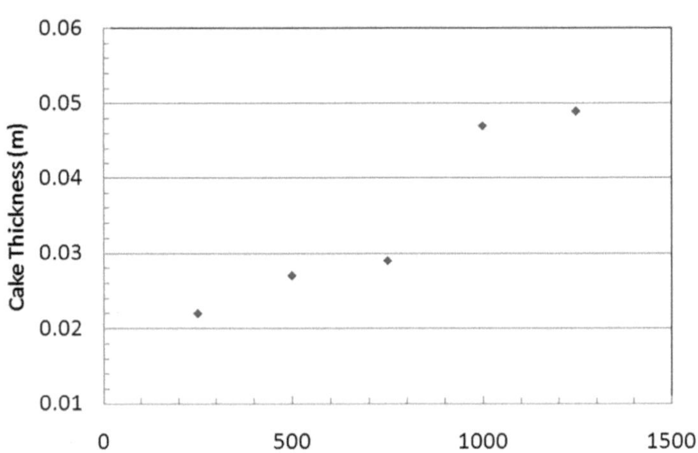

b) Cake thickness over differential pressure.

Fig. 6 **a, b** Effect of overbalance differential pressure on filter cake thickness for deep vertical well. "Reprinted from *Asia-Pac. Journal of Chemical Engineering, 10, 809, 2015-Gamwo and Kabir, with permission from John Wiley and Sons, September 12, 2016*"

3.3 *Effect of Pressure on Flow Pattern and Filter Cake*

To study the effect of overbalance differential pressure on cake thickness, CFD simulations were performed over a range of differential pressures, namely: 250 psi (or 1.72 MPa), 500 psi (or 3.45 MPa), 750 psi (or 5.17 MPa), 1000 psi (or 6.89 MPa), and 1250 psi (or 8.62 MPa). Their results are presented in Fig. 6a, b. Figure 6a qualitatively shows the filter cake pattern at various overbalance differential pressures in the well. As exhibited in the figure, cake thickness increases with increased pressure. The quantitative cake thicknesses were extracted from Fig. 6a and are presented in Fig. 6b, which clearly shows that the cake thickness increases with differential pressure as expected (Fig. 6a, b). Higher pressure forced more fluid into the formation by separating drilling fluid from a large number of solid particles and depositing these particles in the form of thicker cake on the wellbore wall.

Thicker cakes are not desirable because they often lead to drilling difficulties, such as stuck pipe and other drilling-related problems. Yet, extremely thin cake could lead to the loss of drilling fluid into the formation. Therefore, it is necessary to optimize filter cake thickness to achieve more efficient drilling. There is a paucity of experimental results for real-world deep drilling scenarios; therefore, these simulations would provide drilling engineers some guidelines on filter cake pattern and will also help engineers to optimize filter cake.

In addition to overbalance pressure, cake thickness also depends on several other parameters such as drilling fluid properties (density and viscosity), particle size and porosity of the formation.

4 Conclusions

The CFD numerical predictions were performed to mimic the extreme deep drilling process and cake formation in vertical wells located several miles beneath the Earth's surface. The CFD tool has successfully captured cake formation on the vertical subterranean zone during drilling and provided fluid flow patterns and velocity magnitudes of the drilling fluid during a deep drilling process.

Two sensitivity cases studies were carried out, one on the drilling fluid rheology model and one on the overbalance pressure drops. Their impacts on cutting removal performance have been inferred. The conclusions of this study are as follows:

- Two rheological models, the power-law and the Herschel–Bulkley model, were studied and both were able to predict the expected drilling fluid flow pattern at the bottom section of the deep drilling process with descending fluid in the pipe and ascending fluid in the annulus. Deviations were observed for the two models in the annulus ascending drilling fluid. The Herschel–Bulkley model predicted higher fluid and fewer vortices, indicating better removal of cuttings debris for rheological Herschel–Bulkley model fluids.

- The effect of overbalance pressure on filter cake thickness was investigated for five different overbalance pressures from 250 to 1250 psi. Cake thickness was found to increase linearly with the pressure drop. Based on this result, it is necessary to optimize the overbalance pressure when drilling for oil and gas at extremely high-temperature and high-pressure conditions, like those encountered miles underneath the Earth's surface. Such optimization of the overbalance pressure should improve the cuttings removal performance.

Acknowledgements This research was supported in part by an appointment to the National Energy Technology Laboratory Research Participation Program, sponsored by the U.S. Department of Energy and administrated by the Oak Ridge Institute for Science and Education.

References

1. Ali, S. (2006). Reversible drilling-fluid emulsions for improved well performance. *Oilfield Review.*
2. Arthur, K. G., & Peden, J. M. (1988). The evaluation of drilling fluid filter cake properties and their influence on fluid loss. *SPE Journal.*
3. Berry, J. H. (2009). Drilling fluid properties & function, CETCO drilling products. Retrieved from http://www.getco.com.
4. Blkoor, S. O., & Fattah, K. A. (2013). The influence of xc-polymer on drilling fluid filter cake properties and formation damage. *Journal of Petroleum & Environmental Engineering.*
5. Cerasi, P., & Soga, K. (2001). Failure modes of drilling fluid filter cake. *Geotechnique, 51*(9), 777–785.
6. Cornelissena, J. T., Toghipour, F., Escudiéa, R., Ellisa, N., & Gracea, J. R. (2007). CFD modeling of a liquid–solid fluidized bed. *Chemical Engineering Science, 62,* 6334–6348.
7. Delhommer, H. J., & Walker, C. O. (1987). *Method for controlling lost circulation of drilling fluids with hydrocarbon absorbent polymer.* US Patent Number, 4:633,950.
8. Ding, J., & Gidaspow, D. (1990). A bubbling fluidization model using kinetic theory of granular flow. *Journal of A. I. Chemical Engineering, 36,* 523–538.
9. Fordham, E. J., Ladva, H. K. J., Hall, C., Baret, J. F., & Sherwood, J. D. (1988). Dynamic filtration of bentonite muds under different flow conditions. In *63rd Annual SPE Conference, Houston, Texas.*
10. Fisher, K. A., Wakeman, R. J., Chiu, T. W., & Meuric, O. F. J. (2008). Numerical modeling of cake formation and fluid loss from non-Newtonian mud's during drilling using eccentric/concentric drill strings with/without rotation. *Trans IChemE, 78 Part A,* 707–714.
11. Ferguson, C. K., & Klotz, J. A. (1954). Filtration from mud during drilling. *Trans AIME, 201,* 29–42.
12. Fluent. (2006). *FLUENT 6.3 User's Guide.* F. Inc. Lebanon, NH.
13. Fu, L. F., & Dempsey, B. A. (1998). Modeling the effect of particle size and charge on the structure of the filter cake in ultra-filtration. *Journal of Membrane Science, 149,* 221–240.
14. Gamwo, I. K., & Kabir, M. A. (2015). Impact of drilling fluid rheology and wellbore pressure on rock cuttings removal performance: Numerical investigation. *Asia-Pac. Journal of Chemical Engineering.*
15. Gidaspow, D., Bezburuah, R., & Ding, J. (1992). Hydrodynamics of circulating fluidized beds, kinetic theory approach. In *Fluidization VII, Proceedings of the 7th Engineering Foundation Conference on Fluidization.*

16. Hamed, S. B., & Belhadri, M. (2009). Rheological properties of biopolymers drilling fluids. *Journal of Petroleum Science and Engineering, 67,* 84–90.
17. Ishii, M. (1975). Thermo-fluid dynamic theory of two-phase flow. In *Collection de la Direction des Etudes et Recherches d'Electricite de France 22, Eyrolles. Paris.*
18. Jackson, R. (1997). Locally averaged equations of motion for a mixture of identical spherical particles and a Newtonian fluid. *Chemical Engineering Science, 52*(15), 2457–2469.
19. Johnson, P. C., Nott, P., & Jackson, R. (1990). Frictional–collisional equations of motion for participate flows and their application to chutes. *Journal of Fluid Mechanics, 210,* 501–535.
20. Johnson, P. C., & Jackson, R. (1987). Frictional-collisional constitutive relations for granular materials, with application to plane shearing. *Journal of Fluid Mechanics, 176,* 67–93.
21. Jung, J., & Gamwo, I. K. (2008). Multiphase CFD-based models for chemical looping combustion process: Fuel reactor modeling. *Powder Technology, 183,* 401–409.
22. Kabir, M. A., Khan, M. M. K., & Rasul, M. G. (2004). Flow of a low concentration polyacrylamide fluid solution in a channel with a flat plate obstruction at the entry. *Korea-Australia Rheology Journal, 16*(2), 63–73.
23. Kabir, M. A., Khan, M. M. K., & Rasul, M. G. (2013). Reverse flow phenomena of a polyacrylamide solution in a channel with an obstacle at the entry: Influence of obstruction geometry. *Journal of Chemical Engineering Communications, 199*(4), 1–20.
24. Kabir, M. A., & Gamwo, I. K. (2011). Filter cake formation on the vertical well at high temperature and high pressure: Computational fluid dynamics modeling and simulations. *Journal of Petroleum and Gas Engineering, 2*(7), 146–164.
25. Kelsey, J. R., & Carson, C. C. (1987). Geothermal drilling-drilling for geothermal energy. *Geothermal Sciences and Technology, 1,* 39–61.
26. Lun, C. K. K., Savage, S. B., Jerey, D. J., & Chepurniy, N. (1984). Kinetic theories for granular flow: Inelastic particles in Couette flow and slightly inelastic particles in a general flow field. *Journal of Fluid Mechanics, 140,* 223–256.
27. Maurer Engineering Inc. (1997). *Wellbore thermal simulation model: Theory and user's manual*. Maurer Engineering Inc.
28. Myöhänen, K., Hyppänen, T., & Kyrki-Rajamäki, R. (2006). *CFD modeling of fluidized bed systems*. Finland: SIMS.
29. Outmans, H. D. (1963). Mechanics of static and dynamic filtration in the borehole. *SPE, 228*(236).
30. Peden, J. M., Avalos, M. R., & Arthur, K. G. (1982). The analysis of the dynamic filtration and permeability impairment characteristics of inhibited water based muds. In *SPE Formation Damage Control Symposium, Lafayette.*
31. Parn-anurak, S. (2003). *Modeling of fluid filtration and near-wellbore damage along a horizontal well*. Ph.D. thesis, New Mexico Institute of Mining and Technology, USA.
32. Rogers, H. E., Murray, D. A., & Webb, E. D. (1996). *Apparatus and method for removing gelled drilling fluid and filter cake from the side of a wellbore*. US Patent 5564,500.
33. Saha, H. (2009). *Practical application of filtration theory to the minerals industry*. Ph.D. Thesis, The University of Melbourne, Australia.
34. Sherwood, J. D., Meeten, G. H., Farrow, C. A., & Alderman, N. J. (1991). Squeeze-film rheometry of non-uniform mudcak. *Journal of Non-Newtonian Fluid Mechanics, 39,* 311–334.
35. Sherwood, J. D., Meeton, G. H., Farrow, C. A., & Alderman, N. J. (1991). Concentration profile within non-uniform mudcakes. *Journal* of the *Chemical Society, Faraday* Transactions, *84*(4), 611.
36. Spooner, K. M., Bilbo, D., & McNeil, B. (2004). The application of high temperature polymer drilling fluid on Smackover operations in Mississippi. In *AADE-2004 Drilling Fluids Conference, Houston, Texas.*
37. Usher, S. P., Kretser, R. G., & Scales, P. J. (2001). Validation of a new filtration technique for de-waterability characterization. *AIChE Journal, 47*(7), 1561–1570.

38. Vaussard, A., Martin, M., Konirsch, O., & Patroni, J. M. (1986). An experimental study of drilling fluids dynamic filtration. In *SPE 61st Annual Technical Conference, New Orleans.*
39. Warren, B. K., Smith, T. R., & Ravi, K. M. (1993). Static and dynamic fluid-loss characteristics of drilling fluids in a full-scale wellbore. *SPE Journal.*
40. Wikipedia. (2010). Oil well. Retrieved from http://en.wikipedia.org/wiki/oil_well.well.

Optimising Pyrolysis Conditions for Thermal Conversion of Beauty Leaf Tree (*Calophyllum inophyllum* L.) Press Cake

Nanjappa Ashwath, Hyungseok Nam and Sergio C. Capareda

Abstract Beauty leaf tree (BLT) has been recognised as one of the potential species for biodiesel production in the tropics as it can produce up to 3800 L of non-edible oil which can be converted to biodiesel. The BLT is also resilient to stress conditions so it can be cultivated on degraded lands such as salt- and drought-affected soils. Biodiesel production from BLT, however, generates wastes such as the husk, press cake and glycerol. These wastes will increase the cost of producing biodiesel, and they can also add waste management issues. The current study investigated conversion of BLT biodiesel production wastes into other forms of biofuels. Oven-dried press cake samples were pyrolysed using a batch reactor at 300, 400 or 500 °C, with a residence time of 30, 60 or 90 min. The gas generated from this process was condensed to produce bioliquor and biooil, and the uncondensed gas was quantified as syngas. The pyrolysed biomass residue was collected as biochar and quantified. Energy content of these four products was determined, and the results showed that more than 90% of the energy contained in the BLT press cake can be recovered as other forms of biofuel. It was also found that the temperature had greater influence on the conversion process than on residence time. Furthermore, the biochar yield decreased with an increase in temperature, in contrast to biooil and syngas yields. The optimum conditions for thermal conversion of BLT press cake were found to be 500 °C, with a residence time of 30 min. This study demonstrates that the wastes resulting from biodiesel production process can be used as the feedstocks for producing other forms of biofuels. This approach will not only solve the environmental issues, but it will also improve economic viability of BLT biodiesel production process. Based on these results and the additional tests, a portable and continuous feeding auger pyrolysis reactor is recommended for converting BLT whole fruits, press cake or husks into biofuels.

N. Ashwath (✉)
School of Medical and Applied Sciences, Central Queensland University, Rockhampton, QLD 4702, Australia
e-mail: n.ashwath@cqu.edu.au

N. Ashwath · H. Nam · S.C. Capareda
Biological and Agricultural Engineering Department, Texas A&M University (TAMU), College Station, TX 77843, USA

M.M.K. Khan et al. (eds.), *Application of Thermo-fluid Processes in Energy Systems*, Green Energy and Technology, https://doi.org/10.1007/978-981-10-0697-5_12

1 Introduction

The search for renewable energy is continuing, with the hope of finding a source or a process that will produce fuel at a reasonable cost, and is environmentally friendly and sustainable. Amongst the available sources of renewable energy (solar, wind, tide, biomass, geothermal), the biomass energy has attracted the attention of scientists, industrialists and policy makers, as it is practically achievable with low technology and has the potential to stop or even reverse environmental pollution resulting from greenhouse gas emission [1].

Biomass is derived from plant (primary) or animal (secondary) sources. The plant biomass is produced via photosynthesis, wherein CO_2 and water are converted into simple sugars in the leaf in the presence of sunlight. The simple sugars will then form the building blocks for biomass synthesis. In biomass energy, the biomass is converted into various products via thermal conversion or anaerobic digestion. In thermal conversion, high temperature, nitrogen, pressure and catalysts are used. In anaerobic digestion, microbes are used to convert the biomass (cellulose, hemicellulose, lignin, lipids, protein) into gasses such as methane, CO, H_2, CO_2 and sludge. In thermal conversion, the long chained molecules of cellulose, hemicellulose and lignin are broken down into short chained molecules in the form of condensable gasses (to produce biooil and bioliquor), non-condensable gases (syngas; e.g. CO, H_2, CO_2, CH_4) and solid biochar.

Plants accumulate either vegetative (timber, leaves, roots) or reproductive (e.g. fruits and seeds) tissues. The reproductive tissues contain energy-rich materials such as lipids or sugars. The plants that contain sugars are used as feedstocks to produce ethanol [2] and those that contain lipids are used to produce biodiesel [3]. Currently, the bioethanol and biodiesel are the dominant types of biofuels being explored.

In thermal conversion, usually the vegetative parts of plants (e.g. stems, branches, roots; usually in the form of wastes) are used as feedstocks. The reproductive parts (fruits and seeds) are generally used in biodiesel or bioethanol production. The fruits, seeds and other parts that are used in biodiesel or ethanol production still contain residues or unconverted biomass. This biomass could constitute as high as 70% of the fruit or seed biomass [3]. This biomass is known as the press cake or oil cake, and is generally used as animal feed supplement, if the seeds are free toxins.

The press cakes of some biodiesel feedstocks, for example, BLT, pongamia or jatropha contain toxins; they will not be used as animal feed, and hence they will be discarded as a waste. These wastes can serve as excellent feedstocks for thermal conversion, as they contain high energy value (25 MJ/kg) compared to the vegetative material (e.g. stem 12 MJ/kg). A review of the literature shows that the press cakes have been used in animal feed supplements, composting [4], pyrolysis or hydrogen production [5]. The press cakes can also be used to produce activated carbons [6, 7], or they are fermented to produce edible fungus and enzymes [8, 9].

Beauty leaf tree is an Australian native species (Fig. 1) and grows in the coastal areas of tropical Australia. It is also widespread in the Pacific islands. It is a medium to large tree and bears fruits twice a year [3]. The fruits of BLT contain non-edible oil, and the oil content of the kernel (seed; Fig. 2) varies from 20 to 70% Ashwath [3, 10]. Has compared the oil production potential of BLT with a large number of other native species of Australia. Based on this study, he has ranked BLT as number one species for oil production, as it can produce up to 3800 L of oil per ha per year. Since its recognition as a highly potential biodiesel feedstock, a number of studies have been undertaken to further assess BLT's capability to serve as the second-generation biodiesel feedstock. Bhuiya et al. [11] have developed procedures for converting BLT oil into biodiesel; Hathurusingha [12] tested the BLT biodiesel and found that the BLT biodiesel will meet the ASTM standards. Hathurusingha and Ashwath [13] reported that the BLT biodiesel could act as an antiknocking agent. Since the BLT grows in varied climatic conditions, studies have also been undertaken to test its variability in kernel oil content (N. Ashwath; unpublished). The BLT research is continuing with the view to selecting trees that can grow well in degraded soils and also possess high kernel oil content.

The BLT oil is comprised of four common fatty acids such as oleic acid, linoleic acid, palmitic acid and stearic acid, with small proportions (<1%) of linolenic acid, arachidic acid and behemic acid [3, 14]. The BLT kernel oil contains high proportion of free fatty acids [14], and as a result, its conversion to biodiesel has been

Fig. 1 Mature beauty leaf tree grown in a Parkland in Queensland, and the abundant production of fruits even in such a marginal soil demonstrating the potential of this species to serve as a biodiesel/biofuel feedstock

Fig. 2 Cross section of a BLT fruit showing the husk (*right*) and kernel (*left*)

fraught with difficulties. A two-stage acid catalysed-transesterification process has been proposed, but this process requires the use of excessive amounts of methanol [15]. The BLT-derived biodiesel has been tested for its performance, both for engine performance and gas emission. These tests have shown that the BLT biodiesel yields marginally less power than petroleum diesel, but is highly superior in reducing gas emission (M. Bhuiya; personal communication).

Biodiesel production from BLT involves a number of steps, such as (i) harvesting and transportation of fruits, (ii) separation of kernels from the fruits, (iii) extraction of oil, (iv) oil refining, (v) conversion of the oil to biodiesel and (vi) washing and drying of biodiesel before it is being sold. These steps are not only time consuming but also require different types of chemicals, all leading to increased costs of production. As a result, the BLT biodiesel could potentially become more expensive than petroleum diesel. In addition, biodiesel production process results in the loss of up to 70% of the fruit weight in the form of husk, press cake and glycerol, and managing these wastes incur additional costs.

The current research was therefore carried out to test the potential of converting BLT biodiesel by-products into other forms of biofuels via pyrolysis. The study also investigated the effects of temperature and residence time on pyrolysis process, and on the products resulting from this process.

2 Materials and Methods

2.1 *Press Cake Preparation*

The air-dried fruits of BLT were cracked open to separate kernels (Fig. 2). The kernels were then blended using a heavy duty kitchen blender to produce 5–10 mm size fragments. The ground material was dried initially in a fan-forced oven at around 30–35 °C for about three days and then at 50 °C until the moisture content of the kernels reached between 15 and 20% (4–8 days). The oven-dried kernels were used to extract oil using a screw press. The press cake obtained from the above process was dried at 60 °C for 24 h before being used in the pyrolysis experiments.

2.2 *Pyrolysis of Press Cake*

A batch pyrolyser was employed to test thermal conversion of the BLT press cake (Fig. 3).

The batch reactor represents slow (or intermediate) pyrolysis process according to Worasuwannarak et al. [16]. The BLT cake (500 g) was loaded into the reactor chamber, and the chamber was flushed with nitrogen gas for about 15 min. The BLT cake was then heated to 300, 400 or 500 °C, with a gradual increase in temperature (5 °C/min). When the desired temperature was reached, the samples were exposed to predetermined times of 30, 60 or 90 min in independent runs, and each of the above temperature x time combination was repeated three times and the averages of three replications are presented.

The steps taken in thermal conversion of BLT cake are shown in Fig. 4.

The batch reactor generated volatile matter and gas after the chamber temperature approached ~250 °C. These components were passed through a condenser which was maintained at −4 °C. The uncondensed gas was passed through a two-stage solvent condenser [17]. The gas that escaped the second condenser was passed through a gas metre to record the volume. Samples of these gasses were collected and their chemical compositions determined using a gas chromatograph. The gas weight and chemical composition of the gas were used to calculate energy content of the syngas.

2.3 *Sample Collection and Analysis*

On completion of pyrolysis, the reactor was cooled to room temperature (for ~ 10 h), and the biochar, biooil and bioliquor fractions were collected. The weights of each of the above component were determined, and their energy content measured using a bomb calorimeter [17]. The proximate and ultimate analysis of the biochar, biooil and bioliquor was performed according to ASTM methods and

Fig. 3 Batch pyrolyser used in the thermal conversion of BLT press cake

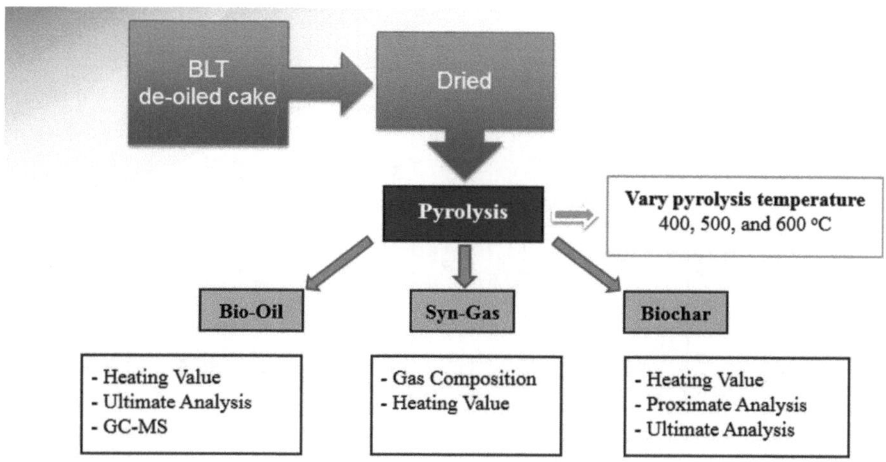

$$\%Mass\ conversion = \frac{W_{product}}{W_{biomass}} \times 100 \qquad \%Energy\ conversion = \frac{W_{product} \times HV_{product}}{W_{biomass} \times HV_{biomass}} \times 100$$

Fig. 4 Procedures used in the pyrolysis of BLT press cake

the compositional analysis was performed based on NREL methods [17]. Weights of the resulting biooil, biochar, bioliquor (vinegar) and the syngas were recorded, and their relative percentages calculated in proportion to the original weights of the BLT cake used. A rotary evaporator was used to separate biooil from bioliquor, and the condensed biooil was mixed with the other biooil fraction.

3 Results and Discussion

Large quantities of press cakes are produced as by-products of oil extraction process. At present, the majority of this cake is used as animal feed supplement. Recent introduction of biofuel crops such as jatropha, pongamia and beauty leaf tree [18–20] will add to the supplies of the press cake. However, these cakes will not be used in animal feed due to the presence of toxic compounds. Consequently, these press cakes will become a burden to biofuel production industries, as they will have to incur additional costs to dispose of these by-products.

Past studies have shown that press cakes, including those containing toxins, can be used to produce other forms of biofuel as in jatropha [21], microalgae [21], pongamia [22], rape seed [23], olive [24], peanut [25], sunflower [26], cotton [26], canola [27], and soybean [28].

A search on Scopus using the terms "oil cake pyrolysis" yielded only 110 research articles. Oil cake production has been occurring for thousands of years; yet, the studies undertaken to utilise these press cakes to produce different forms of biofuels are very much limited, and most of such attempts have occurred in recent

Table 1 Comparison of different species for their responses to pyrolysis using seed cakes

	BLT	Rape seed	Rape seed	Olive	Peanut	Jatropha	Microalgae	Sunflower	Cotton	Canola	Pongamia	Soybean	Mean	SE	Median
Particle size (mm)	5	–	0.2			0.7	0.4	0.2–.4		o.2 −1.8					
Moisture (%)	4.1	–	11.1		5.6	2.7	4.6	5.6	6.1	8.6	0.34	9.9	5.86	1.04	5.6
Calorific value (MJ/kg)	26.6			21.1		20.8	21.1	22.3	17.9	23.6		19.7	21.64	0.93	21.1
Temp (°C)	500	500	500	475	500	440	440	550	550	500	525	500	498	10.08	500
Temp increment rate (°C/min)	5	7	5	10	20				300	10	20	5	47.44	31.92	10
Pyrolyser	Batch	Fixed bed	Fixed bed	Tubular	Semi batch	Fluidised bed	Fluidised bed	Tubular	Tubular	Pressurised FB	Semi batch	Fixed bed			
Carrier gas	N_2	–	N_2				N_2	N_2	N_2	N_2	N_2	N_2			
Exposure time (min)	30	–	180			0.063	0.063		10	1	–	30	35.88	24.54	10
Ash (%)	4.3	–	7.2	4.3	4.8	3.4	7.3	5.9	4.9	6	2.1	5.2	5.04	0.47	4.9
Fixed C (%)	12.6	–	15.1	25.8	6.6	14.1	12.8	14.7	10.3	14.8	24.1	17.6	15.32	1.68	14.7
Volatile matter (%)	83.1	–	66.5	69.9	83	79.8	75.3	73.8	78.7	70.6	73.5	67.4	74.69	1.77	73.8
Biochar (%)	30.6	27.4	6.6	27		28	26.4	18	27	20	35	36.4	25.67	2.51	27
Biooil (%)	9.4	59.7	17.9	34		31.4	41.5	40	35	28	45	14.4	32.39	4.41	34
Bio liquor (%)	37.9	–	40.5	11		14.6	10.3	13	10	13	–	51.5	22.42	5.38	13
Syngas (%)	14.9	12.8	35	29		28	21.8	29	30	33	20	12.2	24.15	2.47	28
Conversion efficiency (%)	92.8	99.9	100	101	0	102	100	100	102	94	100	115	105		
Reference	This study	Ozcimen and Karaosmanoglu [20]	David and Kapac [31]	Derimbas [24]	Agrawalla et al. [22]	Kim et al. [18]	Kim et al. [18]	Yorgun et al. [26]	Ozbay et al. [29]	Ates et al. [27]	Shadangi and Singh [19]	Karagoz [28]			

times. Thus, it may be said that the research on pyrolysis of press cake is at its infancy, and to the best of our knowledge, of this is the first time that the BLT press cake has been explored as a potential feedstock for biofuel production.

A selected number of press cakes were, therefore, considered to compare the results of the current study with the past investigations. In this comparison, the type and conditions of pyrolysis, properties of the feedstock and the properties of the resulting pyrolysis products are listed in Table 1 (see references therein). The results of the current study at 500 °C and 30 min residence time are also listed for comparison.

The majority of the studies on press cake have used <2 mm samples, with a moisture content of <10%. The calorific value of the press cake ranged between 18 and 27 MJ/kg. The selected feedstocks had very high proportion (67−83%) of volatile matter, 10−26% fixed carbon and 2−7% ash. Compared to other press cakes, the BLT press cake had the highest volatile matter (83%), low ash content (4.3%) and moderate levels of fixed carbon (12.6%). Various types of pyrolysis systems, such as the fixed bed, fluidised bed and tubular systems, have been used, and varying heating regimes (500 ± 100 °C) have been employed in the pyrolysis experiments (Table 1). A comparison of the performance of different reactors can be found in Nam et al. [17].

Despite these variations in pyrolysis conditions, and the differences in the compositions of the feedstocks, the pyrolysis reactions have yielded more or less similar proportions of biochar (26%; average of all selected feedstocks), biooil (32%), bioliquor (22%) and syngas (24%). The low standard errors (<6%) for the above parameters again indicate that the products of pyrolysis remain more or less similar irrespective of the feedstock used and the pyrolysis conditions employed, as long as the pyrolysis temperature and the residence times are maintained similar (see Table 1 and the references therein).

The BLT seed cake pyrolysis products, however, differed with the average values that were obtained for the selected 11 other species. Its biochar yield was

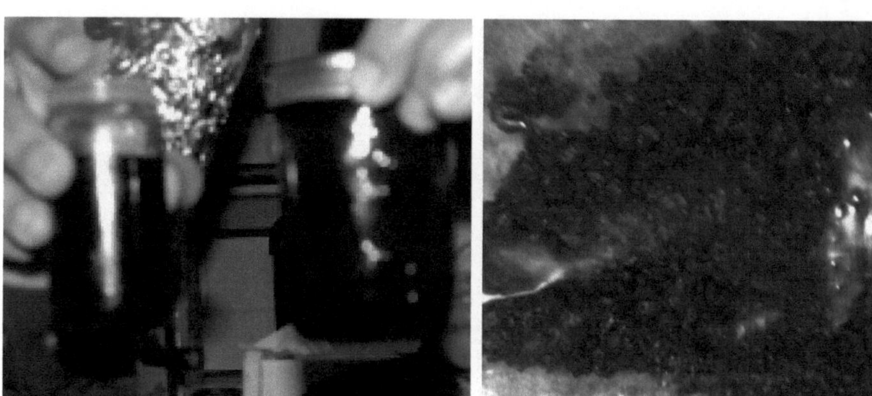

Fig. 5 Biooil (*left*), bioliquor (*middle*) and biochar (*right*) samples from BLT press cake

much higher than that reported for other species (31 vs. 26%), and its bioliquor yield was much higher than the average of other species (38 vs. 22%). In contrast, the biooil and syngas yields were much lower than those reported for other species (Table 1). This discrepancy may partly be explained by the complications associated with the separation of biooil from bioliquor. In the current study, the lower phase in the first condenser was collected as the bioliquor and the upper phase was collected as the biooil. As shown in Fig. 5, this separation by gravity was often not uniform, as some of the biooil tended to mix with the bioliquor during collection.

Overall, the biooil and bioliquor fractions of BLT make up to more than 48% of the pyrolysis products, and this is comparable to 54% of the average of the other 11 other species. The patterns in biooil + bioliquor production by BLT followed similar patterns as those reported for the non-edible species such as jatropha [21] and pongamia [22] but different with the patterns of edible species. This suggests that the non-edible species may respond differently to edible species in their thermal conversion process, and the causes for these differences are yet to be investigated (Table 1 and the references therein).

The pyrolysis products produced from BLT press cake at various combinations of temperature x residence time are shown in Fig. 6. The biochar yield reduced from 70% to 50% and then to 30%, as the pyrolysis temperature increased from 300, to 400 and then to 500 °C, respectively. In contrast, the biooil yield increased from 2.5 to 10% when the temperature was increased from 300 to 500 °C. The yields of syngas and bioliquor (aqueous condensate) also increased with a rise in temperature. The bioliquor yield increased from 300 to 400 °C, but no further increase occurred when the temperature was increased to 500 °C (average of all three residence times).

Responses of the 11 species to pyrolysis show two major patterns in generating pyrolysis products. Coincidentally, the edible species show one pattern and the non-edible species show a different pattern. The effects of residence time were

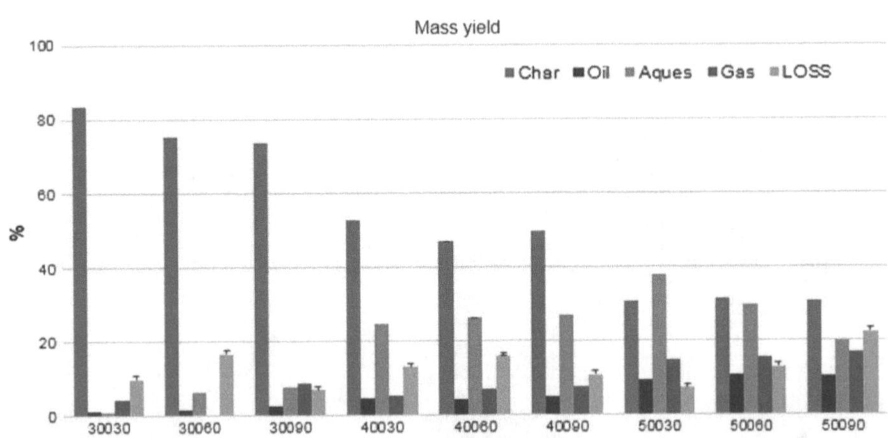

Fig. 6 Mass yields of various pyrolysis products at various combinations of temperature and residence time

highly variable. Chemical compositions of the feedstocks used, heat transfer rate to feedstocks and the complex reactions that take place between the condensable and non-condensable gasses within the reactors, may explain some of the above discrepancies [17]. However, it is interesting to note that the results of this study are comparable to those undertaken with sugarcane bagasse a non-oil producing and cellulose-rich feedstock [30].

Residence time had limited effect on the yields of all four products, except for bioliquor at 500 °C, wherein its yield decreased from 38 to 20%, as the residence time increased from 30 min to 90 min, respectively. Comparison of different product yields at 30, 60 and 90 min and at various temperatures shows no consistent pattern. This may be due to interactive effects of time temperature in releasing different pyrolysis products. Complexities associated with heat transfer processes varying degrees of decomposition of biomass into volatile matters or gasses, and the reactions taking place between volatile matters and the gasses may partly explain these results [17].

The yield losses were calculated based on the original weight of the press cake used, and the weights of the four products obtained from thermal conversion. These losses varied with the temperature and the residence time. At 300 and 400 °C, the yield losses showed no trend with the residence time. However, at 500 °C, the losses increased consistently from 9 to 12% at 30 and 90 min, respectively. The highest loss of 23% was recorded at 500 °C at 90 min (Fig. 5). It was also noted that the yield losses at higher temperatures were exacerbated by longer exposure times. Since the residence time exhibits varying responses at each of the tested temperature, care must be taken in selecting residence times. This is to be considered not only to maximise the target product but also to minimise the electricity consumption during pyrolysis. The current study suggests that a residence time of 30 min seems optimum for BLT press cake, as the longer exposure times have only made a marginal improvement to product yields.

Depending on the needs of the users (farmer vs. city council or an industry), batch pyrolysis (fixed bed; slow pyrolysis) technique can be employed to produce four types of products from the biomass. The gas can be directly injected into a generator to produce electricity, the biooil can be blended with diesel, or it can be upgraded to other fuels or chemicals. The bioliquor could be used as a soil conditioner (to improve soil health), or as an insecticide.

The resulting biochar on the other hand, will have multiple uses. It can be used in the manufacturing of heating briquettes, tyres, iron structures or filtration systems [31]. Most importantly, the biochar can be used as a soil amendment to improve the ability of the soil to retain nutrients and water [31], as this could lead to improved plant productivity [32] and reduced environmental impacts [33].

Biofuel production generates large quantities of wastes (husk, bagasse, press cake, glycerol, mill mud, etc.) which must be managed in a sustainable manner to demonstrate the benefits of biofuels. However, the costs of such management could be so high that they can force certain biofuel plants to shut down on economic viability grounds. The current study has demonstrated that some of those wastes can be transformed into other forms of biofuel. Such an activity will not only minimise the costs of waste management, but it can also generate additional income to the biofuel industry. Most importantly, such an activity will generate products (e.g. biochar) that will help sequester carbon, retain soil nutrients, improve soil health and enhance plant productivity, while also providing a green solution to other environmental issues. Further research is, therefore, needed to utilise pyrolysis products in industrial, agricultural and environmental applications.

4 Conclusions

This study demonstrates that the BLT press cake can serve as a valuable feedstock for thermal conversion, as it can be readily converted into other forms of energy, such as biochar, biooil or syngas.

Temperatures of 300−500 °C could be used depending on the purpose of pyrolysis (biochar, biooil or bioenergy). For example, temperature of approximately 300 °C may be used, if the biochar is the major product, or 500 °C can be maintained for the production of all four products. A temperature of >500 °C may be used for syngas generation.

A residence time of 30 min seems appropriate for BLT press cake, as (i) this time will produce reasonable quantities of biochar, biooil and bioliquor, (ii) the yield losses are low, and most importantly, (iii) the costs of operating a pyrolysis unit will be much lower at 30 than at 60 or 90 minutes.

Acknowledgements NA acknowledges the generosity of Dr. Capareda and Texas A&M University, College Station, Texas, for hosting him as a visiting scientist during his sabbatical leave. He is also grateful to Central Queensland University, Australia (Prof. Grant Stanley and Prof. Hillary Winchester), for granting leave and providing research funds via OSPRO. We also thank Dr. Pramod Shrestha and Mr. Mostafa Bhuiya for supplying some of the feedstocks used in the study, and Dr. Jersson Placido and Dr. Amado Maglinao for providing logistic support in the laboratory.

References

1. Jahirul, M. I., Rasul, M. G., Chowdhury, A. A. & Ashwath, N. (2012). Biofuels Production through Biomass Pyrolysis—A technological review. *Energies: Special Issue on Biomass and Biofuels,* 5, 4952−5001.
2. Rijal, D., Vancov, T., McIntosh, S., Ashwath, N., & Stanley, G. A. (2016). Process options for conversion of *Agave tequilana* leaves into bioethanol. *Industrial Crops Products, 84,* 263–272.
3. Ashwath, N. (2010). *Evaluating biodiesel potential of Australian native and naturalised plant species.* RIRDC publication#10-216, Rural industries research and development corporation, Canberra, 113 p.
4. Varanda, M. G., Pinto, G., & Martins, F. (2011). Life cycle analysis of biodiesel production. *Fuel Processing and Technology, 92,* 1087–1094. doi:10.1016/j.fuproc.2011.01.003.

5. Lopes, S. L., Fragoso, R., Duarte, E., & Marques, P. A. S. S. (2015). Bioconversion of *Jatropha curcas* seed cake to hydrogen by a strain of *Enterobacter aerogenes*. *Fuel, 139,* 715–719. doi:10.1016/j.fuel.2014.09.004.
6. Foo, K. Y., & Hameed, B. H. (2009). Utilization of biodiesel waste as a renewable resource for activated carbon: Application to environmental problems. *Renewable and Sustainable Energy Reviews, 13,* 2495–2504. doi:10.1016/j.rser.2009.06.009.
7. Nunes, A. A., Franca, A. S., & Oliveira, L. S. (2009). Activated carbons from waste biomass: An alternative use for biodiesel production solid residues. *Bioresource Technology, 100,* 1786–1792. doi:10.1016/j.biortech.2008.09.032.
8. Veerabhadrappa, M. B., Shivakumar, S. B., & Devappa, S. (2014). Solid-state fermentation of Jatropha seed cake for optimization of lipase, protease and detoxification of anti-nutrients in Jatropha seed cake using *Aspergillus versicolor* CJS-98. *Journal of Bioscience and Bioengineering, 117,* 208–214. doi:10.1016/j.jbiosc.2013.07.003.
9. da Luz, J. M. R., Paes, S. A., Torres, D. P., Nunes, M. D., da Silva, J. S., Mantovani, H. C., et al. (2013). Production of edible mushroom and degradation of ant nutritional factors in jatropha biodiesel residues. *LWT—Food Science and Technology., 50,* 575–580. doi:10.1016/j.lwt.2012.08.006.
10. Hathurusingha, S., Ashwath, N., & Midmore, D. J. (2011). Provenance variations in seed-related characters and oil content of *Calophyllum inophyllum* L. in northern Australia and Sri Lanka. *New Forests, 41,* 89–94.
11. Bhuiya, M. M. K., Rasul, M. G., Khan, M. M. K., & Ashwath, N. (2016). Comparison of oil refining and biodiesel production process between screw press and n-hexane techniques from beauty leaf feedstock. *International Conference on Mechanical Engineering*: *Proceedings of the 11th International Conference on Mechanical Engineering*, Dec 2015, Dhaka. AIP Conference Processings.1754, 060016 (2016); http://dx.doi.org/10.1063/1.4958457.
12. Hathurusingha, S. (2012). *Potential of Beauty Leaf Tree (Calophyllum inophyllum L.) as a biodiesel feedstock.* Ph.D. thesis, CQ University, Rockhampton, Australia.
13. Hathurusingha. S., & Ashwath, N. (2012). Beauty Leaf Oil Methyl Ester as an Anti-Knocking Agent for Diesel Fuel. In *Proceedings of the XII International Symposium on Environmental Geotechnology, Energy and Global Sustainable Development*, Los Angeles, June 27−29, 2012. http://www.isegnet.org/2012.
14. Crane, S., Aurore, G., Joseph, H., Mouloungui, Z., & Bourgeois, P. (2005). Composition of fatty acids triacylglycerols and unsaponifiable matter in *Calophyllum calaba* L. oil from Guadeloupe. *Phytochemistry, 66,* 1825–1831.
15. Jahirul, M. I., Koh, W., Brown, R. J., Senadera, W., O'Hara, I., & Moghaddam, L. (2014). Biodiesel production from non-edible beauty leaf (*Calophyllum inophyllum*) oil: Process optimization using response surface methodology (RSM). *Energies, 7,* 5317–5331.
16. Worasuwannarak, N., Sonobe, T., & Tanthapanichakoon, W. (2007). Pyrolysis behaviors of rice straw, rice husk, and corncob by TG-MS technique. *J. Anal. Appl. Pyrolysis., 78,* 265–271.
17. Nam, H., Capareda, S. C., Ashwath, N., & Kongkasawan, J. (2015). Experimental investigation of pyrolysis of rice straw using bench-scale auger, batch and fluidized bed reactors. *Energy, 93,* 2384–2394.
18. Hathurusingha, S., Ashwath, N., & Subedi, P. (2011). Variation in oil content and fatty acid profile of *Calophyllum inophyllum* L. with fruit maturity and its implications on resultant biodiesel quality. *Industrial Crops and Products, 33,* 629–632.
19. Jahirul, M. I., Brown, J. R., Senadeera, W., Ashwath, N., Laing, C., Leski-Taylor, J., et al. (2013). Optimisation of bio-oil extraction process from beauty leaf tree (*Calophyllum inophyllum*) oil seed as a second generation biodiesel source. *Procedia Engineering, 56,* 619–624.
20. Bhuiya, M. M. K., Rasul, M. G., Khan, M. M. K., Ashwath, N., Azad, A. K., & Mofijur, M. (2015). Optimisation of oil extraction process from Australian native beauty leaf seed (*Calophyllum inophyllum*). *Energy Proceedia, 75,* 56–61.

21. Kim, S. W., Koo, B. S., & Lee, D. H. (2014). A comparative study of bio-oils from pyrolysis of microalgae and oil seed waste in a fluidized bed. *Bioresource Technology, 162,* 96–102. doi:10.1016/j.biortech.2014.03.136.
22. Shadangi, K. P., & Singh, R. K. (2012). Thermolysis of polanga seed cake to bio-oil using semi batch reactor. *Fuel, 97,* 450–456. doi:10.1016/j.fuel.2012.02.058.
23. Özçimen, D., & Karaosmanoğlu, F. (2004). Production and characterization of bio-oil and biochar from rapeseed cake. *Renewable Energy, 29*(5), 779–787. doi:10.1016/j.renene.2003.09.006.
24. Demirbas, M. F. (2009). Evaluation of olive cake for bio-oil. *Energy Sources, Part A: Recovery, Utilization and Environmental Effects, 31*(14), 1236–1241. doi:10.1080/15567030801952318.
25. Agrawalla, A., Kumar, S., & Singh, R. K. (2011). Pyrolysis of groundnut de-oiled cake and characterization of the liquid product. *Bioresource Technology, 102*(22), 10711–10716. doi:10.1016/j.biortech.2011.08.113.
26. Yorgun, S., Ensöz, S., & Koçkar, Ö. M. (2001). Flash pyrolysis of sunflower oil cake for production of liquid fuels. *Journal of Analytical and Applied Pyrolysis, 60*(1), 1–12. doi:10.1016/S0165-2370(00)00102-9.
27. Ateş, F., Miskolczi, N., & Saricaoğlu, B. (2015). Pressurized pyrolysis of dried distillers grains with solubles and canola seed press cake in a fixed-bed reactor. *Bioresource Technology, 177,* 149–158. doi:10.1016/j.biortech.2014.10.163.
28. Karagöz, S. (2009). Energy production from the pyrolysis of waste biomasses. *International Journal of Energy Research, 33*(6), 576–581. doi:10.1002/er.1493.
29. Ozbay, N., Pütün, A. E., & Pütün, E. (2006). Bio-oil production from rapid pyrolysis of cottonseed cake: Product yields and compositions. *International Journal of Energy Research, 30*(7), 501–510. doi:10.1002/er.1165.
30. Carrier, M., Hugo, T., Gorgens, J., & Knoetze, H. (2011). Comparison of slow and vacuum pyrolysis of sugar cane bagasse. *Journal of Analytical and Applied Pyrolysis, 90,* 18–26.
31. David, E., & Kopac, J. (2014). Activated carbons derived from residual biomass pyrolysis and their CO2 adsorption capacity. *Journal of Analytical and Applied Pyrolysis, 110*(1), 322–332. doi:10.1016/j.jaap.2014.09.021.
32. Pudasaini, K., Walsh, K. B., Ashwath, N., & Bhattarai, T. (2014). Effects of biochar addition on plant available water of a loamy soil and consequences on cowpea growth. *Acta Horticulturae 1112.* doi:10.17660/ActaHortic.2016.1112.48.
33. MacDonald, L. M., Kookana, R. S., Van Zwieten, L., Singh, B. P., Sing, B., & Fareel, M. (2015). Regional considerations for targeted use of biochar in agriculture and remediation in Australia In: M. Guo, Z. He & M. Uchimiya. (Eds.), *Agricultural and environmental applications of biochar: Advances and Barriers.* 63, (pp. 1−30).SSSA Special publications, Madison, USA.

MIX
Papier aus verantwortungsvollen Quellen
Paper from responsible sources
FSC® C105338

If you have any concerns about our products,
you can contact us on
ProductSafety@springernature.com

In case Publisher is established outside the EU,
the EU authorized representative is:
Springer Nature Customer Service Center GmbH
Europaplatz 3, 69115 Heidelberg, Germany

Printed by Libri Plureos GmbH
in Hamburg, Germany